BUILDING CONTROL WITH PASSIVE DAMPERS

Optimal Performance-based Design for Earthquakes

BUILDING CONTROL WITH PASSIVE DAMPERS

Optimal Performance-based Design for Earthquakes

Izuru Takewaki

Kyoto University, Japan

John Wiley & Sons (Asia) Pte. Ltd.

Other Wiley Editorial Offices

John Wiley & Sons, Ltd, The Atrium, Southern Gate, Chichester, West Sussex, PO19 8SQ, UK

John Wiley & Sons Inc., 111 River Street, Hoboken, NJ 07030, USA

Jossey-Bass, 989 Market Street, San Francisco, CA 94103-1741, USA

Wiley-VCH Verlag GmbH, Boschstrasse 12, D-69469 Weinheim, Germany

John Wiley & Sons Australia Ltd, 42 McDougall Street, Milton, Queensland 4064, Australia

John Wiley & Sons Canada Ltd, 5353 Dundas Street West, Suite 400, Toronto, ONT, M9B 6H8, Canada

Wiley also publishes its books in a variety of electronic formats. Some content that appears in print may not be available in electronic books.

Library of Congress Cataloging-in-Publication Data

Takewaki, Izuru.
 Building control with passive dampers / Izuru Takewaki.
   p. cm.
 Includes bibliographical references and index.
  ISBN 978-0-470-82491-7 (cloth)
 1. Earthquake resistant design. 2. Buildings—Earthquake effects. 3. Damping (Mechanics)
 4. Buildings—Vibration. 5. Structural control (Engineering) I. Title.
   TA654.9.T35 2010
   693.8_52—dc22   2009025377

ISBN 978-0-470-82491-7 (HB)

Typeset in 11/13pt Times by Macmillan, Chennai, India.
Printed and bound in Singapore by Markono Print Media Pte Ltd, Singapore.
This book is printed on acid-free paper responsibly manufactured from sustainable forestry in which at least two trees are planted for each one used for paper production.

Contents

Preface

The concept of performance-based design is well accepted in the current structural design practice of buildings. In earthquake-prone countries, the philosophy of earthquake-resistant design to resist ground shaking with sufficient stiffness and strength of the building itself has also been accepted as a relevant structural design concept for many years. On the other hand, a new strategy based on the concept of active and passive structural control has been introduced rather recently in order to provide structural designers with powerful tools for performance-based design (Kobori, 1993; Housner *et al.*, 1994, 1997; Soong and Dargush, 1997; Kobori *et al.*, 1998; Hanson and Soong, 2001; Casciati, 2002; Johnson and Smyth, 2006). Passive control systems are often used to meet flexibly the requirements posed by this performance-based design. However, the structural engineers do not appear to have the tools for the optimal selection and placement of these passive control systems.

Although the use of structural control devices may reduce the overall earthquake response, it is often the case that some local structural responses are amplified; for example, member forces around the control devices. These phenomena should be resolved in the actual structural design. In addition, it should be kept in mind that earthquake ground motions have a lot of uncertainties; for example, see Drenick (1970), Anderson and Bertero (1987), and Takewaki (2006). In order to tackle this difficult problem, smart installation of passive control devices or passive energy dissipation systems is expected to play a vital role.

In this book, optimality criteria-based and optimal sensitivity-based design algorithms are explained in detail for passive control and energy dissipation systems in building structures. Displacement, acceleration, and earthquake input energy are regarded as three major target indices for performance-based design. It is shown that, once the building frames and passive control and energy dissipation systems are modeled appropriately, the optimal quantity and placement of passive control and energy dissipation systems are determined automatically and simultaneously.

The structural systems treated here are fixed-base shear buildings, fixed-base moment-resisting frames, fixed-base three-dimensional buildings, shear buildings with and without tuned mass dampers on surface ground, and bending-shear buildings on surface ground. Since it is well known that the ground or soil under buildings influences the seismic behavior of buildings with and without supplemental dampers, the model including the soil–structure interaction effect has been desired. Both the transfer functions and earthquake responses are introduced as the performance indices.

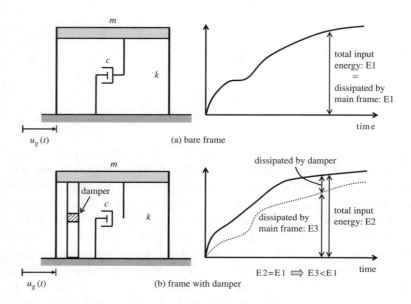

Figure 0.1 Theoretical backbone of effectiveness of supplemental dampers

The design implication of dampers with load and model uncertainties is also included, and the theoretical backbones of the effectiveness of passive control and energy dissipation systems are finally explained from the viewpoint of earthquake input energy. If the earthquake input energy criterion holds even approximately regardless of the existence of supplemental dampers and the supplemental passive dampers can absorb the earthquake input energy as greatly as possible, the input energy to the frame can be reduced drastically (see Figure 0.1). Although main frames are usually designed so as to remain elastic in the case of using passive energy dissipation systems, inelastic dynamic responses of building structures with viscous or hysteretic dampers are also discussed from the viewpoint of the effectiveness of viscous and hysteretic dampers.

The special character of this book may be described as follows. The aim of this book is not merely the presentation of optimization results, but the detailed description of how the optimal size and location are obtained simultaneously. The details of the optimization processes are explained step by step. Therefore, structural engineers can conduct programming by themselves to obtain the optimal size and placement. Most aspects of the techniques explained have been recognized by the leading international academic journals and are reliable. Moreover, the techniques explained are sometimes used as reference key techniques in optimization of passive control systems, and the comparison of these techniques with others is not difficult.

I hope that this book can provide structural designers and engineers with powerful tools and guides for optimal and smart passive structural control.

                                                                                    Izuru Takewaki

References

Anderson, J.C. and Bertero, V.V. (1987) Uncertainties in establishing design earthquakes. *Journal of Structural Engineering*, **113** (8), 1709–1724.

Casciati, F. (ed.) (2002) *Proceedings of 3rd World Conference on Structural Control*, John Wiley & Sons, Ltd, Chichester.

Drenick, R.F. (1970) Model-free design of aseismic structures. *Journal of Engineering Mechanics Division*, **96** (EM4), 483–493.

Hanson, R.D. and Soong, T.T. (2001) *Seismic Design with Supplemental Energy Dissipation Devices*, EERI, Oakland, CA.

Housner, G., Bergmann, L.A., Caughey, T.A., Chassiakos, A.G., Claus, R.O., Masri, S.F., Skelton, R.E., Soong, T.T., Spencer, B.F., and Yao, J.T.P. (1997) Structural control: past, present, and future (special issue). *Journal of Engineering Mechanics*, **123** (9), 897–971.

Housner, G.W., Masri, S.F., and Chassiakos, A.G. (eds) (1994) *Proceedings of 1st World Conference on Structural Control*, IASC, Los Angeles, CA.

Johnson, E., and Smyth, A. (eds) (2006) *Proceedings of 4th World Conference on Structural Control and Monitoring (4WCSCM)*, IASC, San Diego, CA.

Kobori, T. (1993) *Structural Control: Theory and Practice*, Kajima Press (in Japanese).

Kobori, T., Inoue, Y., Seto, K., Iemura, H., and Nishitani, A. (eds) (1998) *Proceedings of 2nd World Conference on Structural Control*, John Wiley & Sons, Ltd, Chichester.

Soong, T.T. and Dargush, G.F. (1997) *Passive Energy Dissipation Systems in Structural Engineering*, John Wiley & Sons, Ltd, Chichester.

Takewaki, I. (2006) *Critical Excitation Methods in Earthquake Engineering*, Elsevier Science, Amsterdam.

1

Introduction

1.1 Background and Review

Structural control has a long and successful history in mechanical and aerospace engineering. However, in the field of civil engineering, it has a different background (Housner *et al.*, 1994, 1997; Kobori, 1996; Soong and Dargush, 1997; Kobori *et al.*, 1998; Rivin, 1999; Srinivasan and McFarland, 2000; Hanson and Soong, 2001; Burns, 2002; Casciati, 2002; Soong and Constantinou, 2002; Christopoulos and Filiatrault, 2006; Johnson and Smyth, 2006; Arora, 2007; de Silva, 2007). Building and civil structures are often subjected to severe earthquake ground motions and wind disturbances with large uncertainties. Therefore, there is a need to take into account these uncertainties in the theory of structural control and its application to actual structures. Professor Soong presented five important areas impacted by structural control in his keynote lecture (Soong, 1998) in the Second World Conference on Structural Control held in Kyoto: (i) a systems approach, (ii) a deepening effect, (iii) a broadening effect, (iv) experimental research, and (v) creative engineering. Among these five areas, the broadening effect includes the effective use of passive dampers in building structures. In this book, this aspect will be explained in detail.

In the early stage of development in passive structural control, the installation itself of supplemental dampers in ordinary buildings was the principal objective. It appears natural that, after extensive developments of various damper systems, another objective was directed to the development of smart and effective installation of supplemental passive dampers.

Although the motivation was inspired and directed to smart and effective installation of supplemental passive dampers, research on optimal passive damper placement has been very limited. The following studies may deal with this subject. Constantinou and Tadjbakhsh (1983) derived the optimum damping coefficient for a damper placed on the first story of a shear building subjected to horizontal random earthquake motions. Gurgoze and Muller (1992) presented a numerical method for finding the

Building Control with Passive Dampers   Izuru Takewaki
© 2009 John Wiley & Sons (Asia) Pte Ltd

optimal placement and the optimal damping coefficient for a single viscous damper in a prescribed linear multidegree-of-freedom (MDOF) system. Zhang and Soong (1992) proposed a seismic design method to find the optimal configuration of viscous dampers for a building with specified story stiffnesses. While their method is based upon an intuitive criterion that an additional damper should be placed sequentially on the story with the maximum interstory drift, it is pioneering. Hahn and Sathi-avageeswaran (1992) performed several parametric studies on the effects of damper distribution on the earthquake response of shear buildings and showed that, for a building with uniform story stiffnesses, dampers should be added to the lower half floors of the building. De Silva (1981) presented a gradient algorithm for the optimal design of discrete passive dampers in the vibration control of a class of flexible systems. Inaudi and Kelly (1993) proposed a procedure for finding the optimal isolation damping for minimum acceleration response of base-isolated structures subjected to stationary random excitation. Tsuji and Nakamura (1996) proposed an algorithm to find both the optimal story stiffness distribution and the optimal damper distribution for a shear building model subjected to a set of spectrum-compatible earthquakes. Connor and Klink (1996) and Connor et al. (1997) introduced an optimal stiffness distribution and restricted the damper distribution to that proportional to the stiffness distribution. They call this distribution a quasi-optimal distribution. Masri et al. (1981) presented a simple yet efficient optimum active control method for reducing the oscillations of distributed parameter systems subjected to arbitrary deterministic or stochastic excitations. While they deal with active control, the result is informative to the development in passive optimal control theories.

Rather recently Takewaki (1997, 1999) opened another door of smart passive damper placement with the help of the concepts of inverse problem approaches and optimal criteria-based design approaches. He solved a problem of optimal passive damper placement by deriving the optimality criteria and then by developing an incremental inverse problem approach. For many years, this research played a role as a pioneering work in this area and many researchers referred this article and compared the results by their methods with the result by Takewaki (1997). Subsequently, Takewaki and Yoshitomi (1998), Takewaki et al. (1999) and Takewaki (2000) introduced a new approach based on the concept of optimal sensitivity. The optimal quantity of passive dampers is obtained automatically together with the optimal placement through this new method.

After these researches, many related studies were developed (Moreschi, 2000; Garcia, 2001; Singh and Moreschi, 2001, 2002; Garcia and Soong, 2002; Liu et al., 2003; Silvestri et al., 2003, 2004, 2006; Uetani et al., 2003; Xu et al., 2003; Asahina et al., 2004; Kiu et al., 2004; Lavan and Levy, 2004, 2005, 2006a, 2006b; Palazzo et al., 2004; Park et al., 2004; Silvestri and Trombetti, 2004, 2007; Trombetti and Silvestri, 2004, 2006, 2007; Wongprasert and Symans, 2004; Xu et al., 2004; Liu et al., 2005; Tan et al., 2005; Levy and Lavan, 2006; Marano et al., 2006; Attard, 2007; Aydin et al., 2007; Cimellaro, 2007; Cimellaro and Retamales, 2007; Wang and Dyke, 2008;

Paola and Navarra, 2009; Viola and Guidi, 2009). Most of them investigated new optimal design methods of supplemental dampers and proposed effective and useful methods.

There are several textbooks dealing with the design of passive dampers. Connor and Klink (1996) introduced a concept of "motion-based design" and provided versatile explanations on various passive and active control systems, namely visco-elastic dampers, viscous dampers, tuned-mass dampers (TMDs), base-isolation systems, and active control systems. Soong and Dargush (1997) explain the fundamental mechanical aspects of passive dampers and present many practical examples of application to realistic buildings. Hanson and Soong (2001) begin with basic concepts of passive dampers and present a few examples of application. Christopoulos and Filiatrault (2006) deal with passive energy dissipation systems and base-isolated buildings. They treat several different systems of supplemental dampers, namely metallic and friction dampers, viscous and visco-elastic dampers, self-centering characteristic dampers, TMDs, and so on. They also explain the energy principle and performance-based design principle. De Silva (2007) collects many useful chapters for passive damper systems and gives an up-to-date review.

1.2 Fundamentals of Passive-damper Installation

The three principal types of passive control system installed in building structures are (i) story-installation-type supplemental passive dampers (viscous damper, visco-elastic damper, hysteretic damper), (ii) TMDs, and (iii) base-isolation systems, as shown in Figure 1.1.

In this book, story-installation-type supplemental passive dampers are principally treated. TMDs are also treated partially. In order to present the fundamental basics for

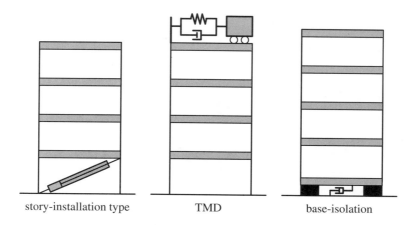

story-installation type   TMD   base-isolation

Figure 1.1 Three principal installation types of passive damper.

mechanical modeling of these story-installation-type supplemental passive dampers, viscous and visco-elastic dampers are taken as examples.

1.2.1 Viscous Dampers

The passive damper system, as shown in Figure 1.2, including a viscous damper can be modeled into two models. One is the dashpot and the other is the dashpot supported by a spring (Maxwell-type model). Let c denote the damping coefficient of the dashpot and k_s denote the stiffness of the supporting spring. This supporting spring represents the stiffness of the viscous damper device itself (e.g., oil damper) or the stiffness of the surrounding supporting system.

As for the Maxwell-type model, the force–displacement relation in the frequency domain can be described by

$$F(\omega) = (K_R + iK_I)U(\omega) = (k_V + i\omega c_V)U(\omega) \tag{1.1}$$

In Equation 1.1, k_V and c_V denote the stiffness of the spring and the damping coefficient of the dashpot of the pseudo Kelvin–Voigt model transformed from the Maxwell-type model.

The complex stiffness in Equation 1.1 may be derived as follows from the formulation of the series model of the dashpot and the supporting spring:

$$K_R + iK_I = \cfrac{1}{\cfrac{1}{i\omega c} + \cfrac{1}{k_s}} \tag{1.2}$$

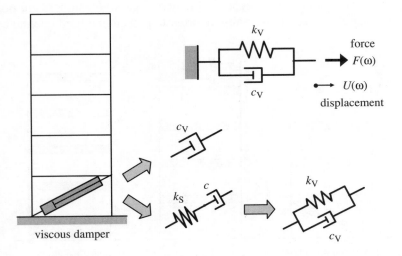

Figure 1.2 Passive damper system including a viscous damper and its modeling into a dashpot model and Maxwell model.

After some manipulation, the real and imaginary parts of the complex stiffness in Equation 1.2 may be expressed by

$$K_R(\omega) = k_V(\omega) = \frac{k_s c^2 \omega^2}{k_s^2 + c^2 \omega^2} \tag{1.3}$$

$$K_I(\omega) = \omega c_V(\omega) = \frac{k_s^2 c \omega}{k_s^2 + c^2 \omega^2} \tag{1.4}$$

It can be understood from Equations 1.3 and 1.4 that k_V and c_V are functions of the excitation frequency.

1.2.2 Visco-elastic Dampers

The passive damper system, as shown in Figure 1.3, including a visco-elastic damper can be modeled into two models. One is the Kelvin–Voigt model and the other is the Kelvin–Voigt model with a support (e.g., Kasai et al., 1998; Fu and Kasai, 1998). Let k denote the stiffness of the visco-elastic damper itself and c denote the damping coefficient of the visco-elastic damper itself. On the other hand, k_s denotes the stiffness of the supporting spring. This supporting spring represents the stiffness of the visco-elastic damper device itself (e.g., steel attachment of the visco-elastic material) or the stiffness of the surrounding supporting system. It is well known that k and c of most of visco-elastic materials depend on frequency, vibration amplitude and temperature, and so on. Therefore, the treatment of visco-elastic damper devices is more difficult than viscous dampers in general.

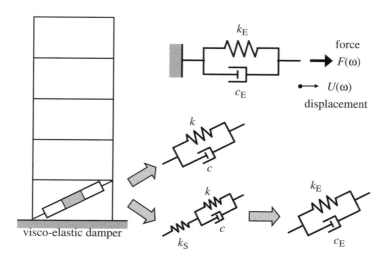

Figure 1.3   Passive damper system including a visco-elastic damper and its modeling into a Kelvin–Voigt model and Kelvin–Voigt model with support.

As for the Kelvin–Voigt model with a support, the force–displacement relation in the frequency domain can be described by

$$F(\omega) = (K_R + iK_I)U(\omega) = (k_E + i\omega c_E)U(\omega) \tag{1.5}$$

In Equation 1.5, k_E and c_E denote the stiffness of the spring and the damping coefficient of the dashpot of the pseudo Kelvin–Voigt model transformed from the Kelvin–Voigt model with a support.

The complex stiffness of this pseudo Kelvin–Voigt model may be derived as follows from the formulation of the series model of the Kelvin–Voigt model and the supporting spring:

$$K_R + iK_I = \cfrac{1}{\cfrac{1}{k + i\omega c} + \cfrac{1}{k_s}} \tag{1.6}$$

After some manipulation, the real and imaginary parts of the complex stiffness in Equation 1.6 may be expressed by

$$K_R(\omega) = k_E(\omega) = \frac{k_s(k^2 + kk_s + c^2\omega^2)}{(k + k_s)^2 + c^2\omega^2} \tag{1.7}$$

$$K_I(\omega) = \omega c_E(\omega) = \frac{\omega k_s^2 c}{(k + k_s)^2 + c^2\omega^2} \tag{1.8}$$

It can be understood from Equations 1.7 and 1.8 that k_E and c_E are functions of the excitation frequency.

1.3 Organization of This Book

In Chapter 1, the background of this book and the fundamentals of passive-damper installation are explained. Furthermore, a comprehensive review in the field of design of passive damper systems, especially story-installation-type systems, is conducted.

In Chapter 2, the optimality criteria-based design is presented for simple mass–spring models (shear building models) which include a single criterion in terms of transfer functions. The ratios of absolute values of the transfer functions evaluated at the undamped fundamental natural frequency of a structural system are taken as controlled quantities together with the undamped fundamental natural frequency. A set of optimality conditions is derived first. Then, in order to derive a solution satisfying the optimality criteria, a formulation of incremental inverse problems is introduced.

Chapter 3 is focused on the optimality criteria-based design including multiple criteria; that is, deformation and acceleration. Under the condition of constant quantity of structural members, deformation and acceleration are optimized. An efficient and systematic method is explained for the *simultaneous optimization* of story stiffness distributions and damping coefficient distributions of dampers. The method is a two-step design method. In the first step, a design is found which satisfies the optimality conditions for a specified set of total story stiffness capacity and total damper capacity. In the second step, the total story stiffness capacity and/or total damper capacity are varied with the optimality conditions satisfied. While deformation is reduced both in the first and second steps, acceleration is reduced only in the second step via increase of total damper capacity.

Chapter 4 introduces a concept of optimal sensitivity-based design of dampers in moment-resisting frames. The damper systems are modeled by a viscous-type model or a Maxwell-type model. The sum of the transfer function amplitudes is treated as the design objective. As the quantity of dampers increases, the optimal placement and quantity are determined automatically based on the optimal sensitivity. A non-monotonic sensitivity case is also treated. A representative schematic diagram for the optimization procedure explained in Chapter 4 is shown in Figure 1.4. While a different algorithm was devised and a variation from a uniform storywise distribution of added dampers was considered in Chapter 2, a variation from the null state is treated here. This treatment helps designers to understand simultaneously which position would be the best and what capacity of dampers would be required to attain a series of desired response performance levels.

In Chapter 5, a method of optimal sensitivity-based design of dampers is explained in three-dimensional structures. As in Chapter 4, the sum of the transfer function

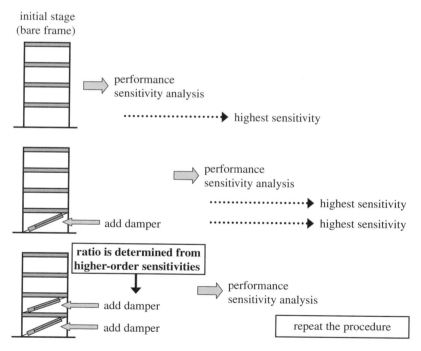

Figure 1.4 A representative schematic diagram of optimization procedures explained in this book.

amplitudes defined for many local interstory drifts is treated as the design objective. The torsional effect of three-dimensional structures is included and the procedure for finding the best placement of dampers is explained. A nonmonotonic path with respect to damper quantity level is also shown in detail.

Chapter 6 deals with an optimal sensitivity-based design of dampers in shear buildings on surface ground under earthquake loading. It is well known that the soil or ground under a structure greatly influences the structural vibration properties. It is important, therefore, to develop an optimization technique for such an interaction model. While some methods of active control have been proposed for such a model, theories for passive control systems were proposed by the present author for the first time and are explained here in detail.

Chapter 7 explains a concept of optimal sensitivity-based design of dampers in bending-shear buildings on surface ground under earthquake loading. It is noted that the bending-shear deformation is important in tall buildings or buildings of large aspect ratios (height/width ratio). It is shown first that the dampers are effective only for shear deformation and a procedure for finding the best damper placement is explained in detail.

In Chapter 8, the optimal sensitivity-based design of dampers is presented in shear buildings with a TMD on surface ground under earthquake loading. While the TMD is very popular as a passive control device, no method of optimal design of combining the interstory-type supplemental viscous dampers with the TMDs has ever been presented. The effect of the TMD system on the optimal placement of interstory-type supplemental viscous dampers will also be explained.

Chapter 9 examines a new perspective of design of dampers in shear buildings with uncertainties. Uncertainty analysis is one of the major subjects recently in the structural design of buildings. The effects of uncertainties in dampers, main structures, and earthquake ground motions on the design of building structures with passive dampers are discussed in detail. In particular, the level of uncertainties in dampers and earthquake ground motion is known in general to be larger than that in main structures, and consideration of the effect of uncertainties in dampers and earthquake ground motions on the design of passive control structures is extremely important in the reliable design of such structures. A critical excitation method (Takewaki, 2006) is introduced first for modeling the uncertainty in earthquake ground motions. Then info-gap uncertainty analysis is introduced to measure the robustness of buildings with passive dampers and a procedure is explained to take into account simultaneously the load and structural model uncertainties.

In Chapter 10, the theoretical backbones of effectiveness of passive control systems are explained using the constant input energy property (see Figure 0.1 in the Preface). The dual formulation in the time and frequency domains will play an important role in the verification of this constant input energy property.

In Chapter 11, the inelastic dynamic critical responses of building structures with viscous or hysteretic dampers are explained and the effectiveness of passive dampers is discussed from the viewpoint of types of passive damper. Resonant sinusoidal waves as approximate critical inputs and response spectrum-compatible waves are employed as input ground motions.

It is very important and useful to remark that, while the present book deals with supplemental viscous dampers only, the methods explained can be applied to supplemental visco-elastic dampers. The procedure shown in Sections 4.6–4.9 is a good example of that extension. Assume that the visco-elastic dampers can be modeled by the Kelvin–Voigt model. When the quantity of visco-elastic dampers changes, the values of the stiffness and the viscous damping coefficient will change simultaneously. In this case, it is appropriate to employ the viscous damping coefficient as the leading (or primary) design variable and regard the stiffness of the Kelvin–Voigt model as the minor (or dependent) variable. The procedure similar to that in Sections 4.6–4.9 can be utilized almost in the same manner.

Figure 1.4 shows a representative schematic diagram for the optimization procedures explained in this book. The optimization procedures in Chapters 4–9 are based on the optimality criteria and related performance sensitivities. The damper placement criteria are derived from these optimality criteria and performance sensitivities. In Figure 1.4, the initial design is a bare frame without supplemental dampers. Sensitivity analysis of the objective function with respect to a design variable (damping coefficient of supplemental damper) is performed first for this bare frame and the highest performance sensitivity is found. Then the damping coefficient of the supplemental damper is added to this story. This implies that the supplemental damper with the highest performance sensitivity can decrease the performance most effectively and the damping coefficient should be added in this supplemental damper. Again, sensitivity analysis is performed next for the frame with a supplemental damper and the highest performance sensitivity is found. If the multiple stories show the highest performance sensitivity, then the damping coefficients of the corresponding supplemental dampers are added. Again, sensitivity analysis is performed next for the frame with supplemental dampers and the procedure explained above is repeated until the required total quantity of supplemental dampers is reached.

References

Arora, J.S. (ed.) (2007) *Optimization of Structural and Mechanical Systems*, World Scientific.

Asahina, D., Bolander, J.E., and Berton, S. (2004) Design optimization of passive devices in multi-degree of freedom structures. Proceedings of 13th World Conference on Earthquake Engineering, paper no. 1600.

Attard, T.L. (2007) Controlling all interstory displacements in highly nonlinear steel buildings using optimal viscous damping. *Journal of Structural Engineering*, **133** (9), 1331–1340.

Aydin, E., Boduroglub, M.H., and Guney, D. (2007) Optimal damper distribution for seismic rehabilitation of planar building structures. *Engineering Structures*, **29**, 176–185.

Burns, S.A. (ed.) (2002) *Recent Advances in Optimal Structural Design*, Structural Engineering Institute Technical Committee on Optimal Structural Design, ASCE, New York.

Casciati, F. (ed.) (2002) *Proceedings of 3rd World Conference on Structural Control*, John Wiley & Sons, Ltd, Chichester.

Christopoulos, C. and Filiatrault, A. (2006) *Principle of Passive Supplemental Damping and Seismic Isolation*, IUSS Press, Pavia.

Cimellaro, G.P. (2007) Simultaneous stiffness-damping optimization of structures with respect to acceleration, displacement and base shear. *Engineering Structures*, **29**, 2853–2870.

Cimellaro, G.P. and Retamales, R. (2007) Optimal softening and damping design for buildings. *Structural Control and Health Monitoring*, **14** (6), 831–857.

Connor, J.J. and Klink, B.S.A. (1996) *Introduction to Motion-Based Design*, WIT Press.

Connor, J.J., Wada, A., Iwata, M., and Huang, Y.H. (1997) Damage-controlled structures, I: preliminary design methodology for seismically active regions. *Journal of Structural Engineering*, **123** (4), 423–431.

Constantinou, M.C. and Tadjbakhsh, I.G. (1983) Optimum design of a first story damping system. *Computers & Structures*, **17** (2), 305–310.

de Silva, C.W. (1981) An algorithm for the optimal design of passive vibration controllers for flexible systems. *Journal of Sound and Vibration*, **75** (4), 495–502.

de Silva, C.W. (ed.) (2007) *Vibration Damping, Control, and Design*, CRC Press.

Fu, Y. and Kasai, K. (1998) Comparative study of frames using viscoelastic and viscous dampers. *Journal of Structural Engineering*, **124** (5), 513–522.

Garcia, D.L. (2001) A simple method for the design of optimal damper configurations in MDOF structures. *Earthquake Spectra*, **17** (3), 387–398.

Garcia, D.L. and Soong, T.T. (2002) Efficiency of a simple approach to damper allocation in MDOF structures. *Journal of Structural Control*, **9** (1), 19–30.

Gurgoze, M. and Muller, P.C. (1992) Optimal positioning of dampers in multi-body systems. *Journal of Sound and Vibration*, **158** (3), 517–530.

Hahn, G.D. and Sathiavageeswaran, K.R. (1992) Effects of added-damper distribution on the seismic response of buildings. *Computers & Structures*, **43** (5), 941–950.

Hanson, R.D. and Soong, T.T. (2001) *Seismic Design with Supplemental Energy Dissipation Devices*, EERI, Oakland, CA.

Housner, G.W., Masri, S.F., and Chassiakos, A.G. (eds) (1994) *Proceedings of 1st World Conference on Structural Control*, IASC, Los Angeles, CA.

Housner, G., Bergmann, L.A., and Caughey, T.A. (1997) Structural control: past, present, and future (special issue). *Journal of Engineering Mechanics*, **123** (9), 897–971.

Inaudi, J.A. and Kelly, J.M. (1993) Optimum damping in linear isolation systems. *Earthquake Engineering & Structural Dynamics*, **22**, 583–598.

Johnson, E. and Smyth, A. (eds) (2006) *Proceedings of 4th World Conference on Structural Control and Monitoring (4WCSCM)*. IASC, San Diego, CA.

Kasai, K., Fu, Y., and Watanebe, A. (1998) Passive control systems for seismic damage mitigation. *Journal of Structural Engineering*, **124** (5), 501–512.

Kiu, W., Tong, M., Wu, Y., and Lee, G. (2004) Optimized damping device configuration design of a steel frame structure based on building performance indices. *Earthquake Spectra*, **20** (1), 67–89.

Kobori, T. (1996) Structural control for large earthquakes. Proceedings of IUTAM Symposium, Kyoto, Japan, pp. 3–28.

Kobori, T., Inoue, Y., Seto, K. *et al.* (eds) (1998) *Proceedings of 2nd World Conference on Structural Control*, John Wiley & Sons, Ltd, Chichester.

Lavan, O. and Levy, R. (2004) Optimal design of supplemental viscous dampers for linear framed structures. Proceedings of 13th World Conference on Earthquake Engineering, paper no. 42.

Lavan, O. and Levy, R. (2005) Optimal design of supplemental viscous dampers for irregular shear-frames in the presence of yielding. *Earthquake Engineering & Structural Dynamics*, **34** (8), 889–907.

Lavan, O. and Levy, R. (2006a) Optimal design of supplemental viscous dampers for linear framed structures. *Earthquake Engineering & Structural Dynamics*, **35** (3), 337–356.

Lavan, O. and Levy, R. (2006b) Optimal peripheral drift control of 3D irregular framed structures using supplemental viscous dampers. *Journal of Earthquake Engineering*, **10** (6), 903–923.

Levy, R. and Lavan, O. (2006) Fully stressed design of passive controllers in framed structures for seismic loadings. *Structural and Multidisciplinary Optimization*, **32** (6), 485–498.

Liu, W., Tong, M., Wu, X., and Lee, G. (2003) Object-oriented modeling of structural analysis and design with application to damping device configuration. *Journal Computing in Civil Engineering*, **17** (2), 113–122.

Liu, W., Tong, M., and Lee, G. (2005) Optimization methodology for damper configuration based on building performance indices. *Journal of Structural Engineering*, **131** (11), 1746–1756.

Marano, G.C., Trentadue, F., and Chiaia, B. (2006) Stochastic reliability based design criteria for linear structure subject to random vibrations. Proceedings of 8th Biennial ASME Conference on Engineering Systems Design and Analysis, ESDA2006.

Masri, S.F., Bekey, G.A., and Caughey, T.K. (1981) Optimum pulse control of flexible structures. *Journal of Applied Mechanics*, **48**, 619–626.

Moreschi, L.M. (2000) Seismic design of energy dissipation systems for optimal structural performance. PhD Dissertation, Virginia Polytechnic Institute and State University.

Palazzo, B., Petti, L., and De Iuliis, M. (2004) A modal approach to optimally place dampers in framed structures. Proceedings of 13th World Conference on Earthquake Engineering, paper no. 769.

Paola, M.D. and Navarra, G. (2009) Stochastic seismic analysis of MDOF structures with nonlinear viscous dampers. *Structural Control and Health Monitoring*, **16**, (3), 303–318.

Park, J.-H., Kim, J., and Min, K.-W. (2004) Optimal design of added viscoelastic dampers and supporting braces. *Earthquake Engineering & Structural Dynamics*, **33** (4), 465–484.

Rivin, E. (1999) *Stiffness and Damping in Mechanical Design*, Marcel Dekker Inc.

Silvestri, S. and Trombetti, T. (2004) Optimal insertion of viscous dampers into shear-type structures: dissipative properties of the MPD system. Proceedings of 13th World Conference on Earthquake Engineering, paper no. 473.

Silvestri, S. and Trombetti, T. (2007) Physical and numerical approaches for the optimal insertion of seismic viscous dampers in shear-type structures. *Journal of Earthquake Engineering*, **11** (5), 787–828.

Silvestri, S., Trombetti, T., and Ceccoli, C. (2003) Inserting the mass proportional damping (MPD) system in a concrete shear-type structure. *International Journal of Structural Engineering and Mechanics*, **16** (2), 177–193.

Silvestri, S., Trombetti, T., and Ceccoli, C. (2004) Optimal insertion of viscous dampers into shear-type structures: performances and applicability of the MPD system. Proceedings of 13th World Conference on Earthquake Engineering, paper no. 477.

Silvestri, S., Trombetti, T., and Gasparini, G. (2006) Effectiveness of inserting dampers between frames and lateral-resisting elements for the mitigation of the seismic effects. Proceedings of the 100th Anniversary Earthquake Conference Commemorating the 1906 San Francisco Earthquake, EERI's Eighth U.S. National Conference on Earthquake Engineering (8NCEE).

Singh, M.P. and Moreschi, L.M. (2001) Optimal seismic response control with dampers. *Earthquake Engineering & Structural Dynamics*, **30** (4), 553–572.

Singh, M.P. and Moreschi, L.M. (2002) Optimal placement of dampers for passive response control. *Earthquake Engineering & Structural Dynamics*, **31** (4), 955–976.

Soong, T.T. (1998) Structural control: impact on structural research in general. *Proceedings of 2nd World Conference on Structural Control*, Kobori, T., Inoue, Y., Seto, K. *et al.* (eds), John Wiley & Sons, Ltd, Chichester 1, 5–14.

Soong, T.T. and Constantinou, M.C. (2002) *Passive and Active Structural Vibration Control in Civil Engineering*, Springer-Verlag.

Soong, T.T. and Dargush, G.F. (1997) *Passive Energy Dissipation Systems in Structural Engineering*, John Wiley & Sons, Ltd, Chichester.

Srinivasan, A.V. and McFarland, D.M. (2000) *Smart Structures: Analysis and Design*, Cambridge University Press.

Takewaki, I. (1997) Optimal damper placement for minimum transfer functions. *Earthquake Engineering & Structural Dynamics*, **26** (11), 1113–1124.

Takewaki, I. (1999) *Dynamic Structural Design: Inverse Problem Approach*, WIT Press.

Takewaki, I. (2000) Optimal damper placement for planar building frames using transfer functions. *Structural and Multidisciplinary Optimization*, **20** (4), 280–287.

Takewaki, I. (2006) *Critical Excitation Methods in Earthquake Engineering*, Elsevier Science, Amsterdam.

Takewaki, I. and Yoshitomi, S. (1998) Effects of support stiffnesses on optimal damper placement for a planar building frame. *Journal of the Structural Design of Tall Buildings*, **7** (4), 323–336.

Takewaki, I., Yoshitomi, S., Uetani, K., and Tsuji, M. (1999) Non-monotonic optimal damper placement via steepest direction search. *Earthquake Engineering & Structural Dynamics*, **28** (6), 655–670.

Tan, P., Dyke, S.J., Richardson, A., and Abdullah, M. (2005) Integrated device placement and control design in civil structures using genetic algorithms. *Journal of Structural Engineering*, **131** (10), 1489–1496.

Trombetti, T. and Silvestri, S. (2004) Added viscous dampers in shear-type structures: the effectiveness of mass proportional damping. *Journal of Earthquake Engineering*, **8** (2), 275–313.

Trombetti, T. and Silvestri, S. (2006) On the modal damping ratios of shear-type structures equipped with Rayleigh damping systems. *Journal of Sound and Vibration*, **292** (1–2), 21–58.

Trombetti, T. and Silvestri, S. (2007) Novel schemes for inserting seismic dampers in shear-type systems based upon the mass proportional component of the Rayleigh damping matrix. *Journal of Sound and Vibration*, **302** (3), 486–526.

Tsuji, M. and Nakamura, T. (1996) Optimum viscous dampers for stiffness design of shear buildings. *Journal of the Structural Design of Tall Buildings* **5**, 217–234.

Uetani, K., Tsuji, M., and Takewaki, I. (2003) Application of optimum design method to practical building frames with viscous dampers and hysteretic dampers. *Engineering Structures*, **25** (5), 579–592.

Viola, E. and Guidi, F. (2009) Influence of the supporting braces on the dynamic control of buildings with added viscous dampers. *Structural Control and Health Monitoring*, **16**, (3), 267–286. DOI: 10.1002/stc.234

Wang, Y. and Dyke, S. (2008) Smart system design for a 3D base-isolated benchmark building. *Structural Control and Health Monitoring*, **30**, 939–957.

Wongprasert, N. and Symans, M.D. (2004) Application of a genetic algorithm for optimal damper distribution within the nonlinear seismic benchmark building. *Journal of Engineering Mechanics*, **130** (4), 401–406.

Xu, Z.D., Shen, Y.P., and Zhao, H.T. (2003) A synthetic optimization analysis method on structures with viscoelastic dampers. *Soil Dynamics and Earthquake Engineering*, **23**, 683–689.

Xu, Z.D., Zhao, H.T., and Li, A.Q. (2004) Optimal analysis and experimental study on structures with viscoelastic dampers. *Journal of Sound and Vibration*, **273** (3), 607–618.

Zhang, R.H. and Soong, T.T. (1992) Seismic design of viscoelastic dampers for structural applications. *Journal of Structural Engineering*, **118** (5), 1375–1392.

2

Optimality Criteria-based Design: Single Criterion in Terms of Transfer Function

2.1 Introduction

Inverse eigenvalue problems and inverse eigenmode problems for undamped structural systems have been investigated extensively and many important results have been accumulated (e.g., Porter, 1970; Barcilon, 1982; Gladwell, 1986, 2004; Nakamura and Yamane, 1986; Boley and Golub, 1987; Ram, 1994; Takewaki and Nakamura, 1995, 1997; Takewaki *et al.*, 1996). It has been demonstrated that sophisticated mathematical treatment can be developed in the inverse problems for undamped structural system and various theories have been proposed. While the inverse problems for undamped structural systems play a fundamental and important role in the development of inverse problems for damped structural systems, damping effects on inversion of system parameters are conspicuous and of great interest recently (Lancaster and Maroulas, 1987; Starek and Inman, 1995). In particular, when the magnitude of damping is fairly large or damping system is nonclassical (nonproportional), the inverse problems for undamped structural systems do not necessarily provide relevant information on design of the corresponding damped structural systems.

In the first part of this chapter, a new analytical procedure is explained for redesign of structural systems with an arbitrary damping system (viscous and/or hysteretic, proportional or nonproportional) for target transfer functions. It may be appropriate to call the present problem an incremental inverse problem. The ratios of absolute values of the transfer functions evaluated at the undamped fundamental natural frequency of a structural system are taken as controlled quantities together with the undamped fundamental natural frequency. To the best of the author's knowledge,

Building Control with Passive Dampers   Izuru Takewaki
© 2009 John Wiley & Sons (Asia) Pte Ltd

this kind of treatment has never been proposed. The following studies may be relevant to this subject. Tsai (1974) proposed a technique to determine approximate modal damping ratios as a classically damped model for a nonclassically damped soil–structure system by matching amplitudes of transfer functions at several natural frequencies in both models and solved a set of simultaneous *nonlinear* equations iteratively. Chen (1992) formulated an optimal design problem for a *classically damped* model subject to constraints on natural frequencies and frequency responses at a fixed frequency. The features of the present formulation are to be able to deal with any damping system (e.g., viscous and/or hysteretic, proportional or nonproportional), to be able to treat any structural system so far as it can be modeled with finite-element (FE) systems and to consist of a systematic algorithm without any indefinite iterative operation.

Since the amplitudes of transfer functions are not necessarily useful from physical points of view until they are transformed into mean squares (statistical quantities) by multiplication with the Fourier transform of a disturbance, only their ratios are specified in this chapter. Then the undamped fundamental natural frequency plays a role for adjusting the level of the amplitudes of transfer functions to a desired level. An undamped fundamental natural frequency is a representative parameter of overall stiffness of a structure, and specification of the undamped fundamental natural frequency is expected to provide a fundamental and useful index for structural design.

Two two-degree-of-freedom (DOF) mass–spring models with viscous or hysteretic dampers are introduced in order to demonstrate the effectiveness of the present technique. Furthermore, for six-DOF mass–spring models with viscous dampers, the effects of damper placement on the level and overall configuration of transfer functions are discussed.

In the second part of this chapter, a new formulation of optimality criteria-based design of passive dampers is presented based on the incremental inverse problem formulation explained in the first part. The problem treated here is to find the optimal damper placement to minimize the sum of amplitudes of the transfer functions evaluated at the undamped fundamental natural frequency of a structural system subject to a constraint on the sum of the damping coefficients of added dampers. For a given shear building model with an arbitrary damping system, an optimal distribution of passive dampers is obtained automatically with the optimality criteria-based sensitivity ratios of amplitudes of transfer functions at the fundamental natural frequency as the target performance indices. The features of the present formulation are to enable one to deal with any damping system (e.g., proportional or nonproportional), to enable one to treat any structural system so far as it can be modeled with FE systems and to be a systematic algorithm without any indefinite iterative operation. It should also be pointed out that, because the present formulation deals with a general dynamical property (i.e. the amplitude of the transfer function), the results are general and are not influenced by characteristics of input motions.

2.2 Incremental Inverse Problem: Simple Example

As a simple example, consider a two-DOF mass–spring–dashpot model including viscous dampers as shown in Figure 2.1(a). Let $\{\overline{m}_1, \overline{m}_2\}$, $\{k_1, k_2\}$, and $\{\overline{c}_1, \overline{c}_2\}$ denote respectively the masses, spring stiffnesses, and damping coefficients of dashpots. It is assumed here that the masses $\{\overline{m}_1, \overline{m}_2\}$ and damping coefficients $\{\overline{c}_1, \overline{c}_2\}$ of dashpots are prescribed. The design variables are the spring stiffnesses $\{k_1, k_2\}$.

Let u_1 and u_2 denote the nodal displacements of masses $\overline{m}_1$ and $\overline{m}_2$ respectively. When this model is subjected to a base acceleration $\ddot{u}_g$, the equations of motion for this model may be written as

$$\begin{bmatrix} k_1 + k_2 & -k_2 \\ -k_2 & k_2 \end{bmatrix} \begin{Bmatrix} u_1 \\ u_2 \end{Bmatrix} + \begin{bmatrix} \overline{c}_1 + \overline{c}_2 & -\overline{c}_2 \\ -\overline{c}_2 & \overline{c}_2 \end{bmatrix} \begin{Bmatrix} \dot{u}_1 \\ \dot{u}_2 \end{Bmatrix}$$

$$+ \begin{bmatrix} \overline{m}_1 & 0 \\ 0 & \overline{m}_2 \end{bmatrix} \begin{Bmatrix} \ddot{u}_1 \\ \ddot{u}_2 \end{Bmatrix} = - \begin{bmatrix} \overline{m}_1 & 0 \\ 0 & \overline{m}_2 \end{bmatrix} \begin{Bmatrix} 1 \\ 1 \end{Bmatrix} \ddot{u}_g \qquad (2.1)$$

It is effective to discuss the equations of motion in the frequency domain. Let $U_1(\omega)$, $U_2(\omega)$, and $\ddot{U}_g(\omega)$ as functions of frequency denote the Fourier transforms of u_1, u_2,

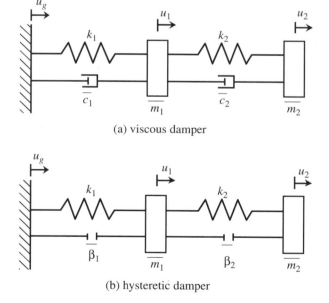

(a) viscous damper

(b) hysteretic damper

Figure 2.1 Two-DOF mass–spring model: (a) viscous damper model; (b) hysteretic damper model. (Originally published in I. Takewaki, "Efficient redesign of damped structural systems for target transfer functions," *Computer Methods in Applied Mechanics and Engineering*, **147**, no. 3–4, 275–286, 1997, Elsevier B.V.).

and $\ddot{u}_g$ respectively, and let ω denote a circular frequency. Fourier transformation of Equation 2.1 may be reduced to the following form:

$$\left(\begin{bmatrix} k_1 + k_2 & -k_2 \\ -k_2 & k_2 \end{bmatrix} + i\omega \begin{bmatrix} \bar{c}_1 + \bar{c}_2 & -\bar{c}_2 \\ -\bar{c}_2 & \bar{c}_2 \end{bmatrix} - \omega^2 \begin{bmatrix} \overline{m}_1 & 0 \\ 0 & \overline{m}_2 \end{bmatrix} \right) \left\{ \begin{array}{c} U_1(\omega) \\ U_2(\omega) \end{array} \right\}$$

$$= - \begin{bmatrix} \overline{m}_1 & 0 \\ 0 & \overline{m}_2 \end{bmatrix} \left\{ \begin{array}{c} 1 \\ 1 \end{array} \right\} \ddot{U}_g(\omega) \qquad (2.2)$$

In Equation 2.2, i is the imaginary unit. The Fourier transforms $\delta_1(\omega)$ and $\delta_2(\omega)$ of the interstory drifts $d_1 = u_1$ and $d_2 = u_2 - u_1$ can be expressed in terms of $U_1(\omega)$ and $U_2(\omega)$ by

$$\left\{ \begin{array}{c} \delta_1(\omega) \\ \delta_2(\omega) \end{array} \right\} = \begin{bmatrix} 1 & 0 \\ -1 & 1 \end{bmatrix} \left\{ \begin{array}{c} U_1(\omega) \\ U_2(\omega) \end{array} \right\} \qquad (2.3)$$

Since the fundamental natural frequency of a structure usually plays an important role, the steady-state response at this fundamental natural frequency is considered here. Let ω_1 denote the undamped fundamental natural circular frequency of the model considered here and let us define new complex-value quantities $\hat{U}_1$ and $\hat{U}_2$ by

$$\hat{U}_1 \equiv \frac{U_1(\omega_1)}{\ddot{U}_g(\omega_1)} \quad \text{and} \quad \hat{U}_2 \equiv \frac{U_2(\omega_1)}{\ddot{U}_g(\omega_1)} \qquad (2.4)$$

In Equation 2.4, $\hat{U}_i$ represents the value such that ω_1 is substituted in the frequency response function obtained as $U_i(\omega)$ after substituting $\ddot{U}_g(\omega) = 1$ in Equation 2.2. This complex-value quantity indicates the resonant amplitude of the floor displacement at the fundamental natural circular frequency ω_1. When a base acceleration input with broad frequency-band is taken, this quantity is regarded as representing the response to this input. In addition, new complex-value quantities $\hat{\delta}_1$ and $\hat{\delta}_2$ are defined by $\hat{\delta}_1 \equiv \hat{U}_1$ and $\hat{\delta}_2 \equiv \hat{U}_2 - \hat{U}_1$. Owing to Equations 2.2 and 2.4, it is concluded that the quantities $\hat{U}_1$ and $\hat{U}_2$ must satisfy the following equation:

$$\mathbf{A} \left\{ \begin{array}{c} \hat{U}_1 \\ \hat{U}_2 \end{array} \right\} = - \left\{ \begin{array}{c} \overline{m}_1 \\ \overline{m}_2 \end{array} \right\} \qquad (2.5)$$

where $\mathbf{A}$ is defined by

$$\mathbf{A} = \begin{bmatrix} (k_1 + k_2) + i\omega_1(\bar{c}_1 + \bar{c}_2) - \omega_1^2 \overline{m}_1 & -k_2 - i\omega_1 \bar{c}_2 \\ -k_2 - i\omega_1 \bar{c}_2 & k_2 + i\omega_1 \bar{c}_2 - \omega_1^2 \overline{m}_2 \end{bmatrix} \qquad (2.6)$$

Differentiation of Equation 2.5 including multiplication of a matrix and a vector with respect to a design variable k_j provides

$$\mathbf{A}_{,j} \left\{ \begin{array}{c} \hat{U}_1 \\ \hat{U}_2 \end{array} \right\} + \mathbf{A} \left\{ \begin{array}{c} \hat{U}_{1,j} \\ \hat{U}_{2,j} \end{array} \right\} = \mathbf{0} \qquad (2.7)$$

Here and in the following, the symbol $(\cdot)_{,j}$ denotes the partial differentiation with respect to the design variable k_j. Since the matrix $\mathbf{A}$ is regular, the first-order sensitivities of the complex-value quantities $\hat{U}_1$ and $\hat{U}_2$ are derived from Equation 2.7 as

$$\left\{ \begin{matrix} \hat{U}_{1,j} \\ \hat{U}_{2,j} \end{matrix} \right\} = -\mathbf{A}^{-1}\mathbf{A}_{,j} \left\{ \begin{matrix} \hat{U}_1 \\ \hat{U}_2 \end{matrix} \right\} \tag{2.8}$$

The first-order sensitivities of the complex-value quantities $\hat{\delta}_1$ and $\hat{\delta}_2$ are derived by substituting the relations $\hat{\delta}_1 \equiv \hat{U}_1$ and $\hat{\delta}_2 \equiv \hat{U}_2 - \hat{U}_1$ into Equation 2.8 as

$$\left\{ \begin{matrix} \hat{\delta}_{1,j} \\ \hat{\delta}_{2,j} \end{matrix} \right\} = - \begin{bmatrix} 1 & 0 \\ -1 & 1 \end{bmatrix} \mathbf{A}^{-1}\mathbf{A}_{,j} \begin{bmatrix} 1 & 0 \\ 1 & 1 \end{bmatrix} \left\{ \begin{matrix} \hat{\delta}_1 \\ \hat{\delta}_2 \end{matrix} \right\} \tag{2.9}$$

The complex-value quantity $\hat{\delta}_i$ may be rewritten symbolically as

$$\hat{\delta}_i = \mathrm{Re}[\hat{\delta}_i] + \mathrm{i}\,\mathrm{Im}[\hat{\delta}_i] \tag{2.10}$$

where the symbols $\mathrm{Re}[\ ]$ and $\mathrm{Im}[\ ]$ indicate the real and imaginary parts respectively of a complex number. The first-order sensitivity of the quantity $\hat{\delta}_i$ may be formally or symbolically expressed as

$$\hat{\delta}_{i,j} = (\mathrm{Re}[\hat{\delta}_i])_{,j} + i(\mathrm{Im}[\hat{\delta}_i])_{,j} \tag{2.11}$$

The absolute value of the complex-value quantity $\hat{\delta}_i$ is defined by

$$|\hat{\delta}_i| = \sqrt{(\mathrm{Re}[\hat{\delta}_i])^2 + (\mathrm{Im}[\hat{\delta}_i])^2} \tag{2.12}$$

The first-order sensitivity of the complex amplitude $|\hat{\delta}_i|$ may be expressed as

$$|\hat{\delta}_i|_{,j} = \frac{1}{|\hat{\delta}_i|} \{ \mathrm{Re}[\hat{\delta}_i](\mathrm{Re}[\hat{\delta}_i])_{,j} + \mathrm{Im}[\hat{\delta}_i](\mathrm{Im}[\hat{\delta}_i])_{,j} \} \tag{2.13}$$

where $(\mathrm{Re}[\hat{\delta}_i])_{,j}$ and $(\mathrm{Im}[\hat{\delta}_i])_{,j}$ can be calculated from Equation 2.9.

Let us define a new quantity as the ratio between two interstory drifts defined by

$$\alpha_1(\mathbf{k}) = \frac{|\hat{\delta}_2(\mathbf{k})|}{|\hat{\delta}_1(\mathbf{k})|} \tag{2.14}$$

Linear increments $\Delta\Omega_1(\mathbf{k})$ and $\Delta\alpha_1(\mathbf{k})$ of the lowest eigenvalue $\Omega_1(\mathbf{k})$ (square of the undamped fundamental natural circular frequency $\omega_1(\mathbf{k})$) and $\alpha_1(\mathbf{k})$ in Equation 2.14 may be described as follows:

$$\Delta\Omega_1(\mathbf{k}) = \frac{\partial\Omega_1(\mathbf{k})}{\partial\mathbf{k}}\Delta\mathbf{k} \tag{2.15a}$$

$$\Delta\alpha_1(\mathbf{k}) = \left(\frac{\partial|\hat{\delta}_2(\mathbf{k})|}{\partial\mathbf{k}}\frac{1}{|\hat{\delta}_1(\mathbf{k})|} - \frac{\partial|\hat{\delta}_1(\mathbf{k})|}{\partial\mathbf{k}}\frac{|\hat{\delta}_2(\mathbf{k})|}{|\hat{\delta}_1(\mathbf{k})|^2}\right)\Delta\mathbf{k}$$

$$= \frac{1}{|\hat{\delta}_1(\mathbf{k})|}\left(\frac{\partial|\hat{\delta}_2(\mathbf{k})|}{\partial\mathbf{k}} - \frac{\partial|\hat{\delta}_1(\mathbf{k})|}{\partial\mathbf{k}}\alpha_1(\mathbf{k})\right)\Delta\mathbf{k} \tag{2.15b}$$

Equations 2.15a and 2.15b may be arranged to the following set of simultaneous linear equations with respect to $\Delta\mathbf{k}$:

$$\begin{bmatrix} \dfrac{\partial\Omega_1(\mathbf{k})}{\partial\mathbf{k}} \\[2ex] \dfrac{\partial|\hat{\delta}_2(\mathbf{k})|}{\partial\mathbf{k}} - \left(\dfrac{\partial|\hat{\delta}_1(\mathbf{k})|}{\partial\mathbf{k}}\right)\alpha_1(\mathbf{k}) \end{bmatrix}\Delta\mathbf{k} = \left\{\begin{array}{c} \Delta\Omega_1(\mathbf{k}) \\[1ex] |\hat{\delta}_1(\mathbf{k})|\Delta\alpha_1(\mathbf{k}) \end{array}\right\} \tag{2.16}$$

Equation 2.16 implies that, once the increments $\Delta\Omega_1(\mathbf{k})$ and $\Delta\alpha_1(\mathbf{k})$ are given and the sensitivities $\partial\Omega_1(\mathbf{k})/\partial\mathbf{k}$ and $\partial|\hat{\delta}_j(\mathbf{k})|/\partial\mathbf{k}$ of these increments are evaluated, the increment $\Delta\mathbf{k}$ of stiffness to be determined can be found. The concept of this procedure can be found in Figure 2.2.

The first-order sensitivity of the lowest eigenvalue $\Omega_1(\mathbf{k})$ is well-known (Fox and Kapoor, 1968) in the field of structural optimization and is expressed as

$$\Omega_1(\mathbf{k})_{,j} = \mathbf{V}^{(1)\mathrm{T}}\mathbf{K}_{,j}\mathbf{V}^{(1)} \tag{2.17}$$

In Equation 2.17, $\mathbf{V}^{(1)}$ is the undamped lowest eigenvector of the present model and is to be normalized by

$$\mathbf{V}^{(1)\mathrm{T}}\mathbf{M}\mathbf{V}^{(1)} = 1 \tag{2.18}$$

In Equation 2.18, $\mathbf{M} = \mathrm{diag}(m_1\ m_2)$. Differentiation of the definition $\omega_1(\mathbf{k})^2 = \Omega_1(\mathbf{k})$ of the lowest eigenvalue with respect to the design variable k_j leads to the following expression of the first-order sensitivity of the undamped fundamental natural circular frequency $\omega_1(\mathbf{k})$:

$$\omega_1(\mathbf{k})_{,j} = \frac{1}{2\omega_1(\mathbf{k})}\mathbf{V}^{(1)\mathrm{T}}\mathbf{K}_{,j}\mathbf{V}^{(1)} \tag{2.19}$$

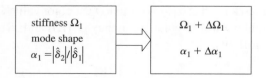

Figure 2.2 Concept of incremental inverse problem.

If the damping system considered is hysteretic as shown in Figure 2.1(b), then the coefficient matrix defined by Equation 2.6 may be modified to the following expression:

$$\mathbf{A_H} = \begin{bmatrix} k_1(1 + 2\overline{\beta}_1 i) + k_2(1 + 2\overline{\beta}_2 i) - \omega_1^2 \overline{m}_1 & -k_2(1 + 2\overline{\beta}_2 i) \\ -k_2(1 + 2\overline{\beta}_2 i) & k_2(1 + 2\overline{\beta}_2 i) - \omega_1^2 \overline{m}_2 \end{bmatrix} \quad (2.20)$$

where $\overline{\beta}_1$ and $\overline{\beta}_2$ are the prescribed hysteretic damping factors.

A technique on how to predetermine the quantities $\Delta\Omega_1(\mathbf{k})$ and $\Delta\alpha_1(\mathbf{k})$ will be shown in the following section.

2.3 Incremental Inverse Problem: General Formulation

For the general treatment of incremental inverse problems described above, consider an n-DOF FE model. The model is not restricted to mass–spring–dashpot models. The design variables in the present model are $\{k_1 \cdots k_m\}$, where m denotes the number of the design variables. After some manipulation of mass, stiffness, and damping matrices, Equation 2.2 may be generalized into the following form:

$$(\mathbf{K} + i\omega\mathbf{C} - \omega^2\mathbf{M})\mathbf{U}(\omega) = -\mathbf{M}\mathbf{r}\ddot{U}_g(\omega) \quad (2.21)$$

where $\mathbf{K}, \mathbf{C}$, and $\mathbf{M}$ are the system stiffness, damping, and mass matrices respectively and $\mathbf{r}$ in the right-hand side of the equation is the influence coefficient vector for a base acceleration input. Furthermore, $\mathbf{U}(\omega) = \{U_1(\omega) \cdots U_n(\omega)\}^T$ represents the nodal displacements in the frequency domain.

By applying a procedure similar to that in the previous section, Equation 2.5 may be generalized into

$$\mathbf{A}\hat{\mathbf{U}} = -\mathbf{M}\mathbf{r} \quad (2.22)$$

where $\hat{\mathbf{U}} = \{\hat{U}_1 \cdots \hat{U}_n\}^T$ are the displacements evaluated at the undamped fundamental natural circular frequency ω_1; that is, $\hat{U}_j = U_j(\omega_1)/\ddot{U}_g(\omega_1)$. The coefficient matrix $\mathbf{A}$ is defined by

$$\mathbf{A} = \mathbf{K} + i\omega_1\mathbf{C} - \omega_1^2\mathbf{M} \quad (2.23)$$

The Fourier transforms $\hat{\boldsymbol{\delta}} = \{\hat{\delta}_1 \cdots \hat{\delta}_m\}^T$ of the deformation quantities evaluated at the undamped fundamental natural circular frequency are assumed to be derived from $\hat{\mathbf{U}}$ by

$$\hat{\boldsymbol{\delta}} = \mathbf{T}\hat{\mathbf{U}} \quad (2.24)$$

Here, $\mathbf{T}$ indicates the deformation–displacement transformation matrix. Depending on whether $\mathbf{T}$ is a square matrix or not, $\hat{\boldsymbol{\delta}}$ has to be chosen carefully. The procedure on how to treat a square matrix $\mathbf{T}$ is discussed by Takewaki and Nakamura (1995). By applying a procedure similar to that in the previous section, Equation 2.7 may be generalized into

$$\mathbf{A}_{,j}\hat{\mathbf{U}} + \mathbf{A}\hat{\mathbf{U}}_{,j} = \mathbf{0} \quad (2.25)$$

From Equation 2.25, $\hat{\mathbf{U}}_{,j}$ can be expressed in terms of the displacement quantities $\hat{\mathbf{U}}$:

$$\hat{\mathbf{U}}_{,j} = -\mathbf{A}^{-1}\mathbf{A}_{,j}\hat{\mathbf{U}} \tag{2.26}$$

An explicit expression of $\mathbf{A}^{-1}$ can be derived for a mass–spring model having a tri-diagonal coefficient matrix $\mathbf{A}$ (Takewaki and Nakamura, 1995). However, in the models without this property, numerical manipulation is required. From Equations 2.24 and 2.26, the sensitivity $\hat{\boldsymbol{\delta}}_{,j}$ of deformation is expressed in terms of the displacement quantities $\hat{\mathbf{U}}$:

$$\hat{\boldsymbol{\delta}}_{,j} = -\mathbf{T}\mathbf{A}^{-1}\mathbf{A}_{,j}\hat{\mathbf{U}} \tag{2.27}$$

The linear increment of the absolute value $|\hat{\delta}_k(\mathbf{k})|$ of deformation may be expressed as

$$\Delta|\hat{\delta}_k(\mathbf{k})| = \frac{\partial|\hat{\delta}_k(\mathbf{k})|}{\partial\mathbf{k}}\Delta\mathbf{k} \quad (k = 1, \cdots, m) \tag{2.28}$$

A new quantity $\alpha_k(\mathbf{k})$ as a ratio of deformations is defined by

$$\alpha_k(\mathbf{k}) = \frac{|\hat{\delta}_{k+1}(\mathbf{k})|}{|\hat{\delta}_1(\mathbf{k})|} \quad (k = 1, \cdots, m-1) \tag{2.29}$$

This quantity will be referred to as the transfer function ratio. The linear increment of this ratio $\alpha_k(\mathbf{k})$ may be expressed as

$$\Delta\alpha_k(\mathbf{k}) = \frac{\Delta|\hat{\delta}_{k+1}(\mathbf{k})|}{|\hat{\delta}_1(\mathbf{k})|} - \frac{|\hat{\delta}_{k+1}(\mathbf{k})|}{|\hat{\delta}_1(\mathbf{k})|^2}\Delta|\hat{\delta}_1(\mathbf{k})| = \frac{\Delta|\hat{\delta}_{k+1}(\mathbf{k})|}{|\hat{\delta}_1(\mathbf{k})|} - \frac{\alpha_k(\mathbf{k})}{|\hat{\delta}_1(\mathbf{k})|}\Delta|\hat{\delta}_1(\mathbf{k})| \tag{2.30}$$

The quantity $\Delta|\hat{\delta}_1(\mathbf{k})|$ in Equation 2.30 is expressed in terms of $\Delta\mathbf{k}$ in Equation 2.28. Then, substitution of Equation 2.28 into Equation 2.30 provides the following set of simultaneous linear equations of $\Delta\mathbf{k}$:

$$\left\{\frac{\partial|\hat{\delta}_{k+1}(\mathbf{k})|}{\partial\mathbf{k}} - \frac{\partial|\hat{\delta}_1(\mathbf{k})|}{\partial\mathbf{k}}\alpha_k(\mathbf{k})\right\}\Delta\mathbf{k} = |\hat{\delta}_1(\mathbf{k})|\Delta\alpha_k(\mathbf{k}) \quad (k = 1, \cdots, m-1) \tag{2.31}$$

Combination of Equation 2.31 with the definition $\{\partial\Omega_1(\mathbf{k})/\partial\mathbf{k}\}\Delta\mathbf{k} = \Delta\Omega_1(\mathbf{k})$ of the linear increment of the lowest eigenvalue yields the following set of simultaneous linear equations with respect to $\Delta\mathbf{k}$:

$$\begin{bmatrix} \dfrac{\partial\Omega_1(\mathbf{k})}{\partial\mathbf{k}} \\ \dfrac{\partial|\hat{\delta}_2(\mathbf{k})|}{\partial\mathbf{k}} - \dfrac{\partial|\hat{\delta}_1(\mathbf{k})|}{\partial\mathbf{k}}\alpha_1(\mathbf{k}) \\ \vdots \\ \dfrac{\partial|\hat{\delta}_m(\mathbf{k})|}{\partial\mathbf{k}} - \dfrac{\partial|\hat{\delta}_1(\mathbf{k})|}{\partial\mathbf{k}}\alpha_{m-1}(\mathbf{k}) \end{bmatrix}\Delta\mathbf{k} = \left\{\begin{array}{c} \Delta\Omega_1(\mathbf{k}) \\ |\hat{\delta}_1(\mathbf{k})|\Delta\alpha_1(\mathbf{k}) \\ \vdots \\ |\hat{\delta}_1(\mathbf{k})|\Delta\alpha_{m-1}(\mathbf{k}) \end{array}\right\} \tag{2.32}$$

The first-order sensitivity of the absolute value $|\hat{\delta}_i(\mathbf{k})|$ of deformation is derived from Equations 2.13 and 2.27.

Let α_0 and α_F denote the initial values of the transfer function ratios and the corresponding target values. The linear increments of α to be specified in this formulation are given as follows:

$$\Delta\alpha = \frac{1}{N}(\alpha_F - \alpha_0) \tag{2.33}$$

As for the lowest eigenvalue, let $\Omega_{1(0)}$ and Ω_{1F} denote the initial value of the lowest eigenvalue of the undamped model and the corresponding target value. The linear increment of Ω_1 to be specified in this formulation is given as

$$\Delta\Omega_1 = \frac{1}{N}(\Omega_{1F} - \Omega_{1(0)}) \tag{2.34}$$

Since Equation 2.32 constitutes a set of simultaneous linear equations with respect to the stiffness increments $\Delta\mathbf{k}$, regular repetitive application of Equation 2.32 in the range (α_0, α_F) of the transfer function ratios and $(\Omega_{1(0)}, \Omega_{1F})$ of the lowest eigenvalue of the undamped model will provide a definite target design.

2.4 Numerical Examples I

2.4.1 Viscous Damping Model

A viscous damping model is discussed first. Consider the two-DOF mass–spring–dashpot model shown in Figure 2.1(a). In the models to be considered in three examples, the terminology "story" is used to indicate the position between two adjacent masses. The masses and viscous damping coefficients of dashpots are prescribed as $\bar{m}_1 = 1.0 \times 10^5$ kg, $\bar{m}_2 = 0.8 \times 10^5$ kg, and $\bar{c}_1 = \bar{c}_2 = 2.0 \times 10^5$ N s/m. The design variables in this case are the spring stiffnesses k_1 and k_2. The partial derivatives of the coefficient matrix $\mathbf{A}$ defined in Equation 2.6 with respect to design variables, k_1 and k_2, are expressed as

$$\mathbf{A}_{,1} = \begin{bmatrix} 1 & 0 \\ 0 & 0 \end{bmatrix} + i\omega_{1,1}\begin{bmatrix} \bar{c}_1 + \bar{c}_2 & -\bar{c}_2 \\ -\bar{c}_2 & \bar{c}_2 \end{bmatrix} - \Omega_{1,1}\begin{bmatrix} \bar{m}_1 & 0 \\ 0 & \bar{m}_2 \end{bmatrix} \tag{2.35a}$$

$$\mathbf{A}_{,2} = \begin{bmatrix} 1 & -1 \\ -1 & 1 \end{bmatrix} + i\omega_{1,2}\begin{bmatrix} \bar{c}_1 + \bar{c}_2 & -\bar{c}_2 \\ -\bar{c}_2 & \bar{c}_2 \end{bmatrix} - \Omega_{1,2}\begin{bmatrix} \bar{m}_1 & 0 \\ 0 & \bar{m}_2 \end{bmatrix} \tag{2.35b}$$

The initial values of spring stiffnesses are taken as $k_1 = k_2 = 4.0 \times 10^7$ N/m. In addition, the initial lowest eigenvalue of the undamped model is taken as $\Omega_{1(0)} = 178$ rad^2/s^2 and the initial value for the transfer function ratio is taken as $\alpha_{01} = 0.554$. These values have been obtained from the initial values of spring stiffnesses. The target values of the lowest eigenvalue and the transfer function ratio are specified as $\Omega_{1F} = 178$ rad^2/s^2 and $\alpha_{F1} = 1.0$. The increment $\Delta\Omega_1$ of the lowest eigenvalue of the undamped model is specified as zero in this example. The number of steps in the redesign process is prescribed as $N = 50$. In order to verify the accuracy of this procedure for redesign, the eigenvalue analysis has been done and the transfer function ratio has been calculated from Equation 2.5 for the model with the final design variables. The maximum discrepancies (differences) of the lowest eigenvalue and the transfer function ratio from the target values turned out to be 0.12% and 0.37% respectively. This supports the validity of the present procedure for redesign.

Figure 2.3(a) shows the amplitudes of transfer functions defined by $\delta_1(\omega)/\ddot{U}_g(\omega)$ and $\delta_2(\omega)/\ddot{U}_g(\omega)$ for the initial design and Figure 2.3(b) presents those for the target

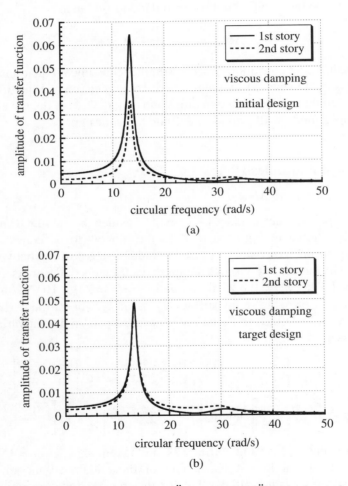

(a)

(b)

Figure 2.3 Amplitudes of transfer functions $\delta_1(\omega)/\ddot{U}_g(\omega)$ and $\delta_2(\omega)/\ddot{U}_g(\omega)$ in viscous damper model: (a) initial design; (b) target design. (Originally published in I. Takewaki, "Efficient redesign of damped structural systems for target transfer functions," *Computer Methods in Applied Mechanics and Engineering*, **147**, no. 3–4, 275–286, 1997, Elsevier B.V.).

design. It can be seen from Figure 2.3(b) that the amplitudes of transfer functions $\delta_1(\omega)/\ddot{U}_g(\omega)$ and $\delta_2(\omega)/\ddot{U}_g(\omega)$ at the undamped fundamental natural circular frequency (13.34 rad/s) in the target design attain almost the same value by the constraint $\alpha_{F1} = 1.0$. This clearly shows the validity of the present procedure.

2.4.2 Hysteretic Damping Model

A hysteretic damping model may be another representative model of damping. Consider the two-DOF mass–spring model with hysteretic damping systems shown in Figure 2.1(b). The nodal masses are prescribed as $\overline{m}_1 = 1.0 \times 10^5$ kg and $\overline{m}_2 = 0.8 \times 10^5$ kg. The coefficients of the hysteretic damping system are prescribed as $\overline{\beta}_1 = \overline{\beta}_2 = 0.02$. The hysteretic damping coefficients can be introduced by replacing $k_1 + i\omega\overline{c}_1$ and $k_2 + i\omega\overline{c}_2$ in Equation 2.2 by $k_1(1 + 2\overline{\beta}_1 i)$ and $k_2(1 + 2\overline{\beta}_2 i)$ respectively. The design variables in this case are the spring stiffnesses k_1 and k_2 in this example also. The partial derivatives of the coefficient matrix $\mathbf{A}$, derived by replacing $k_1 + i\omega\overline{c}_1$ and $k_2 + i\omega\overline{c}_2$ in Equation 2.6 by $k_1(1 + 2\overline{\beta}_1 i)$ and $k_2(1 + 2\overline{\beta}_2 i)$, with respect to design variables can be expressed as

$$\mathbf{A}_{,1} = \begin{bmatrix} 1 & 0 \\ 0 & 0 \end{bmatrix} + i \begin{bmatrix} 2\overline{\beta}_1 & 0 \\ 0 & 0 \end{bmatrix} - \Omega_{1,1} \begin{bmatrix} \overline{m}_1 & 0 \\ 0 & \overline{m}_2 \end{bmatrix} \tag{2.36a}$$

$$\mathbf{A}_{,2} = \begin{bmatrix} 1 & -1 \\ -1 & 1 \end{bmatrix} + i \begin{bmatrix} 2\overline{\beta}_2 & -2\overline{\beta}_2 \\ -2\overline{\beta}_2 & 2\overline{\beta}_2 \end{bmatrix} - \Omega_{1,2} \begin{bmatrix} \overline{m}_1 & 0 \\ 0 & \overline{m}_2 \end{bmatrix} \tag{2.36b}$$

The initial values of spring stiffnesses are taken as $k_1 = k_2 = 4.0 \times 10^7$ N/m. The initial lowest eigenvalue of the undamped model is taken as $\Omega_{1(0)} = 178$ rad^2/s^2 and the initial value for the transfer function ratio is taken as $\alpha_{01} = 0.554$. These are the same as in the previous example for viscous damping. The targets of the lowest eigenvalue and the transfer function ratio are specified as $\Omega_{1F} = 178$ rad^2/s^2 and $\alpha_{F1} = 1.0$. The number of steps in the redesign process is assumed to be $N = 50$. An accuracy check demonstrates that the maximum discrepancies (differences) of the lowest eigenvalue and the transfer function ratio from the target values are 0.12% and 0.37% respectively in this example also. This supports the validity of the present procedure for redesign.

Figure 2.4(a) presents the amplitudes of transfer functions defined by $\delta_1(\omega)/\ddot{U}_g(\omega)$ and $\delta_2(\omega)/\ddot{U}_g(\omega)$ for the initial design and Figure 2.4(b) shows those for the target design. It can be seen that the amplitudes of transfer functions $\delta_1(\omega)/\ddot{U}_g(\omega)$ and $\delta_2(\omega)/\ddot{U}_g(\omega)$ at the undamped fundamental natural circular frequency in the target design attain almost the same value by the constraint $\alpha_{F1} = 1.0$. This shows the validity of the present procedure. Furthermore, it can be observed that the overall configuration of the amplitude of the transfer function is different from that for the model with viscous damping systems and that the relative value of the amplitude of the transfer function at the undamped second natural circular frequency to that at

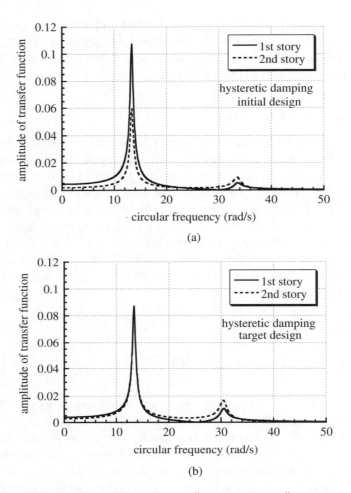

Figure 2.4 Amplitudes of transfer functions $\delta_1(\omega)/\ddot{U}_g(\omega)$ and $\delta_2(\omega)/\ddot{U}_g(\omega)$ in hysteretic damper model: (a) initial design; (b) target design. (Originally published in I. Takewaki, "Efficient redesign of damped structural systems for target transfer functions," *Computer Methods in Applied Mechanics and Engineering*, **147**, no. 3–4, 275–286, 1997, Elsevier B.V.).

the undamped fundamental natural frequency is larger than that for the model with viscous damping systems. This indicates that damping effects in higher modes do not decay rapidly in a model with hysteretic damping systems compared with a model with viscous damping systems.

2.4.3 Six-DOF Models with Various Possibilities of Damper Placement

As an example of MDOF models, consider the six-DOF model shown in Figure 2.5. All the nodal masses are assumed to be prescribed as $\overline{m}_1 = \cdots = \overline{m}_6 = 0.8 \times 10^5$ kg.

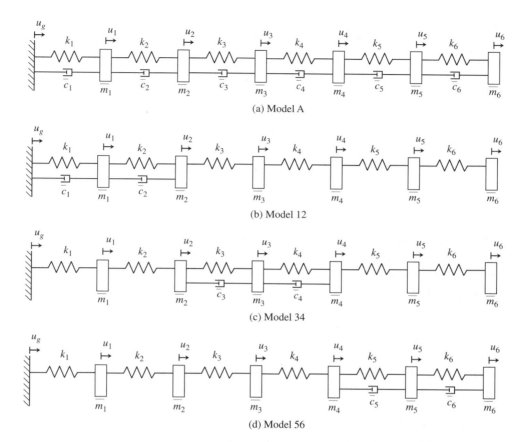

Figure 2.5 Six-DOF model with various damper placements. (Originally published in I. Takewaki, "Efficient redesign of damped structural systems for target transfer functions," *Computer Methods in Applied Mechanics and Engineering*, **147**, no. 3–4, 275–286, 1997, Elsevier B.V.).

Four different models, called Model A, Model 12, Model 34, and Model 56, are considered with various possibilities of damper placement. Every viscous damper is to have the same prescribed value of damping coefficient: 1.5×10^6 N s/m. Model A has the viscous dampers in all stories; that is, $\bar{c}_1 = \cdots = \bar{c}_6 = 1.5 \times 10^6$ N s/m. On the other hand, Model 12 has the viscous dampers only in the first and second stories; that is, $\bar{c}_1 = \bar{c}_2 = 1.5 \times 10^6$ N s/m. As another model, Model 34 has the viscous dampers only in the third and fourth stories; that is, $\bar{c}_3 = \bar{c}_4 = 1.5 \times 10^6$ N s/m. Finally, Model 56 has the viscous dampers only in the fifth and sixth stories; that is, $\bar{c}_5 = \bar{c}_6 = 1.5 \times 10^6$ N s/m.

The design variables are now the spring stiffnesses $k_1, \ldots, k_6$. The initial values of spring stiffnesses are taken as $k_1 = \cdots = k_6 = 4.0 \times 10^7$ N/m. The initial value of the lowest eigenvalue of the undamped model is taken as $\Omega_{1(0)} = 29.1$ rad^2/s^2 and the

Table 2.1 Initial values of transfer function ratios $\{\alpha_j\}$ for Models A, 12, 34, and 56. (Originally published in I. Takewaki, "Efficient redesign of damped structural systems for target transfer functions," *Computer Methods in Applied Mechanics and Engineering*, **147**, no. 3–4, 275–286, 1997, Elsevier B.V.).

	Model A	Model 12	Model 34	Model 56
α_{01}	0.936	0.940	0.939	0.941
α_{02}	0.820	0.844	0.807	0.828
α_{03}	0.659	0.680	0.649	0.666
α_{04}	0.461	0.477	0.464	0.457
α_{05}	0.237	0.245	0.239	0.235

initial values of the transfer function ratios for Model A, Model 12, Model 34, and Model 56 are given in Table 2.1. These values have been obtained from the initial values of spring stiffnesses. The targets of the lowest eigenvalue and the transfer function ratios are specified as $\Omega_{1F} = 29.1\,\text{rad}^2/\text{s}^2$ and $\alpha_{F1} = \cdots = \alpha_{F5} = 1.0$. The same target value of the lowest eigenvalue enables one to compare four models with the same overall stiffness. The number of steps in the redesign process is assumed to be $N = 50$.

Figure 2.6 presents the spring stiffness distributions of the target designs obtained for Model A, Model 12, Model 34, and Model 56. It can be observed and reasoned that the spring stiffnesses in the lower stories in Model 12 become smaller than those in Model A due to the existence of viscous dampers only in the lower stories and the spring stiffnesses in the upper stories become larger in compensation for those. Figure 2.7 shows the variation of spring stiffnesses in Model A. It is further understood that the variation of the spring stiffnesses in the upper stories exhibits higher nonlinearities with respect to the step number. Figure 2.8(a) indicates the amplitudes of the transfer functions of the target design for Model A. It can be confirmed that the amplitude of the transfer function in every story at the undamped fundamental natural circular frequency attains almost the same value as desired and predicted. Figure 2.8(b) shows the amplitudes of the transfer functions of the target design obtained for Model 12. Figure 2.9 presents the amplitudes of the transfer functions of the target designs obtained for Model 34 and Model 56. It can be observed that, while the amplitudes of the transfer functions in all stories at the undamped fundamental natural circular frequency in Models 12, 34, and 56 attain almost the same value, the level is almost three times that for Model A. This fact just corresponds to the fact that the damping ratios in the lowest eigenvibration of Models 12, 34, and 56 are almost one-third that of Model A (see Table 2.2) and that the amplitude of a transfer function at the resonance frequency is inversely proportional to the damping ratio in a single-DOF (SDOF) model. Furthermore, it can be observed that the damping effects in higher modes (i.e., the amplitudes of transfer functions at the higher natural frequencies) are quite sensitive to damper placement.

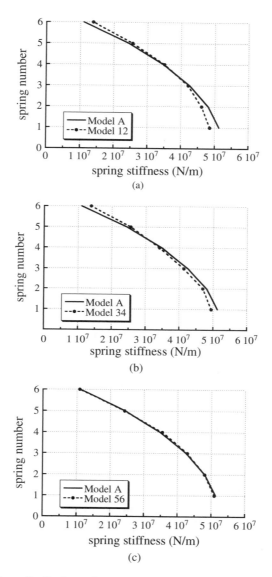

Figure 2.6 Spring stiffness distributions of target designs for Model A, Model 12, Model 34, and Model 56. (Originally published in I. Takewaki, "Efficient redesign of damped structural systems for target transfer functions," *Computer Methods in Applied Mechanics and Engineering*, **147**, no. 3–4, 275–286, 1997, Elsevier B.V.).

2.5 Optimality Criteria-based Design of Dampers: Simple Example

Consider, as an illustrative example, the two-story shear building model with added viscous dampers shown in Figure 2.10. Let $\{\overline{m}_1, \overline{m}_2\}$ and $\{\overline{k}_1, \overline{k}_2\}$ denote the masses

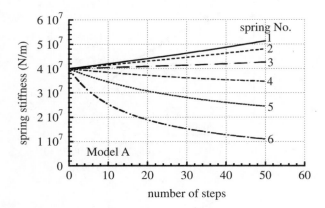

Figure 2.7 Variation of spring stiffnesses in Model A. (Originally published in I. Takewaki, "Efficient redesign of damped structural systems for target transfer functions," *Computer Methods in Applied Mechanics and Engineering*, **147**, no. 3–4, 275–286, 1997, Elsevier B.V.).

and story stiffnesses respectively of the shear building model and let $\{c_1, c_2\}$ denote the damping coefficients of the added dampers. Assume that $\{\overline{m}_1, \overline{m}_2\}$ and $\{\overline{k}_1, \overline{k}_2\}$ are prescribed. The design variables are the damping coefficients $\mathbf{c} = \{c_1, c_2\}$ of added dampers. It is also assumed here that the original structural damping is negligible compared with the damping of the added dampers. This assumption can be removed, if necessary, without difficulty. Let u_1 and u_2 denote the displacements of masses $\overline{m}_1$ and $\overline{m}_2$ respectively. When this model is subjected to a horizontal base acceleration $\ddot{u}_g$, the equations of motion for this model may be written as

$$\begin{bmatrix} \overline{k}_1 + \overline{k}_2 & -\overline{k}_2 \\ -\overline{k}_2 & \overline{k}_2 \end{bmatrix} \begin{Bmatrix} u_1 \\ u_2 \end{Bmatrix} + \begin{bmatrix} c_1 + c_2 & -c_2 \\ -c_2 & c_2 \end{bmatrix} \begin{Bmatrix} \dot{u}_1 \\ \dot{u}_2 \end{Bmatrix} + \begin{bmatrix} \overline{m}_1 & 0 \\ 0 & \overline{m}_2 \end{bmatrix} \begin{Bmatrix} \ddot{u}_1 \\ \ddot{u}_2 \end{Bmatrix}$$

$$= -\begin{bmatrix} \overline{m}_1 & 0 \\ 0 & \overline{m}_2 \end{bmatrix} \begin{Bmatrix} 1 \\ 1 \end{Bmatrix} \ddot{u}_g \tag{2.37}$$

Let $U_1(\omega)$, $U_2(\omega)$, and $\ddot{U}_g(\omega)$ denote the Fourier transforms of u_1, u_2, and $\ddot{u}_g$ respectively, and let ω denote a circular frequency of excitation. Fourier transformation of Equation 2.37 may be reduced to the following form:

$$\left(\begin{bmatrix} \overline{k}_1 + \overline{k}_2 & -\overline{k}_2 \\ -\overline{k}_2 & \overline{k}_2 \end{bmatrix} + i\omega \begin{bmatrix} c_1 + c_2 & -c_2 \\ -c_2 & c_2 \end{bmatrix} - \omega^2 \begin{bmatrix} \overline{m}_1 & 0 \\ 0 & \overline{m}_2 \end{bmatrix} \right) \begin{Bmatrix} U_1(\omega) \\ U_2(\omega) \end{Bmatrix}$$

$$= -\begin{bmatrix} \overline{m}_1 & 0 \\ 0 & \overline{m}_2 \end{bmatrix} \begin{Bmatrix} 1 \\ 1 \end{Bmatrix} \ddot{U}_g(\omega) \tag{2.38}$$

where i is the imaginary unit. The Fourier transforms $\delta_1(\omega)$ and $\delta_2(\omega)$ of the interstory drifts $d_1 = u_1$ and $d_2 = u_2 - u_1$ are related to $U_1(\omega)$ and $U_2(\omega)$ by

$$\begin{Bmatrix} \delta_1(\omega) \\ \delta_2(\omega) \end{Bmatrix} = \mathbf{T} \begin{Bmatrix} U_1(\omega) \\ U_2(\omega) \end{Bmatrix} \tag{2.39}$$

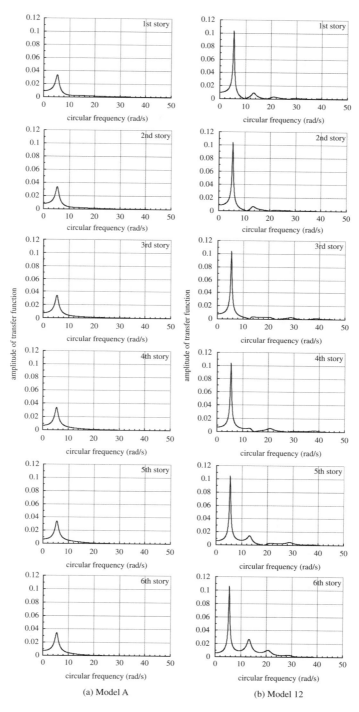

Figure 2.8 Amplitudes of transfer functions in target design: (a) Model A; (b) Model 12. (Originally published in I. Takewaki, "Efficient redesign of damped structural systems for target transfer functions," *Computer Methods in Applied Mechanics and Engineering*, **147**, no. 3–4, 275–286, 1997, Elsevier B.V.).

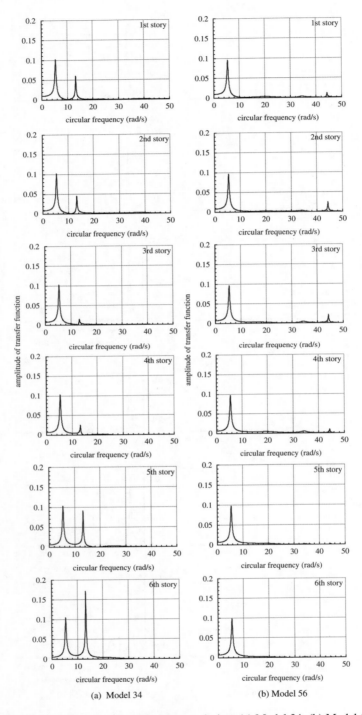

Figure 2.9 Amplitudes of transfer functions in target design: (a) Model 34; (b) Model 56. (Originally published in I. Takewaki, "Efficient redesign of damped structural systems for target transfer functions," *Computer Methods in Applied Mechanics and Engineering*, **147**, no. 3–4, 275–286, 1997, Elsevier B.V.).

Table 2.2 Fundamental natural circular frequencies and damping ratios in the lowest eigenvibration due to complex eigenvalue analysis of target designs for Models A, 12, 34, and 56. (Originally published in I. Takewaki, "Efficient redesign of damped structural systems for target transfer functions," *Computer Methods in Applied Mechanics and Engineering*, **147**, no. 3–4, 275–286, 1997, Elsevier B.V.).

	Model A	Model 12	Model 34	Model 56
ω_1^*(rad/s)	5.45	5.41	5.42	5.51
$h^{(1)}$	0.120	0.0384	0.0390	0.0422

ω_1^*: fundamental natural circular frequency due to complex eigenvalue analysis; $h^{(1)}$: damping ratios in the lowest eigenvibration due to complex eigenvalue analysis.

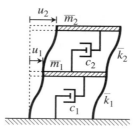

Figure 2.10 Two-story shear building model with supplemental viscous dampers. (I. Takewaki, "Optimal Damper Placement for Minimum Transfer Functions," *Earthquake Engineering and Structural Dynamics*, Vol.26, No.11. © 1997 John Wiley & Sons, Ltd).

where

$$\mathbf{T} = \begin{bmatrix} 1 & 0 \\ -1 & 1 \end{bmatrix} \tag{2.40}$$

Let ω_1 denote the undamped fundamental natural circular frequency of the model and let us define new quantities $\hat{U}_1$ and $\hat{U}_2$ by

$$\hat{U}_1 \equiv \frac{U_1(\omega_1)}{\ddot{U}_g(\omega_1)} \quad \text{and} \quad \hat{U}_2 \equiv \frac{U_2(\omega_1)}{\ddot{U}_g(\omega_1)} \tag{2.41}$$

As stated in Equation 2.4, $\hat{U}_i$ indicates the value such that ω_1 is substituted in the frequency response function obtained as $U_i(\omega)$ after substituting $\ddot{U}_g(\omega) = 1$ in Equation 2.38. Because $\{\bar{m}_1, \bar{m}_2\}$ and $\{\bar{k}_1, \bar{k}_2\}$ are prescribed, ω_1 is a given value. It may be convenient to introduce new quantities $\hat{\delta}_1$ and $\hat{\delta}_2$ defined by $\hat{\delta}_1 \equiv \hat{U}_1$ and $\hat{\delta}_2 \equiv \hat{U}_2 - \hat{U}_1$ (see Figure 2.11). Judging from Equation 2.38, after substituting $\omega = \omega_1$ and Equation 2.41, $\hat{U}_1$ and $\hat{U}_2$ must satisfy the following equation:

$$\mathbf{A} \left\{ \begin{matrix} \hat{U}_1 \\ \hat{U}_2 \end{matrix} \right\} = - \left\{ \begin{matrix} \bar{m}_1 \\ \bar{m}_2 \end{matrix} \right\} \tag{2.42}$$

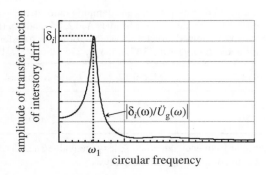

Figure 2.11 Amplitudes of new quantities $\hat{\delta}_1$ and $\hat{\delta}_2$ as transfer functions of interstory drifts defined by $\hat{\delta}_1 \equiv \hat{U}_1$ and $\hat{\delta}_2 \equiv \hat{U}_2 - \hat{U}_1$ ($\hat{U}_1$, $\hat{U}_2$: transfer functions of horizontal displacements). (I. Takewaki, "Optimal Damper Placement for Minimum Transfer Functions," *Earthquake Engineering and Structural Dynamics*, Vol.26, No.11. © 1997 John Wiley & Sons, Ltd).

where

$$\mathbf{A} = \begin{bmatrix} (\overline{k}_1 + \overline{k}_2) + i\omega_1(c_1 + c_2) - \omega_1^2 \overline{m}_1 & -\overline{k}_2 - i\omega_1 c_2 \\ -\overline{k}_2 - i\omega_1 c_2 & \overline{k}_2 + i\omega_1 c_2 - \omega_1^2 \overline{m}_2 \end{bmatrix} \quad (2.43)$$

It should be remarked here that the squares of the amplitudes of transfer functions are useful from physical points of view because they can be transformed into response mean squares (statistical quantities) after multiplication with the power spectral density (PSD) function of a disturbance and integration in the frequency range. Since the amplitude of the transfer function of an interstory drift evaluated at the undamped fundamental natural circular frequency can be related to the level of this response mean square, these amplitudes of transfer functions are treated here as controlled quantities.

The problem of optimal damper placement (PODP) for a fixed shear building model may be described as follows.

Problem 2.1 PODP Find the damping coefficients of added dampers minimizing the sum of amplitudes of the transfer functions of interstory drifts evaluated at the undamped fundamental natural circular frequency

$$V = \sum_{i=1}^{2} |\hat{\delta}_i(\mathbf{c})| \quad (2.44)$$

subject to a constraint on the sum of the damper damping coefficients

$$\sum_{i=1}^{2} c_i = \overline{W} \quad (\overline{W} : \text{specified value}) \quad (2.45)$$

The quantity V defined in Equation 2.44 represents the global flexibility and its minimization is preferred from the view point of performance-based design.

The Lagrangian L for Problem 2.1 may be expressed in terms of a Lagrange multiplier λ:

$$L(\mathbf{c}, \lambda) = \sum_{i=1}^{2} |\hat{\delta}_i(\mathbf{c})| + \lambda \left(\sum_{i=1}^{2} c_i - \overline{W} \right) \qquad (2.46)$$

For simplicity of expression, the argument $(\mathbf{c})$ will be omitted hereafter.

2.5.1 Optimality Criteria

The optimality criteria for Problem 2.1 may be derived from the stationarity conditions of $L(\mathbf{c}, \lambda)$ with respect to $\mathbf{c}$ and λ:

$$\left(\sum_{i=1}^{2} |\hat{\delta}_i| \right)_{,j} + \lambda = 0 \quad (j = 1, 2) \qquad (2.47)$$

$$\sum_{i=1}^{2} c_i - \overline{W} = 0 \qquad (2.48)$$

Here, and in the following, $(\cdot)_{,j}$ denotes the partial differentiation with respect to c_j. It should be noted that the damping coefficient of every added damper must be a nonnegative value. If $c_j = 0$, then Equation 2.47 must be modified into the following form:

$$\left(\sum_{i=1}^{2} |\hat{\delta}_i| \right)_{,j} + \lambda \geq 0 \qquad (2.49)$$

The optimality criteria, Equation 2.47, include a parameter λ. It may be convenient from the viewpoint of construction of a systematic solution algorithm to express the optimality criteria without this parameter. Let us define a new quantity defined by

$$\gamma_1 = \frac{\left(\sum_{i=1}^{2} |\hat{\delta}_i| \right)_{,2}}{\left(\sum_{i=1}^{2} |\hat{\delta}_i| \right)_{,1}} \qquad (2.50)$$

The alternative expression of the optimality criteria, Equation 2.47, may then be reduced to $\gamma_1 = 1$ by eliminating λ. If $c_1 = 0$, then $\gamma_1 \geq 1$. If $c_2 = 0$, then $\gamma_1 \leq 1$.

2.5.2 Solution Algorithm

Differentiation of Equation 2.42 with respect to a design variable c_j provides

$$\mathbf{A}_{,j}\left\{\begin{matrix} \hat{U}_1 \\ \hat{U}_2 \end{matrix}\right\} + \mathbf{A}\left\{\begin{matrix} \hat{U}_{1,j} \\ \hat{U}_{2,j} \end{matrix}\right\} = \mathbf{0} \qquad (2.51)$$

$\mathbf{A}_{,j}$ in Equation 2.51 can be expressed as

$$\mathbf{A}_{,1} = \mathrm{i}\omega_1\begin{bmatrix} 1 & 0 \\ 0 & 0 \end{bmatrix} \qquad \mathbf{A}_{,2} = \mathrm{i}\omega_1\begin{bmatrix} 1 & -1 \\ -1 & 1 \end{bmatrix} \qquad (2.52)$$

Since $\mathbf{A}$ defined by Equation 2.43 is regular, the first-order sensitivities of $\hat{U}_1$ and $\hat{U}_2$ are derived from Equation 2.51 as

$$\left\{\begin{matrix} \hat{U}_{1,j} \\ \hat{U}_{2,j} \end{matrix}\right\} = -\mathbf{A}^{-1}\mathbf{A}_{,j}\left\{\begin{matrix} \hat{U}_1 \\ \hat{U}_2 \end{matrix}\right\} \qquad (2.53)$$

The first-order sensitivities of $\hat{\delta}_1$ and $\hat{\delta}_2$ are then derived by differentiating $\hat{\delta}_1 \equiv \hat{U}_1$ and $\hat{\delta}_2 \equiv \hat{U}_2 - \hat{U}_1$ (or Equation 2.39 after substituting $\omega = \omega_1$) and using Equation 2.53 as

$$\left\{\begin{matrix} \hat{\delta}_{1,j} \\ \hat{\delta}_{2,j} \end{matrix}\right\} = -\mathbf{T}\mathbf{A}^{-1}\mathbf{A}_{,j}\,\mathbf{T}^{-1}\left\{\begin{matrix} \hat{\delta}_1 \\ \hat{\delta}_2 \end{matrix}\right\} \qquad (2.54)$$

The quantity $\hat{\delta}_i$ as a complex number may be rewritten symbolically as

$$\hat{\delta}_i = \mathrm{Re}[\hat{\delta}_i] + \mathrm{i}\mathrm{Im}[\hat{\delta}_i] \qquad (2.55)$$

where Re[ ] and Im[ ] indicate the real and imaginary parts respectively of a complex number. The first-order sensitivity of $\hat{\delta}_i$ with respect to c_j may then be formally expressed as

$$\hat{\delta}_{i,j} = (\mathrm{Re}[\hat{\delta}_i])_{,j} + \mathrm{i}(\mathrm{Im}[\hat{\delta}_i])_{,j} \qquad (2.56)$$

The absolute value of $\hat{\delta}_i$ is defined by

$$|\hat{\delta}_i| = \sqrt{(\mathrm{Re}[\hat{\delta}_i])^2 + (\mathrm{Im}[\hat{\delta}_i])^2}. \qquad (2.57)$$

Then, the first-order sensitivity of $|\hat{\delta}_i|$ with respect to c_j may be expressed as

$$|\hat{\delta}_i|_{,j} = \frac{1}{|\hat{\delta}_i|}\{\mathrm{Re}[\hat{\delta}_i](\mathrm{Re}[\hat{\delta}_i])_{,j} + \mathrm{Im}[\hat{\delta}_i](\mathrm{Im}[\hat{\delta}_i])_{,j}\} \qquad (2.58)$$

where $(\mathrm{Re}[\hat{\delta}_i])_{,j}(=\mathrm{Re}[\hat{\delta}_{i,j}])$ and $(\mathrm{Im}[\hat{\delta}_i])_{,j}(=\mathrm{Im}[\hat{\delta}_{i,j}])$ are calculated from Equation 2.54.

A numerical example of this first-order sensitivity of $|\hat{\delta}_i|$ with respect to c_j for a two-story shear building model is presented in Appendix 2.A. Interested readers should examine the results shown in Appendix 2.A, which will be useful in understanding the damping sensitivity.

Consider next the variation of γ_1 defined in Equation 2.50. The linear increment $\Delta\gamma_1$ of γ_1 may be described as follows:

$$\Delta\gamma_1 = \left(\frac{1}{B_1}\frac{\partial B_2}{\partial \mathbf{c}} - \frac{B_2}{B_1^2}\frac{\partial B_1}{\partial \mathbf{c}}\right)\Delta\mathbf{c} = \frac{1}{B_1}\left(\frac{\partial B_2}{\partial \mathbf{c}} - \frac{\partial B_1}{\partial \mathbf{c}}\gamma_1\right)\Delta\mathbf{c} \qquad (2.59)$$

where

$$B_1 = \left(\sum_{i=1}^{2}|\hat{\delta}_i|\right)_{,1}, \quad B_2 = \left(\sum_{i=1}^{2}|\hat{\delta}_i|\right)_{,2} \qquad (2.60)$$

The increments $\Delta\mathbf{c}$ must satisfy the following relation due to the constraint (2.48):

$$\sum_{i=1}^{2}\Delta c_i = 0 \qquad (2.61)$$

Equations 2.59 and 2.61 lead to the following set of simultaneous linear equations with respect to $\Delta\mathbf{c}$:

$$\begin{bmatrix}\frac{1}{B_1}\left(\frac{\partial B_2}{\partial \mathbf{c}} - \frac{\partial B_1}{\partial \mathbf{c}}\gamma_1\right) \\ 1 \quad 1\end{bmatrix}\Delta\mathbf{c} = \left\{\begin{matrix}\Delta\gamma_1 \\ 0\end{matrix}\right\} \qquad (2.62)$$

Note that $(\partial B_2/\partial \mathbf{c} - \gamma_1\partial B_1/\partial \mathbf{c})/B_1$ in the left-hand side of Equation 2.62 is a 1×2 vector. Equation 2.62 implies that, once $\Delta\gamma_1$ is given and $\partial B_1/\partial \mathbf{c}$, $\partial B_2/\partial \mathbf{c}$ are evaluated, $\Delta\mathbf{c}$ can be found. Let γ_{01} denote the initial value of γ_1. The increment $\Delta\gamma_1$ is given here as $\Delta\gamma_1 = (1 - \gamma_{01})/N$, where N is the number of steps. It should be remarked that, if either one of c_1 or c_2 vanishes, the following relation must be satisfied. If $c_1 = 0$, then $\gamma_1 \geq 1$. If $c_2 = 0$, then $\gamma_1 \leq 1$.

The derivatives $\partial B_1/\partial \mathbf{c}$ and $\partial B_2/\partial \mathbf{c}$ can be evaluated in the following manner. Differentiation of Equation 2.58 with respect to c_k leads to

$$|\hat{\delta}_i|_{,jk} = \frac{1}{|\hat{\delta}_i|^2}(|\hat{\delta}_i|\{(\text{Re}[\hat{\delta}_i])_{,k}(\text{Re}[\hat{\delta}_i])_{,j} + \text{Re}[\hat{\delta}_i](\text{Re}[\hat{\delta}_i])_{,jk} + (\text{Im}[\hat{\delta}_i])_{,k}(\text{Im}[\hat{\delta}_i])_{,j}$$

$$+ \text{Im}[\hat{\delta}_i](\text{Im}[\hat{\delta}_i])_{,jk}\} - |\hat{\delta}_i|_{,k}\{\text{Re}[\hat{\delta}_i](\text{Re}[\hat{\delta}_i])_{,j} + \text{Im}[\hat{\delta}_i](\text{Im}[\hat{\delta}_i])_{,j}\}) \quad (2.63)$$

$\partial B_1/\partial \mathbf{c}$ and $\partial B_2/\partial \mathbf{c}$ may then be expressed as follows:

$$\frac{\partial B_1}{\partial \mathbf{c}} = \{\,|\hat{\delta}_1|_{,11} + |\hat{\delta}_2|_{,11} \quad |\hat{\delta}_1|_{,12} + |\hat{\delta}_2|_{,12}\,\} \qquad (2.64a)$$

$$\frac{\partial B_2}{\partial \mathbf{c}} = \{\,|\hat{\delta}_1|_{,21} + |\hat{\delta}_2|_{,21} \quad |\hat{\delta}_1|_{,22} + |\hat{\delta}_2|_{,22}\,\} \qquad (2.64b)$$

$(\mathrm{Re}[\hat{\delta}_i])_{,jk} (= \mathrm{Re}[\hat{\delta}_{i,jk}])$ and $(\mathrm{Im}[\hat{\delta}_i])_{,jk} (= \mathrm{Im}[\hat{\delta}_{i,jk}])$ in Equation 2.63 are found from

$$\begin{Bmatrix} \hat{\delta}_{1,jk} \\ \hat{\delta}_{2,jk} \end{Bmatrix} = \mathbf{TA}^{-1}\mathbf{A}_{,k}\,\mathbf{A}^{-1}\mathbf{A}_{,j}\,\mathbf{T}^{-1} \begin{Bmatrix} \hat{\delta}_1 \\ \hat{\delta}_2 \end{Bmatrix} - \mathbf{TA}^{-1}\mathbf{A}_{,j}\,\mathbf{T}^{-1} \begin{Bmatrix} \hat{\delta}_{1,k} \\ \hat{\delta}_{2,k} \end{Bmatrix} \qquad (2.65)$$

Equation 2.65 is derived by differentiating Equation 2.54 with respect to c_k and using the relation $\mathbf{A}_{,k}^{-1} = -\mathbf{A}^{-1}\mathbf{A}_{,k}\,\mathbf{A}^{-1}$. It should be noted here that, since the components in the matrix $\mathbf{A}$ are linear functions of $\mathbf{c}$, $\mathbf{A}_{,jk}$ becomes a null matrix for all j and k.

2.6 Optimality Criteria-based Design of Dampers: General Formulation

Consider an n-story shear building model. The design variables are the damping coefficients $\{c_1 \cdots c_n\}$ of added dampers. The problem of optimal damper placement may be stated almost in the same manner as in Problem 2.1 by replacing 2 by n.

Equation 2.38 may be generalized into the following form:

$$(\mathbf{K} + i\omega\mathbf{C} - \omega^2\mathbf{M})\mathbf{U}(\omega) = -\mathbf{Mr}\ddot{U}_g(\omega) \qquad (2.66)$$

where $\mathbf{K}$, $\mathbf{C}$, and $\mathbf{M}$ are the system stiffness, damping, and mass matrices respectively and $\mathbf{r} = \{1 \ \cdots \ 1\}^T$ is the influence coefficient vector for a horizontal base input. Furthermore, $\mathbf{U}(\omega)$ is defined by $\mathbf{U}(\omega) = \{U_1(\omega) \cdots U_n(\omega)\}^T$.

The optimality criteria are now expressed as follows. If $c_1 \neq 0$ and $c_j \neq 0$ $(j \neq 1)$, then $\gamma_{j-1} = 1$. If $c_1 \neq 0$ and $c_j = 0$ $(j \neq 1)$, then $\gamma_{j-1} \leq 1$. If $c_1 = 0$ and $c_j \neq 0$, then $\gamma_{j-1} \geq 1$. If $c_1 = 0$ and $c_j = 0$, then γ_{j-1} is arbitrary.

Equation 2.42 is generalized into

$$\mathbf{A}\hat{\mathbf{U}} = -\mathbf{Mr} \qquad (2.67)$$

where $\hat{\mathbf{U}} = \{\hat{U}_1 \cdots \hat{U}_n\}^T$ $(\hat{U}_j = U_j(\omega_1)/\ddot{U}_g(\omega_1))$ and

$$\mathbf{A} = \mathbf{K} + i\omega_1\mathbf{C} - \omega_1^2\mathbf{M} \qquad (2.68)$$

The transfer functions $\hat{\boldsymbol{\delta}} = \{\hat{\delta}_1 \cdots \hat{\delta}_m\}^T$ of interstory drifts evaluated at the undamped fundamental natural circular frequency are related to $\hat{\mathbf{U}}$ by

$$\hat{\boldsymbol{\delta}} = \mathbf{T}\hat{\mathbf{U}} \qquad (2.69)$$

where $\mathbf{T}$ is the deformation–displacement transformation matrix (generalized version of Equation 2.40). Equation 2.51 may be generalized into

$$\mathbf{A}_{,j}\hat{\mathbf{U}} + \mathbf{A}\hat{\mathbf{U}}_{,j} = \mathbf{0} \qquad (2.70)$$

From Equation 2.70, $\hat{\mathbf{U}}_{,j}$ can be expressed in terms of $\hat{\mathbf{U}}$:

$$\hat{\mathbf{U}}_{,j} = -\mathbf{A}^{-1}\mathbf{A}_{,j}\hat{\mathbf{U}} \qquad (2.71)$$

It should be noted that an explicit expression of $\mathbf{A}^{-1}$ can be derived for a mass–spring model (shear building model) having a tri-diagonal matrix $\mathbf{A}$ (Takewaki and Nakamura, 1995; Takewaki *et al.*, 1996). From Equations 2.69 and 2.71, $\hat{\boldsymbol{\delta}}_{,j}$ is expressed in terms of $\hat{\boldsymbol{\delta}}$:

$$\hat{\boldsymbol{\delta}}_{,j} = -\mathbf{T}\mathbf{A}^{-1}\mathbf{A}_{,j}\mathbf{T}^{-1}\hat{\boldsymbol{\delta}} \tag{2.72}$$

Now let us define a new quantity γ_j by

$$\gamma_j = \frac{\left(\sum_{i=1}^{n} |\hat{\delta}_i|\right)_{,j+1}}{\left(\sum_{i=1}^{n} |\hat{\delta}_i|\right)_{,1}} \quad (j = 1, \cdots, n-1) \tag{2.73}$$

The linear increment of γ_j may be expressed as

$$\Delta\gamma_j = \left(\frac{1}{B_1}\frac{\partial B_{j+1}}{\partial \mathbf{c}} - \frac{B_{j+1}}{B_1^2}\frac{\partial B_1}{\partial \mathbf{c}}\right)\Delta\mathbf{c} = \frac{1}{B_1}\left(\frac{\partial B_{j+1}}{\partial \mathbf{c}} - \frac{\partial B_1}{\partial \mathbf{c}}\gamma_j\right)\Delta\mathbf{c} \tag{2.74}$$

where

$$B_j = \left(\sum_{i=1}^{n} |\hat{\delta}_i|\right)_{,j} \tag{2.75}$$

Combination of Equations 2.74 with $\sum_{i=1}^{n} \Delta c_i = 0$ yields the following set of simultaneous linear equations with respect to $\Delta\mathbf{c}$:

$$\begin{bmatrix} \frac{1}{B_1}\left(\frac{\partial B_2}{\partial \mathbf{c}} - \frac{\partial B_1}{\partial \mathbf{c}}\gamma_1\right) \\ \vdots \\ \frac{1}{B_1}\left(\frac{\partial B_n}{\partial \mathbf{c}} - \frac{\partial B_1}{\partial \mathbf{c}}\gamma_{n-1}\right) \\ 1 \cdots 1 \end{bmatrix}\Delta\mathbf{c} = \begin{Bmatrix} \Delta\gamma_1 \\ \vdots \\ \Delta\gamma_{n-1} \\ 0 \end{Bmatrix} \tag{2.76}$$

The derivatives $|\hat{\delta}_i|_{,jk}$ are derived from Equation 2.63. $(\text{Re}[\hat{\delta}_i])_{,j}$ and $(\text{Im}[\hat{\delta}_i])_{,j}$ in Equation 2.63 are calculated from Equation 2.54 and $(\text{Re}[\hat{\delta}_i])_{,jk}$ and $(\text{Im}[\hat{\delta}_i])_{,jk}$ in Equation 2.63 are found from Equation 2.65.

Let $\boldsymbol{\gamma}_0$ and $\boldsymbol{\gamma}_F$ denote the initial values of the quantities defined in Equation 2.73 and their target values. It should be noted that $\boldsymbol{\gamma}_F = \{1 \cdots 1\}^T$. The linear increments $\Delta\boldsymbol{\gamma} = \{\Delta\gamma_1 \cdots \Delta\gamma_{n-1}\}^T$ of $\boldsymbol{\gamma}$ to be specified are given as follows:

$$\Delta\boldsymbol{\gamma} = \frac{1}{N}(\boldsymbol{\gamma}_F - \boldsymbol{\gamma}_0) \tag{2.77}$$

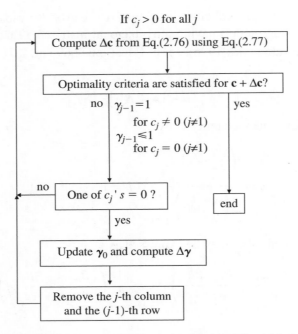

If $c_j > 0$ for all j

Compute $\Delta \mathbf{c}$ from Eq.(2.76) using Eq.(2.77)

Optimality criteria are satisfied for $\mathbf{c} + \Delta \mathbf{c}$?

no $\quad \gamma_{j-1} = 1$
    for $c_j \neq 0$ $(j \neq 1)$
$\gamma_{j-1} \leq 1$
    for $c_j = 0$ $(j \neq 1)$

yes

no   One of c_j's $= 0$?

end

yes

Update $\boldsymbol{\gamma}_0$ and compute $\Delta \boldsymbol{\gamma}$

Remove the j-th column
and the (j-1)-th row

Figure 2.12 Flowchart of the procedure for optimality criteria-based design of dampers.

The solution algorithm may then be summarized as follows:

Step 1 If $c_j > 0$ for all j, compute $\Delta \mathbf{c}$ from Equation 2.76 using Equation 2.77.
Step 2 If one of c_j values vanishes, check whether the ratio γ_{j-1} corresponding to $c_j = 0$ satisfies the condition $\gamma_{j-1} \leq 1$.
Step 3 Update $\boldsymbol{\gamma}_0$ for the model in (Step 2) and compute $\Delta \boldsymbol{\gamma}$ by Equation 2.77.
Step 4 Remove the jth column in the coefficient matrix in the left-hand side in Equation 2.76 and the $(j-1)$th row in the same coefficient matrix corresponding to $c_j = 0$. The other Δc_j values (and c_j values) are then computed sequentially from the resulting reduced set of simultaneous linear equations.
Step 5 Repeat from Step 2 to Step 4 until all the optimality criteria are satisfied.

If $c_1 = 0$, the denominator of Equation 2.73 should be changed to B_j for another story; for example, the story with the maximum value of B_j. Since the present solution algorithm does not include any indefinite iterative operation, it will provide a definite target design (optimal design). Figure 2.12 shows the flowchart of this solution procedure and Figure 2.13 explains the outline of the present optimization procedure.

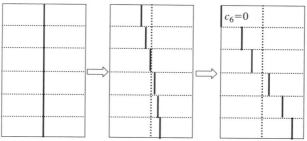

Distribution of damping coefficients

Figure 2.13   Outline of present optimization procedure.

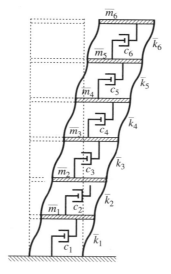

Figure 2.14   Six-story shear building model with supplemental passive viscous dampers. (I. Takewaki, "Optimal Damper Placement for Minimum Transfer Functions," *Earthquake Engineering and Structural Dynamics*, Vol.26, No.11. © 1997 John Wiley & Sons, Ltd).

2.7 Numerical Examples II

2.7.1 Example 1: Model with a Uniform Distribution of Story Stiffnesses

Consider a six-story shear building model as shown in Figure 2.14. All the masses are assumed to be prescribed as $\overline{m}_1 = \cdots = \overline{m}_6 = 0.8 \times 10^5$ kg. Every story stiffness is to have the same prescribed value: $\overline{k}_1 = \cdots = \overline{k}_6 = 4.0 \times 10^7$ N/m. The design variables are now the damping coefficients $c_1, \ldots, c_6$ of added viscous dampers. Every viscous damper is to have the same initial damping coefficient: 1.5×10^6 N s/m. The undamped fundamental natural circular frequency of the model is $\omega_1 = 5.39$ rad/s and the initial

Table 2.3 Initial values of $\{\gamma_j\}$ defined in Equation 2.73. (I. Takewaki, "Optimal Damper Placement for Minimum Transfer Functions," *Earthquake Engineering and Structural Dynamics*, Vol.26, No.11. © 1997 John Wiley & Sons, Ltd).

j	1	2	3	4	5
γ_{0j}	0.8795	0.6798	0.4438	0.2222	0.0627

Table 2.4 Final values of $\{\gamma_j\}$ attained in optimal design. (I. Takewaki, "Optimal Damper Placement for Minimum Transfer Functions," *Earthquake Engineering and Structural Dynamics*, Vol.26, No.11. © 1997 John Wiley & Sons, Ltd).

j	1	2	3	4	5
γ_{Fj}	1.001	0.8548	0.5550	0.2726	0.0723

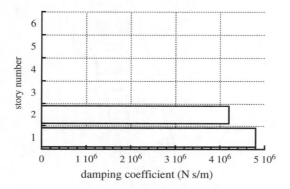

Figure 2.15 Distribution of optimal damping coefficients obtained via the present procedure. (I. Takewaki, "Optimal Damper Placement for Minimum Transfer Functions," *Earthquake Engineering and Structural Dynamics*, Vol.26, No.11. © 1997 John Wiley & Sons, Ltd).

values of γ are shown in Table 2.3. The target values of γ are $\gamma_{Fj} = 1$ (for all j) and the final values of γ attained in the optimal design are shown in Table 2.4. The number of steps in the redesign process is $N = 50$.

The distribution of the optimal damping coefficients obtained via the present procedure is plotted in Figure 2.15. It can be observed that, in the optimal design, the dampers are concentrated in the stories where the largest interstory drifts are attained in the initial design (uniform distribution of damping coefficients) (see Figure 2.16). This fact just corresponds to the conclusions in previous studies (Zhang and Soong, 1992; Tsuji and Nakamura, 1996). It has been disclosed automatically through the solution algorithm explained in the previous section that unnecessary added dampers should be removed downward from the top.

Figure 2.16 shows the amplitudes of the transfer functions for the initial design and the optimal design. It can be observed that the amplitudes of transfer functions of

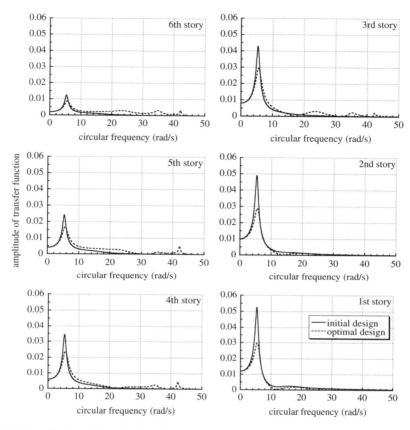

Figure 2.16 Amplitudes of transfer functions for initial and optimal designs. (I. Takewaki, "Optimal Damper Placement for Minimum Transfer Functions," *Earthquake Engineering and Structural Dynamics*, Vol.26, No.11. © 1997 John Wiley & Sons, Ltd).

interstory drifts have been reduced significantly, especially in the lower stories. The objective function V has been reduced from 0.2139 (initial design) to 0.1351 (optimal design).

2.7.2 Example 2: Model with a Uniform Distribution of Amplitudes of Transfer Functions

Another shear building model with a different distribution of story stiffnesses is considered as the second example. The distributions of masses and initial damping coefficients of added viscous dampers are the same as in the previous example. The story stiffnesses of the shear building model have been determined so that it has an undamped fundamental natural circular frequency $\omega_1 = 5.39$ rad/s and a uniform distribution of the amplitudes $\{|\hat{\delta}_j|\}$ of transfer functions of interstory drifts. This procedure can be conducted efficiently using an algorithm for an incremental inverse problem (Takewaki, 1997a; Takewaki, 1997b) explained in Sections 2.2–2.4. The distribution of story stiffnesses is shown in Table 2.5 and the initial values of $\boldsymbol{\gamma}$ are shown in Table 2.6. For this model, the solution algorithm explained in the previous section has been applied. The number of steps in the redesign process is $N = 500$. Figure 2.17 shows the distribution of the optimal damping coefficients. Different from the previous example with a uniform distribution of story stiffnesses, the dampers are not concentrated in the specific stories. This may result from the fact that the initial values of $\boldsymbol{\gamma}$ in Table 2.6 indicate nearly optimal values in this model and drastic reduction of the objective function cannot be expected by the redistribution of added dampers. Actually, the objective function has been slightly reduced from 0.2033 to 0.2027.

Table 2.5 Story stiffnesses $\{\bar{k}_j\}$ in the second example. (I. Takewaki, "Optimal Damper Placement for Minimum Transfer Functions," *Earthquake Engineering and Structural Dynamics*, Vol.26, No.11. © 1997 John Wiley & Sons, Ltd).

j	1	2	3	4	5	6
$\bar{k}_j(\times 10^7 \text{N/m})$	5.131	4.810	4.260	3.476	2.444	1.100

Table 2.6 Initial values of $\{\gamma_j\}$ in the second example. (I. Takewaki, "Optimal Damper Placement for Minimum Transfer Functions," *Earthquake Engineering and Structural Dynamics*, Vol.26, No.11. © 1997 John Wiley & Sons, Ltd).

j	1	2	3	4	5
γ_{0j}	1.021	1.042	1.062	1.071	0.8999

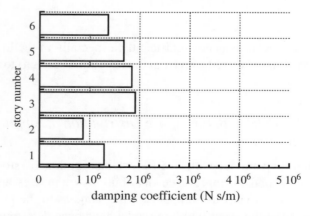

Figure 2.17 Distribution of optimal damping coefficients. (I. Takewaki, "Optimal Damper Placement for Minimum Transfer Functions," *Earthquake Engineering and Structural Dynamics*, Vol.26, No.11. © 1997 John Wiley & Sons, Ltd).

2.8 Comparison with Other Methods

2.8.1 Method of Lopez Garcia

Lopez Garcia (2001) developed an efficient and practical method (the Simplified Sequential Search Algorithm (SSSA)) of optimal damper placement and compared the result by his approach with that of Takewaki (1997c) explained in Sections 2.5–2.7. Figure 2.18 shows this comparison for four recorded ground motions. He concluded that the efficiency of the damper configurations given by the SSSA is similar to the efficiency of the damper configuration given by the optimal placement for minimum transfer functions by Takewaki (1997c). This fact strongly supports the reliability of the method by Takewaki (1997c) explained in Sections 2.5–2.7.

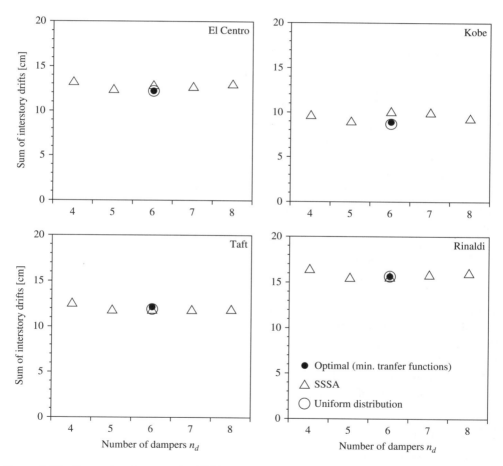

Figure 2.18 Comparison between the SSSA and optimal placement for minimum transfer functions: sum of interstory drifts for six-story structure (Reproduced with permission from D. L. Garcia, "A simple method for the design of optimal damper figurations in MDOF structures," *Earthquake Spectra*, **17**, no. 3, 387–398, 2001. © 2001 EERI).

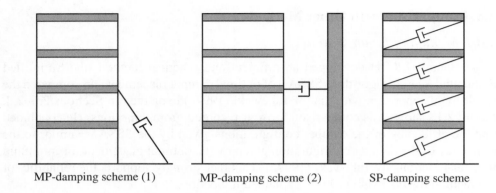

MP-damping scheme (1)   MP-damping scheme (2)   SP-damping scheme

Figure 2.19   MP damping scheme and SP damping scheme.

2.8.2 *Method of Trombetti and Silvestri*

Trombetti and Silvestri (2004) developed an efficient mass-proportional (MP) damp-ing system, as shown in Figure 2.19, and compared their system with the Takewaki (1997c) approach developed for a stiffness-proportional (SP) system and explained in Sections 2.5–2.7. They presented the comparison in terms of the maximum structural displacement and the maximum damper forces for a six-story shear-type structure subjected to three recorded ground motions: Imperial Valley, 1940, El Centro record, NS component; Kern County, 1952, Taft Lincoln School record, EW component; and Hyogoken-Nanbu, 1995, Kobe University record, NS component. It is noted that the good performance of the MP damping scheme, within the class of Rayleigh damping systems, applies to some nonclassical damping systems, as the top-story mean-square response of the structure damped with an MP damping scheme is much smaller than that of the same structure with added viscous dampers placed and sized according to the optimization criteria proposed by Takewaki (1997c). It should be noted that the optimization criteria proposed by Takewaki (1997c) only consider dampers placed between adjacent stories (as is the case of the stiffness-proportional (SP) damping scheme), while the MP damping scheme proposed by Trombetti and Silvestri (2004) is characterized by dampers placed so that they connect each story to a fixed point.

Careful treatment should be made on how this MP damping scheme is possible in the practical situation of building construction.

2.9 Summary

The results are summarized as follows.

1. An analytical procedure has been explained for redesign of structural sys-tems with an arbitrary damping system (viscous or hysteretic, proportional or

nonproportional) for target transfer functions. Since the target transfer functions as response parameters are specified, this problem is a kind of incremental inverse problem. The ratios of absolute values of the transfer functions evaluated at the undamped fundamental natural frequency of a structural system have been taken as controlled quantities together with the undamped fundamental natural frequency. Because amplitudes of transfer functions are not necessarily useful from physical points of view, their ratios have been specified. An undamped fundamental natural frequency is a representative parameter of the overall stiffness of a structure and specification of the undamped fundamental natural frequency is expected to provide a fundamental and useful index for structural design. The features of the present formulation are (i) possible treatment of any damping system stated above, (ii) possible treatment of any structural system so far as it can be modeled with FE systems and (iii) realization of a systematic algorithm without any indefinite iterative operation.

2. An efficient and systematic procedure has been shown for finding the optimal damper placement in structural systems with an arbitrary damping system (e.g., proportional or nonproportional). This problem is aimed at minimizing the sum of amplitudes of the transfer functions evaluated at the undamped fundamental natural frequency of a structural system subject to a constraint on the sum of the damping coefficients of added dampers. The optimal damper placement is determined based upon the newly derived optimality criteria. The optimal location and size of dampers have been obtained automatically via an incremental inverse problem. The features of the present formulation are to be able to deal with any damping system stated above, to be able to treat any structural system so far as it can be modeled with FE systems and to consist of a systematic algorithm without any indefinite iterative operation. It is also interesting to point out that, owing to the employment of a general dynamical property (i.e., the amplitude of a transfer function), the results are general and are not influenced by characteristics of input motions. The efficiency and reliability of the present procedure have been demonstrated through two examples of a six-story shear building model.

3. In the present formulation of optimal damper placement, an upper bound of the damping coefficient of the added dampers in each story is not given as a design constraint. However, that upper bound may be necessary because the maximum power of a damper and the number of dampers that can be added to each floor are limited in a realistic situation. In this case, the optimality conditions (i.e., Equation 2.47) must be modified. The present formulation is expected to be applicable to such a case almost in the same manner. It should also be pointed out that, when optimizing damper placement, the nonstationary nature of earthquake ground motions may lead to results different from those obtained on the basis of the assumption of stationary excitations.

4. The present formulation based on the optimality criteria and the incremental inverse problem treatment exhibits a good performance compared with other methods.

Appendix 2.A: Numerical Example of Damping Sensitivity for a Two-story Shear Building Model

A simple numerical example is presented here for a two-story shear building model. While structural damping was ignored in the formulation of Section 2.6, it is assumed in this example that the damping system consists of the structural damping $\bar{c}_1, \bar{c}_2$ and the supplemental damping c_1, c_2. Then, Equation 2.37 may be modified to

$$
\begin{bmatrix} \bar{k}_1 + \bar{k}_2 & -\bar{k}_2 \\ -\bar{k}_2 & \bar{k}_2 \end{bmatrix} \begin{Bmatrix} u_1 \\ u_2 \end{Bmatrix} + \begin{bmatrix} \bar{c}_1 + \bar{c}_2 + c_1 + c_2 & -\bar{c}_2 - c_2 \\ -\bar{c}_2 - c_2 & \bar{c}_2 + c_2 \end{bmatrix} \begin{Bmatrix} \dot{u}_1 \\ \dot{u}_2 \end{Bmatrix} + \begin{bmatrix} \bar{m}_1 & 0 \\ 0 & \bar{m}_2 \end{bmatrix} \begin{Bmatrix} \ddot{u}_1 \\ \ddot{u}_2 \end{Bmatrix}
$$

$$
= -\begin{bmatrix} \bar{m}_1 & 0 \\ 0 & \bar{m}_2 \end{bmatrix} \begin{Bmatrix} 1 \\ 1 \end{Bmatrix} \ddot{u}_g \tag{A2.1}
$$

The coefficient matrix $\mathbf{A}$ in Equation 2.43 may be expressed as

$$
\mathbf{A} = \begin{bmatrix} (\bar{k}_1 + \bar{k}_2) + \mathrm{i}\omega_1(\bar{c}_1 + \bar{c}_2 + c_1 + c_2) - \omega_1^2 \bar{m}_1 & -\bar{k}_2 - \mathrm{i}\omega_1(\bar{c}_2 + c_2) \\ -\bar{k}_2 - \mathrm{i}\omega_1(\bar{c}_2 + c_2) & \bar{k}_2 + \mathrm{i}\omega_1(\bar{c}_2 + c_2) - \omega_1^2 \bar{m}_2 \end{bmatrix} \tag{A2.2}
$$

Let us consider the first-order derivatives $\partial V / \partial c_1$ and $\partial V / \partial c_2$ at $c_1 = c_2 = 0$ (no supplemental damper) of the objective function V defined in Equation 2.44. The structural parameters are $\bar{m}_1 = 1.0 \times 10^5$ kg, $\bar{m}_2 = 0.8 \times 10^5$ kg, $\bar{k}_1 = \bar{k}_2 = 4.0 \times 10^7$ N/m, $\bar{c}_1 = \bar{c}_2 = 2.0 \times 10^5$ N s/m, and $\omega_1 = \sqrt{178}$ rad/s (undamped fundamental natural circular frequency). The coefficient matrix $\mathbf{A}$ for $c_1 = c_2 = 0$ can be obtained numerically as

$$
\mathbf{A} = 1.00 \times 10^7 \begin{bmatrix} 6.22 + 0.534\mathrm{i} & -4.00 - 0.267\mathrm{i} \\ -4.00 - 0.267\mathrm{i} & 2.58 + 0.267\mathrm{i} \end{bmatrix} \tag{A2.3}
$$

Here, and in the following, the units are omitted for simplificity. The inverse $\mathbf{A}^{-1}$ is derived as

$$
\mathbf{A}^{-1} = 1.00 \times 10^{-6} \begin{bmatrix} 0.0142 - 0.287\mathrm{i} & 0.00570 - 0.445\mathrm{i} \\ 0.00570 - 0.445\mathrm{i} & 0.0220 - 0.693\mathrm{i} \end{bmatrix} \tag{A2.4}
$$

The transfer functions of displacements at $\omega = \omega_1$ defined by Equation 2.41 can be computed as

$$
\begin{Bmatrix} \hat{U}_1 \\ \hat{U}_2 \end{Bmatrix} = \begin{Bmatrix} -0.00190 + 0.0643\mathrm{i} \\ -0.00230 + 0.0999\mathrm{i} \end{Bmatrix} \tag{A2.5}
$$

The first-order derivatives (damping sensitivity) of the transfer functions of displacements can be derived from Equation 2.53 as

$$\left\{ \begin{array}{c} \hat{U}_{1,1} \\ \hat{U}_{2,1} \end{array} \right\} = 1.00 \times 10^{-6} \left\{ \begin{array}{c} 0.0193 - 0.246i \\ 0.0160 - 0.382i \end{array} \right\} \tag{A2.6a}$$

$$\left\{ \begin{array}{c} \hat{U}_{1,2} \\ \hat{U}_{2,2} \end{array} \right\} = 1.00 \times 10^{-6} \left\{ \begin{array}{c} -0.00310 - 0.0750i \\ 0.00920 - 0.118i \end{array} \right\} \tag{A2.6b}$$

In this model, the transfer functions of interstory drifts can be obtained from Equation A2.5 as

$$\left\{ \begin{array}{c} \hat{\delta}_1 \\ \hat{\delta}_2 \end{array} \right\} = \left\{ \begin{array}{c} -0.001\,90 + 0.0643i \\ -0.000\,500 + 0.0356i \end{array} \right\} \tag{A2.7}$$

The first-order derivatives of the interstory drifts can be computed from Equation 2.54 as

$$\left\{ \begin{array}{c} \hat{\delta}_{1,1} \\ \hat{\delta}_{2,1} \end{array} \right\} = 1.00 \times 10^{-6} \left\{ \begin{array}{c} 0.0193 - 0.246i \\ -0.003\,30 - 0.136i \end{array} \right\} \tag{A2.8a}$$

$$\left\{ \begin{array}{c} \hat{\delta}_{1,2} \\ \hat{\delta}_{2,2} \end{array} \right\} = 1.00 \times 10^{-7} \left\{ \begin{array}{c} -0.0308 - 0.750i \\ 0.123 - 0.425i \end{array} \right\} \tag{A2.8b}$$

Finally, the first-order derivatives $\partial V / \partial c_1$ and $\partial V / \partial c_2$ at $c_1 = c_2 = 0$ of the objective function can be derived from Equation 2.58 as follows:

$$\partial V / \partial c_1 = -3.82 \times 10^{-7} \tag{A2.9a}$$

$$\partial V / \partial c_2 = -1.18 \times 10^{-7} \tag{A2.9b}$$

This implies that the increase of the supplemental damper c_1 in the first story is more effective than that of the damper c_2 in the second story. A schematic diagram is shown in Figure 2.20 of the relation of the gradient vector of the objective function (performance) with the constraint on total damper quantity.

It should be remarked that the above-mentioned first-order derivatives $\partial V / \partial c_1$ and $\partial V / \partial c_2$ of the objective function are equal to those at the added damping $c_1 = c_2 = 2.0 \times 10^5$ N s/m in the formulation of Section 2.5 without structural damping.

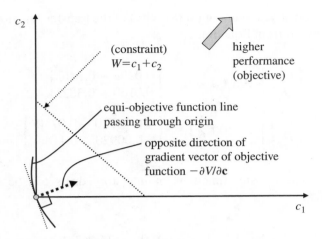

Figure 2.20 Schematic diagram of the relation of the gradient vector of the objective function (performance) with the constraint on damper quantity.

References

Barcilon, V. (1982) Inverse problem for the vibrating beam in the free-clamped configuration. *Philosophical Transactions of the Royal Society, London, Series A*, **304**, 211–251.

Boley, D. and Golub, G.H. (1987) A survey of matrix inverse eigenvalue problems. *Inverse Problems*, **3**, 595–622.

Chen, T.-Y. (1992) Optimum design of structures with both natural frequency and frequency response constraints. *International Journal of Numerical Methods in Engineering*, **33**, 1927–1940.

Fox, R.L. and Kapoor, M.P. (1968) Rates of change of eigenvalues and eigenvectors. *AIAA Journal*, **6** (12), 2426–2429.

Garcia, D.L. (2001) A simple method for the design of optimal damper configurations in MDOF structures. *Earthquake Spectra*, **17** (3), 387–398.

Gladwell, G.M.L. (1986) *Inverse Problems in Vibration*, Martinus Nijhoff Publishers, Dordrecht.

Gladwell, G.M.L. (2004) *Inverse Problems in Vibration*, 2nd edn, Springer-Verlag.

Lancaster, P. and Maroulas, J. (1987) Inverse eigenvalue problems for damped vibrating systems. *Journal of Mathematical Analysis and Applications*, **123** (1), 238–261.

Nakamura, T. and Yamane, T. (1986) Optimum design and earthquake-response constrained design of elastic shear buildings. *Earthquake Engineering & Structural Dynamics*, **14** (5), 797–815.

Porter, B. (1970) Synthesis of linear lumped-parameter vibrating systems by an inverse Holzer technique. *Journal of Mechanical Engineering Science*, **12** (1), 17–19.

Ram, Y.M. (1994) Inverse mode problems for the discrete model of a vibrating beam. *Journal of Sound and Vibration*, **169** (2), 239–252.

Starek, L. and Inman, D.J. (1995) A symmetric inverse vibration problem with overdamped modes. *Journal of Sound and Vibration*, **181** (5), 893–903.

Takewaki, I. (1997a) Incremental inverse eigenmode problem for performance-based structural redesign. *Finite Elements in Analysis and Design*, **27** (2), 175–191.

Takewaki, I. (1997b) Efficient redesign of damped structural systems for target transfer functions. *Computer Methods in Applied Mechanics and Engineering*, **147** (3/4), 275–286.

Takewaki, I. (1997c) Optimal damper placement for minimum transfer functions. *Earthquake Engineering & Structural Dynamics*, **26** (11), 1113–1124.

Takewaki, I. and Nakamura, T. (1995) Hybrid inverse mode problems for FEM-shear models. *Journal of Engineering Mechanics*, **121** (8), 873–880.

Takewaki, I. and Nakamura, T. (1997) Hybrid inverse mode problem for a structure–foundation system. *Journal of Engineering Mechanics*, **123** (4), 312–321.

Takewaki, I., Nakamura, T., and Arita, Y. (1996) A hybrid inverse mode problem for fixed-fixed mass–spring models. *Journal of Vibration and Acoustics*, **118** (4), 641–648.

Trombetti, T. and Silvestri, S. (2004) Added viscous dampers in shear-type structures: The effectiveness of mass proportional damping. *Journal of Earthquake Engineering*, **8** (2), 275–313.

Tsai, N.-C. (1974) Modal damping for soil–structure interaction. *Journal of the Engineering Mechanics Division*, **100** (EM2), 323–341.

Tsuji, M. and Nakamura, T. (1996) Optimum viscous dampers for stiffness design of shear buildings. *Journal of the Structural Design of Tall Buildings* **5**, 217–234.

Zhang, R.H. and Soong, T.T. (1992) Seismic design of viscoelastic dampers for structural applications. *Journal of Structural Engineering*, **118** (5), 1375–1392.

3

Optimality Criteria-based Design: Multiple Criteria in Terms of Seismic Responses

3.1 Introduction

In Chapter 2, the problem of optimal damper placement was discussed for a shear building model with specified story stiffnesses. The amplitude of the transfer function at the fundamental natural frequency has been treated as the performance to be minimized (Takewaki, 1997a; Takewaki, 1997b; Takewaki, 1998). The problem treated in this chapter is to find simultaneously the optimal story stiffness distribution and the optimal passive damper placement in a shear building model by minimizing the sum of mean-square interstory drifts of the shear building model to stationary random excitations or by maximizing the mean-square top-floor absolute acceleration from the viewpoint of stiffness design. While the maximization of acceleration may give an impression somewhat contradictory to the concept of performance-based design, the minimization of acceleration will also be discussed by increasing the total quantity of supplemental dampers.

While research on active and passive control has been developed extensively (e.g., Housner *et al.*, 1994, 1997; Kobori, 1996; Soong and Dargush, 1997; Kobori *et al.*, 1998; Casciati, 2002; Johnson and Smyth, 2006), research on optimal passive damper placement has been limited. In particular, research on simultaneous optimization of stiffness and damping is very limited. Since the stiffness distribution and the damping coefficient distribution interact with each other very sensitively, simultaneous optimization of both distributions is often complicated.

An efficient and systematic method is explained in this chapter for the *simultaneous optimization* of story stiffness distributions and damping coefficient distributions of

Building Control with Passive Dampers   Izuru Takewaki
© 2009 John Wiley & Sons (Asia) Pte Ltd

supplemental dampers. The method is a two-step design method. In the first step, a design is found which satisfies the optimality conditions for a specified set of total story stiffness capacity and total damper capacity. In the second step, the total story stiffness capacity and/or total damper capacity are varied with the optimality conditions satisfied. While deformation is reduced both in the first and second steps, acceleration is reduced only in the second step via an increase of total damper capacity. It is also shown numerically that the deformation minimization and acceleration maximization are almost equivalent. The features of the present formulation are (i) possible treatment of any damping system (e.g., proportional or nonproportional, viscous type, Kelvin–Voigt type or Maxwell type), (ii) possible treatment of any structural system so far as it can be modeled with FE systems, (iii) possible treatment of any random excitations (any configuration of spectral density functions) if stationary, and (iv) realization of a systematic algorithm without any indefinite iterative operation.

3.2 Illustrative Example

Consider the two-story damped shear building model shown in Figure 3.1. It is assumed here that structural damping is negligible compared with added viscous damping. Let $\{m_1, m_2\}$, $\{k_1, k_2\}$, and $\{c_1, c_2\}$ denote the floor masses, story stiffnesses, and story damping coefficients respectively. Assume that $\{m_1, m_2\}$ is prescribed. While only the damping coefficients were the design variables in Chapter 2, the design variables are $\{k_1, k_2\}$ and $\{c_1, c_2\}$ in this chapter.

Let u_1 and u_2 denote the displacements of masses m_1 and m_2 respectively. When this model is subjected to a stationary random base acceleration $\ddot{u}_g$ with zero mean, the equations of motion for this model can be written as

$$
\begin{bmatrix} k_1 + k_2 & -k_2 \\ -k_2 & k_2 \end{bmatrix} \begin{Bmatrix} u_1 \\ u_2 \end{Bmatrix} + \begin{bmatrix} c_1 + c_2 & -c_2 \\ -c_2 & c_2 \end{bmatrix} \begin{Bmatrix} \dot{u}_1 \\ \dot{u}_2 \end{Bmatrix} + \begin{bmatrix} m_1 & 0 \\ 0 & m_2 \end{bmatrix} \begin{Bmatrix} \ddot{u}_1 \\ \ddot{u}_2 \end{Bmatrix}
$$
$$
= - \begin{bmatrix} m_1 & 0 \\ 0 & m_2 \end{bmatrix} \begin{Bmatrix} 1 \\ 1 \end{Bmatrix} \ddot{u}_g \tag{3.1}
$$

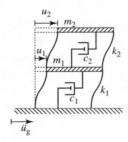

Figure 3.1 Two-story damped shear building model. (I. Takewaki, "Displacement-Acceleration Control via Stiffness-Damping Collaboration," *Earthquake Engineering and Structural Dynamics*, Vol.28, No.12. © 1999 John Wiley & Sons, Ltd).

Let $U_1(\omega)$, $U_2(\omega)$ and $\ddot{U}_g(\omega)$ denote the Fourier transforms of u_1, u_2, and $\ddot{u}_g$ respectively, and let ω denote a circular frequency. Fourier transformation of Equation 3.1 may be reduced to the following form:

$$\mathbf{A}\mathbf{U}(\omega) = \mathbf{B}\ddot{U}_g(\omega) \tag{3.2}$$

where

$$\mathbf{A} = \begin{bmatrix} k_1 + k_2 + i\omega(c_1 + c_2) - \omega^2 m_1 & -k_2 - i\omega c_2 \\ -k_2 - i\omega c_2 & k_2 + i\omega c_2 - \omega^2 m_2 \end{bmatrix} \tag{3.3a}$$

$$\mathbf{B} = -\begin{Bmatrix} m_1 \\ m_2 \end{Bmatrix} \tag{3.3b}$$

$$\mathbf{U}(\omega) = \begin{Bmatrix} U_1(\omega) \\ U_2(\omega) \end{Bmatrix} \tag{3.3c}$$

The notation i denotes the imaginary unit.

The Fourier transforms $\Delta_1(\omega)$ and $\Delta_2(\omega)$ of the interstory drifts $d_1 = u_1$ and $d_2 = u_2 - u_1$ are related to $\mathbf{U}(\omega) = \{U_1(\omega)\; U_2(\omega)\}^T$ by

$$\begin{Bmatrix} \Delta_1(\omega) \\ \Delta_2(\omega) \end{Bmatrix} = \begin{bmatrix} 1 & 0 \\ -1 & 1 \end{bmatrix} \begin{Bmatrix} U_1(\omega) \\ U_2(\omega) \end{Bmatrix} \equiv \mathbf{T}\mathbf{U}(\omega) \tag{3.4a}$$

$$\mathbf{T} = \begin{bmatrix} 1 & 0 \\ -1 & 1 \end{bmatrix} \tag{3.4b}$$

Let $\mathbf{T}_i$ denote the ith row of the matrix $\mathbf{T}$. $\Delta_i(\omega)$ may then be described as

$$\Delta_i(\omega) = \mathbf{T}_i \mathbf{A}^{-1} \mathbf{B} \ddot{U}_g(\omega) \equiv H_{\Delta_i}(\omega)\ddot{U}_g(\omega) \tag{3.5}$$

On the other hand, the Fourier transforms $\ddot{\mathbf{U}}(\omega) = \{\ddot{U}_1(\omega)\;\; \ddot{U}_2(\omega)\}^T$ of the floor accelerations $\ddot{u}_1$ and $\ddot{u}_2$ relative to the base are related to $\ddot{U}_g(\omega)$ by

$$\ddot{\mathbf{U}}(\omega) = -\omega^2 \mathbf{U}(\omega) = -\omega^2 \mathbf{A}^{-1}\mathbf{B}\ddot{U}_g(\omega) \tag{3.6}$$

The Fourier transforms $\ddot{\mathbf{U}}_A(\omega)$ of the absolute floor accelerations are then expressed as

$$\ddot{\mathbf{U}}_A(\omega) = \ddot{\mathbf{U}}(\omega) + \mathbf{1}\ddot{U}_g(\omega) = (\mathbf{1} - \omega^2 \mathbf{A}^{-1}\mathbf{B})\ddot{U}_g(\omega) \equiv \mathbf{H}_A(\omega)\ddot{U}_g(\omega) \tag{3.7}$$

where $\mathbf{1} = \{1\; 1\}^T$.

Let $S_g(\omega)$ denote the PSD function of the input $\ddot{u}_g(t)$. Using the random vibration theory, the mean-square response of the ith interstory drift d_i can be computed from

$$\sigma_{\Delta_i}^2 = \int_{-\infty}^{\infty} |H_{\Delta_i}(\omega)|^2 S_g(\omega) d\omega = \int_{-\infty}^{\infty} H_{\Delta_i}(\omega) H_{\Delta_i}^*(\omega) S_g(\omega) d\omega \tag{3.8}$$

where $(\ )^*$ denotes the complex conjugate. On the other hand, the mean-square response of the ith-floor acceleration may be evaluated by

$$\sigma_{A_i}^2 = \int_{-\infty}^{\infty} \left| H_{A_i}(\omega) \right|^2 S_g(\omega) d\omega = \int_{-\infty}^{\infty} H_{A_i}(\omega) H_{A_i}^*(\omega) S_g(\omega) d\omega \tag{3.9}$$

where $H_{A_i}(\omega)$ is the ith component of $\mathbf{H}_A(\omega)$ defined in Equation 3.7.

The first problem of displacement control may be described thus:

Problem 3.1 Displacement Control   Find the story stiffnesses $\{k_1, k_2\}$ and damper damping coefficients $\{c_1, c_2\}$ of the model which minimize $f = \sum_{i=1}^{2} \sigma_{\Delta_i}^2$ subject to $\sum_{i=1}^{2} k_i = \overline{W}_K$, $\sum_{i=1}^{2} c_i = \overline{W}_C$, and $0 \le k_i \le \overline{k}_i$ $(i = 1, 2)$, $0 \le c_i \le \overline{c}_i$ $(i = 1, 2)$, where $\overline{W}_K$, $\overline{W}_C$, $\overline{k}_i$, and $\overline{c}_i$ are the prescribed parameters.

The second problem of acceleration control may be stated thus:

Problem 3.2 Acceleration Control   Find the story stiffnesses $\{k_1, k_2\}$ and damper damping coefficients $\{c_1, c_2\}$ of the model which maximize $f = \sigma_{A_2}^2$ subject to $\sum_{i=1}^{2} k_i = \overline{W}_K$ and $\sum_{i=1}^{2} c_i = \overline{W}_C$ and $0 \le k_i \le \overline{k}_i (i = 1, 2)$, $0 \le c_i \le \overline{c}_i (i = 1, 2)$.

It is well known that, as a structure becomes stiffer, the response acceleration to seismic excitations with wide-band frequency contents becomes larger in general. This property is utilized in Problem 3.2 within the context of stiffness design (see Figure 3.9 shown later). Since smaller acceleration is preferred in the structural design in general, such a direction should also be discussed carefully. This problem will be investigated in Section 3.6. The correspondence of Problems 3.1 and 3.2 will also be discussed later in numerical examples.

3.3 General Problem

Consider an n-story damped shear building model subjected to a stationary random horizontal base acceleration $\ddot{u}_g$ with zero mean. The problem of finding the optimal stiffness-damping positioning and sizing for displacement–acceleration (SDDA) simultaneous control may be described as follows

Problem 3.3 SDDA   Find the story stiffnesses $\mathbf{k} = \{k_i\}$ and damper damping coefficients $\mathbf{c} = \{c_i\}$ which minimize the weighted sum of mean-square interstory drifts and a mean-square top-floor absolute acceleration

$$f = a \left(\sum_{i=1}^{n} \sigma_{\Delta_i}^2 \right) + b(D_0 \sigma_{A_n}^2) \tag{3.10}$$

subject to a constraint on the sum of the story stiffnesses

$$\sum_{i=1}^{n} k_i = \overline{W}_K \qquad (\overline{W}_K\text{: specified value}) \tag{3.11a}$$

a constraint on the sum of the damper damping coefficients

$$\sum_{i=1}^{n} c_i = \overline{W}_C \qquad (\overline{W}_C\text{: specified value}) \tag{3.11b}$$

and to bounding constraints on the story stiffnesses and damper damping coefficients

$$0 \le k_i \le \overline{k}_i \quad (i = 1, \cdots, n) \tag{3.12a}$$

$$0 \le c_i \le \overline{c}_i \quad (i = 1, \cdots, n) \tag{3.12b}$$

In this problem, $\overline{k}_i$ is the upper bound of the story stiffness and $\overline{c}_i$ is that of the damper damping coefficient. The parameters a and b are weighting parameters on deformation and acceleration respectively, and D_0 is a parameter for adjusting dimensions of deformation and acceleration.

Problem 3.1 corresponds to the parameter set $a = 1$, $b = 0$ and Problem 3.2 corresponds to $a = 0$, $b = -1$. As shown in the later numerical examples, Problems 3.1 and 3.2 are almost equivalent. Therefore, the problem with the parameter set of nonzero a and b may not lead to meaningful solutions in the present model. However, treatment of objective performances in Problem 3.3 is expected to lead to a unified approach for other structural systems. For example, an objective function $f = \sum_{i=2}^{n} \sigma_{\Delta_i}^2 + D_0 \sum_{i=1}^{n} \sigma_{A_i}^2$ may lead to generation of base-isolated structures. Problems 3.1 and 3.2 are treated hereafter and Problem 3.3 with the parameter set of nonzero a and b is not dealt with directly. Acceleration control may be considered in Problem 3.1 via an increase of total damper capacity. Since the optimality conditions for Problems 3.1–3.3 can be derived in a unified manner, the derivation is shown in the following.

The generalized Lagrangian L for Problems 3.1–3.3 may be expressed in terms of Lagrange multipliers λ_K, λ_C, $\boldsymbol{\alpha} = \{\alpha_i\}$, $\boldsymbol{\beta} = \{\beta_i\}$, $\boldsymbol{\mu} = \{\mu_i\}$, and $\boldsymbol{\nu} = \{\nu_i\}$.

$$L(\mathbf{k}, \mathbf{c}, \lambda_K, \lambda_C, \boldsymbol{\alpha}, \boldsymbol{\beta}, \boldsymbol{\mu}, \boldsymbol{\nu}) = f + \lambda_K \left(\sum_{i=1}^{n} k_i - \overline{W}_K \right) + \lambda_C \left(\sum_{i=1}^{n} c_i - \overline{W}_C \right)$$

$$+ \sum_{i=1}^{n} \alpha_i (0 - k_i) + \sum_{i=1}^{n} \beta_i (k_i - \overline{k}_i)$$

$$+ \sum_{i=1}^{n} \mu_i (0 - c_i) + \sum_{i=1}^{n} \nu_i (c_i - \overline{c}_i) \tag{3.13}$$

For simplicity of expression, the partial differentiation with respect to design variables is hereafter denoted by $(\ )^{\cdot j} \equiv \partial(\ )/\partial k_j$ and $(\ )_{,j} \equiv \partial(\ )/\partial c_j$.

3.4 Optimality Criteria

The principal optimality criteria for Problems 3.1–3.3 without active upper and lower bound constraints on story stiffnesses and damping coefficients may be derived from the stationarity conditions of L ($\alpha = \beta = \mu = \nu = 0$) with respect to $\mathbf{k}$, $\mathbf{c}$, λ_K, and λ_C. The optimality conditions can be described as

$$f^{\cdot j} + \lambda_K = 0 \quad \text{for } 0 < k_j < \overline{k}_j \quad (j = 1, \cdots, n) \tag{3.14a}$$

$$f_{,j} + \lambda_C = 0 \quad \text{for } 0 < c_j < \overline{c}_j \quad (j = 1, \cdots, n) \tag{3.14b}$$

$$\sum_{i=1}^{n} k_i - \overline{W}_K = 0 \tag{3.15a}$$

$$\sum_{i=1}^{n} c_i - \overline{W}_C = 0 \tag{3.15b}$$

If either one of the story stiffnesses and damper damping coefficients attains its limit, then the optimality conditions should be modified to

$$f^{\cdot j} + \lambda_K \geq 0 \quad \text{for } k_j = 0 \tag{3.16a}$$

$$f^{\cdot j} + \lambda_K \leq 0 \quad \text{for } k_j = \overline{k}_j \tag{3.16b}$$

$$f_{,j} + \lambda_C \geq 0 \quad \text{for } c_j = 0 \tag{3.17a}$$

$$f_{,j} + \lambda_C \leq 0 \quad \text{for } c_j = \overline{c}_j \tag{3.17b}$$

3.5 Solution Algorithm

The present method consists of two design stages; that is, stage (i) and stage (ii) (see Figure 3.2). In the first stage, an initial model with uniform story stiffnesses and uniform damping coefficients is introduced for the specified sets of total stiffness capacity and total damping capacity and an optimal design is found at the end of redesign. In the second stage, a series of optimal designs is obtained sequentially for various stiffness and damping capacity levels. It may be convenient to introduce new parameters $s_i = f^{\cdot i+1}/f^{\cdot 1}$ and $t_i = f_{,i+1}/f_{,1}$ to express the optimality conditions. The optimality conditions (Equations 3.14a and 3.14b) corresponding to $0 < k_j < \overline{k}_j$ and $0 < c_j < \overline{c}_j$ for all j can then be described as $s_i = t_i = 1$ $(i = 1, \cdots, n - 1)$.

The linear increments of s_i and t_i may be expressed as

$$\Delta s_i = \frac{1}{f^{,1}}\left(\sum_{j=1}^{n}f^{,(i+1)j}\Delta k_j + \sum_{j=1}^{n}f_{,j}^{i+1}\Delta c_j\right) - \frac{f^{,i+1}}{(f^{,1})^2}\left(\sum_{j=1}^{n}f^{,1j}\Delta k_j + \sum_{j=1}^{n}f_{,j}^{1}\Delta c_j\right)$$

(3.18a)

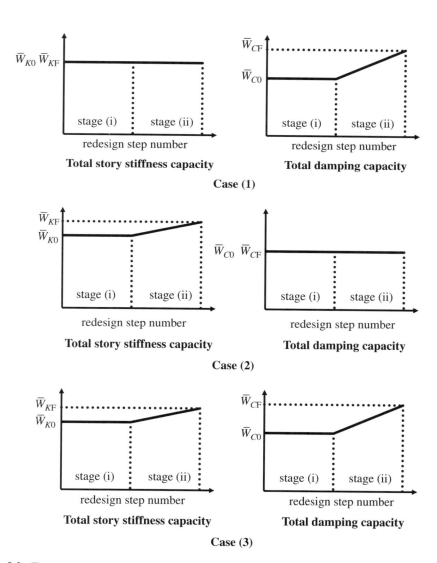

Figure 3.2 Two-stage design procedure: Cases (1), (2), and (3). (I. Takewaki, "Displacement-Acceleration Control via Stiffness-Damping Collaboration," *Earthquake Engineering and Structural Dynamics*, Vol.28, No.12. © 1999 John Wiley & Sons, Ltd).

$$\Delta t_i = \frac{1}{f_{,1}}\left(\sum_{j=1}^{n}f_{,i+1}^{,j}\Delta k_j + \sum_{j=1}^{n}f_{,(i+1)j}\Delta c_j\right) - \frac{f_{,i+1}}{(f_{,1})^2}\left(\sum_{j=1}^{n}f_{,1}^{,j}\Delta k_j + \sum_{j=1}^{n}f_{,1j}\Delta c_j\right)$$

(3.18b)

The linear increments of Equations 3.15a and 3.15b can be expressed as

$$\sum_{i=1}^{n}\Delta k_i - \Delta \overline{W}_K = 0 \qquad (3.19a)$$

$$\sum_{i=1}^{n}\Delta c_i - \Delta \overline{W}_C = 0 \qquad (3.19b)$$

Note that the constraints (Equations 3.11a and 3.11b) may be nonlinear functions of $\{k_i\}$ and $\{c_i\}$ so far as their linear increments can be expressed mathematically. Combination of Equations 3.18a and 3.18b with Equations 3.19a and 3.19b leads to a set of simultaneous linear equations in terms of $\{\Delta k_i\}$ and $\{\Delta c_i\}$. An example for a two-story model can be expressed as

$$\begin{bmatrix} f^{,1}f^{,21} - f^{,2}f^{,11} & f^{,1}f^{,22} - f^{,2}f^{,12} & f^{,1}f_{,1}^{,2} - f^{,2}f_{,1}^{,1} & f^{,1}f_{,2}^{,2} - f^{,2}f_{,2}^{,1} \\ 1 & 1 & 0 & 0 \\ f_{,1}f_{,2}^{,1} - f_{,2}f_{,1}^{,1} & f_{,1}f_{,2}^{,2} - f_{,2}f_{,1}^{,2} & f_{,1}f_{,21} - f_{,2}f_{,11} & f_{,1}f_{,22} - f_{,2}f_{,12} \\ 0 & 0 & 1 & 1 \end{bmatrix}\begin{Bmatrix}\Delta k_1 \\ \Delta k_2 \\ \Delta c_1 \\ \Delta c_2\end{Bmatrix}$$
$$= \begin{Bmatrix}\Delta s_1(f^{,1})^2 \\ \Delta \overline{W}_K \\ \Delta t_1(f_{,1})^2 \\ \Delta \overline{W}_C\end{Bmatrix}$$

(3.20)

In the first stage, the model with uniform story stiffnesses and uniform damping coefficients is employed as the initial model. The solution algorithm in the case of $k_j < \overline{k}_j$, $c_j < \overline{c}_j$ for all j may be summarized as follows:

Stage (i)

Step 0 Compute initial uniform story stiffnesses $k_j = \hat{k}$ $(j = 1, \cdots, n)$ and initial uniform damper damping coefficients $c_j = \hat{c}$ $(j = 1, \cdots, n)$ for the initial sums $\overline{W}_{K0}$ and $\overline{W}_{C0}$ of stiffness and damping coefficients by setting $\hat{k} = \overline{W}_{K0}/n$ and $\hat{c} = \overline{W}_{C0}/n$.

Step 1 Calculate the parameters $\{s_{0i}\}$ and $\{t_{0i}\}$ for the initial model and obtain the incremental parameters $\{\Delta s_i\}$ and $\{\Delta t_i\}$ by setting $\Delta s_i = (1 - s_{0i})/N_1$ and $\Delta t_i = (1 - t_{0i})/N_1$, where N_1 is the number of redesign steps in stage (i).

Step 2 Find the optimal story stiffnesses and damper damping coefficients by sequential application of Equations 3.20 with $\Delta \overline{W}_K = 0$ and $\Delta \overline{W}_C = 0$.

Stage (ii)

Step 3 Set the second sums $\overline{W}_{KF}$ and $\overline{W}_{CF}$ of the stiffness and damping coefficients and calculate $\Delta\overline{W}_K$ and $\Delta\overline{W}_C$ by $\Delta\overline{W}_K = (\overline{W}_{KF} - \overline{W}_{K0})/N_2$ and $\Delta\overline{W}_C = (\overline{W}_{CF} - \overline{W}_{C0})/N_2$, where N_2 is the number of redesign steps in stage (ii).

Step 4 Find the optimal story stiffnesses and damper damping coefficients by sequential application of Equations 3.20 with $\Delta s_i = 0$ and $\Delta t_i = 0$ (for all i).

The flowchart of this solution procedure is shown in Figure 3.3(a) and the corresponding schematic diagram is presented in Figure 3.3(b). The concept in stage (i) may be somewhat similar to the concept of incremental inverse problems in the first part of Chapter 2. However, in the present problem, optimality criteria are included and higher order design sensitivities are required as in the second part of Chapter 2. In the present algorithm, the first- and second-order design sensitivities of the objective function are required. Those expressions for deformation control may be derived as follows.

First-order sensitivities:

$$(\sigma^2_{\Delta_i})^{,j} = \int_{-\infty}^{\infty} \{H_{\Delta_i}(\omega)\}^{,j} H^*_{\Delta_i}(\omega) S_g(\omega) d\omega + \int_{-\infty}^{\infty} H_{\Delta_i}(\omega) \{H^*_{\Delta_i}(\omega)\}^{,j} S_g(\omega) d\omega$$

$$(3.21a)$$

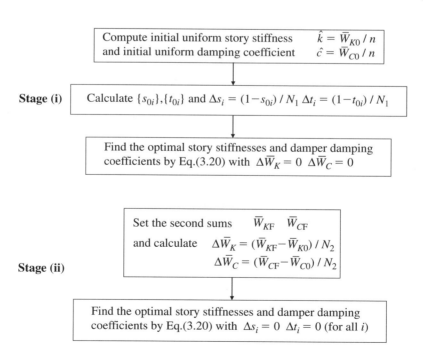

Figure 3.3 (a) Flowchart of solution procedure in two-stage design method; (b) schematic diagram of solution procedure in two-stage design method.

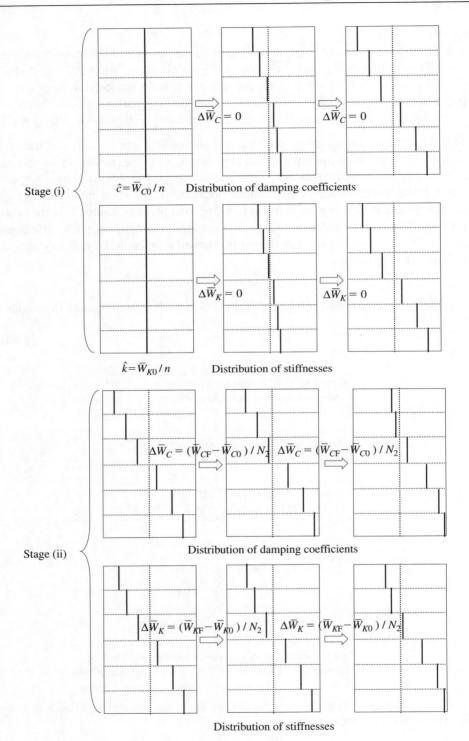

Figure 3.3 (*Continued*)

$$(\sigma_{\Delta_i}^2)_{,j} = \int_{-\infty}^{\infty} \{H_{\Delta_i}(\omega)\}_{,j} H_{\Delta_i}^*(\omega) S_g(\omega) d\omega + \int_{-\infty}^{\infty} H_{\Delta_i}(\omega)\{H_{\Delta_i}^*(\omega)\}_{,j} S_g(\omega) d\omega$$

(3.21b)

Second-order sensitivities:

$$(\sigma_{\Delta_i}^2)^{,jl} = \int_{-\infty}^{\infty} \{H_{\Delta_i}(\omega)\}^{,j} \{H_{\Delta_i}^*(\omega)\}^{,l} S_g(\omega) d\omega + \int_{-\infty}^{\infty} \{H_{\Delta_i}(\omega)\}^{,l} \{H_{\Delta_i}^*(\omega)\}^{,j} S_g(\omega) d\omega$$

$$+ \int_{-\infty}^{\infty} \{H_{\Delta_i}(\omega)\}^{,jl} H_{\Delta_i}^*(\omega) S_g(\omega) d\omega + \int_{-\infty}^{\infty} H_{\Delta_i}(\omega)\{H_{\Delta_i}^*(\omega)\}^{,jl} S_g(\omega) d\omega$$

(3.22a)

$$(\sigma_{\Delta_i}^2)_{,l}^{,j} = \int_{-\infty}^{\infty} \{H_{\Delta_i}(\omega)\}^{,j} \{H_{\Delta_i}^*(\omega)\}_{,l} S_g(\omega) d\omega + \int_{-\infty}^{\infty} \{H_{\Delta_i}(\omega)\}_{,l} \{H_{\Delta_i}^*(\omega)\}^{,j} S_g(\omega) d\omega$$

$$+ \int_{-\infty}^{\infty} \{H_{\Delta_i}(\omega)\}_{,l}^{,j} H_{\Delta_i}^*(\omega) S_g(\omega) d\omega + \int_{-\infty}^{\infty} H_{\Delta_i}(\omega)\{H_{\Delta_i}^*(\omega)\}_{,l}^{,j} S_g(\omega) d\omega$$

(3.22b)

$$(\sigma_{\Delta_i}^2)_{,j}^{,l} = \int_{-\infty}^{\infty} \{H_{\Delta_i}(\omega)\}_{,j} \{H_{\Delta_i}^*(\omega)\}^{,l} S_g(\omega) d\omega + \int_{-\infty}^{\infty} \{H_{\Delta_i}(\omega)\}^{,l} \{H_{\Delta_i}^*(\omega)\}_{,j} S_g(\omega) d\omega$$

$$+ \int_{-\infty}^{\infty} \{H_{\Delta_i}(\omega)\}_{,j}^{,l} H_{\Delta_i}^*(\omega) S_g(\omega) d\omega + \int_{-\infty}^{\infty} H_{\Delta_i}(\omega)\{H_{\Delta_i}^*(\omega)\}_{,j}^{,l} S_g(\omega) d\omega$$

(3.22c)

$$(\sigma_{\Delta_i}^2)_{,jl} = \int_{-\infty}^{\infty} \{H_{\Delta_i}(\omega)\}_{,j} \{H_{\Delta_i}^*(\omega)\}_{,l} S_g(\omega) d\omega + \int_{-\infty}^{\infty} \{H_{\Delta_i}(\omega)\}_{,l} \{H_{\Delta_i}^*(\omega)\}_{,j} S_g(\omega) d\omega$$

$$+ \int_{-\infty}^{\infty} \{H_{\Delta_i}(\omega)\}_{,jl} H_{\Delta_i}^*(\omega) S_g(\omega) d\omega + \int_{-\infty}^{\infty} H_{\Delta_i}(\omega)\{H_{\Delta_i}^*(\omega)\}_{,jl} S_g(\omega) d\omega$$

(3.22d)

In Equations 3.21a, 3.21b and 3.22a–3.22d, the sensitivities of the transfer functions are expressed as follows:

$$\{H_{\Delta_i}(\omega)\}^{,j} = \mathbf{T}_i (\mathbf{A}^{-1})^{,j} \mathbf{B} \qquad (3.23a)$$

$$\{H_{\Delta_i}(\omega)\}_{,j} = \mathbf{T}_i (\mathbf{A}^{-1})_{,j} \mathbf{B} \qquad (3.23b)$$

$$\{H_{\Delta_i}(\omega)\}^{,jl} = \mathbf{T}_i (\mathbf{A}^{-1})^{,jl} \mathbf{B} \qquad (3.23c)$$

$$\{H_{\Delta_i}(\omega)\}_{,l}^{,j} = \mathbf{T}_i (\mathbf{A}^{-1})_{,l}^{,j} \mathbf{B} \qquad (3.23d)$$

$$\{H_{\Delta_i}(\omega)\}_{,j}^{,l} = \mathbf{T}_i (\mathbf{A}^{-1})_{,j}^{,l} \mathbf{B} \qquad (3.23e)$$

$$\{H_{\Delta_i}(\omega)\}_{,jl} = \mathbf{T}_i (\mathbf{A}^{-1})_{,jl} \mathbf{B} \qquad (3.23f)$$

$$\{H^*_{\Delta_i}(\omega)\}^{\cdot j} = \mathbf{T}_i(\mathbf{A}^{-1*})^{\cdot j}\mathbf{B} \tag{3.24a}$$

$$\{H^*_{\Delta_i}(\omega)\}_{,j} = \mathbf{T}_i(\mathbf{A}^{-1*})_{,j}\mathbf{B} \tag{3.24b}$$

$$\{H^*_{\Delta_i}(\omega)\}^{\cdot jl} = \mathbf{T}_i(\mathbf{A}^{-1*})^{\cdot jl}\mathbf{B} \tag{3.24c}$$

$$\{H^*_{\Delta_i}(\omega)\}^{\cdot j}_{,l} = \mathbf{T}_i(\mathbf{A}^{-1*})^{\cdot j}_{,l}\mathbf{B} \tag{3.24d}$$

$$\{H^*_{\Delta_i}(\omega)\}^{\cdot l}_{,j} = \mathbf{T}_i(\mathbf{A}^{-1*})^{\cdot l}_{,j}\mathbf{B} \tag{3.24e}$$

$$\{H^*_{\Delta_i}(\omega)\}_{,jl} = \mathbf{T}_i(\mathbf{A}^{-1*})_{,jl}\mathbf{B} \tag{3.24f}$$

Note that operations of complex conjugate and partial differentiation are exchangeable. Furthermore, the first- and second-order sensitivities of $\mathbf{A}^{-1}$ may be described as follows:

$$(\mathbf{A}^{-1})^{\cdot j} = -\mathbf{A}^{-1}\mathbf{A}^{\cdot j}\mathbf{A}^{-1} \tag{3.25a}$$

$$(\mathbf{A}^{-1})_{,j} = -\mathbf{A}^{-1}\mathbf{A}_{,j}\mathbf{A}^{-1} \tag{3.25b}$$

$$(\mathbf{A}^{-1})^{\cdot jl} = \mathbf{A}^{-1}(\mathbf{A}^{\cdot l}\mathbf{A}^{-1}\mathbf{A}^{\cdot j} + \mathbf{A}^{\cdot j}\mathbf{A}^{-1}\mathbf{A}^{\cdot l})\mathbf{A}^{-1} \tag{3.25c}$$

$$(\mathbf{A}^{-1})^{\cdot j}_{,l} = \mathbf{A}^{-1}(\mathbf{A}_{,l}\mathbf{A}^{-1}\mathbf{A}^{\cdot j} + \mathbf{A}^{\cdot j}\mathbf{A}^{-1}\mathbf{A}_{,l})\mathbf{A}^{-1} \tag{3.25d}$$

$$(\mathbf{A}^{-1})^{\cdot l}_{,j} = \mathbf{A}^{-1}(\mathbf{A}^{\cdot l}\mathbf{A}^{-1}\mathbf{A}_{,j} + \mathbf{A}_{,j}\mathbf{A}^{-1}\mathbf{A}^{\cdot l})\mathbf{A}^{-1} \tag{3.25e}$$

$$(\mathbf{A}^{-1})_{,jl} = \mathbf{A}^{-1}(\mathbf{A}_{,l}\mathbf{A}^{-1}\mathbf{A}_{,j} + \mathbf{A}_{,j}\mathbf{A}^{-1}\mathbf{A}_{,l})\mathbf{A}^{-1} \tag{3.25f}$$

When the sensitivities of $\sigma^2_{A_i}$ are computed, the transfer function $H_{\Delta_i}(\omega)$ must be replaced by $H_{A_i}(\omega)$ defined in Equation 3.7. Furthermore, the sensitivities of the transfer functions $H_{A_i}(\omega)$ must be computed independently of Equations 3.23a–3.23f and 3.24a–3.24f. As an example, $\{H_{A_i}(\omega)\}^{\cdot j}$ and $\{H_{A_i}(\omega)\}^{\cdot jl}$ can be obtained from $\{H_{A_i}(\omega)\}^{\cdot j} = -\omega^2\{H_{U_i}(\omega)\}^{\cdot j}$ and $\{H_{A_i}(\omega)\}^{\cdot jl} = -\omega^2\{H_{U_i}(\omega)\}^{\cdot jl}$, where $\{H_{U_i}(\omega)\}$ is the ith component of $\mathbf{H}_U(\omega) = \mathbf{A}^{-1}\mathbf{B}$.

For a two-story model, $\mathbf{A}^{\cdot j}$ and $\mathbf{A}_{,j}$ are expressed as

$$\mathbf{A}^{\cdot 1} = \begin{bmatrix} 1 & 0 \\ 0 & 0 \end{bmatrix} \quad \mathbf{A}^{\cdot 2} = \begin{bmatrix} 1 & -1 \\ -1 & 1 \end{bmatrix} \quad \mathbf{A}_{,1} = i\omega\begin{bmatrix} 1 & 0 \\ 0 & 0 \end{bmatrix} \quad \mathbf{A}_{,2} = i\omega\begin{bmatrix} 1 & -1 \\ -1 & 1 \end{bmatrix} \tag{3.26}$$

It is interesting to note that, if the matrix $\mathbf{A}$ is a tri-diagonal matrix as in the present model, then the inverse of the matrix $\mathbf{A}$ can be expressed in closed form (Takewaki and Nakamura, 1995) and the computational efficiency of the present method for models with many degrees of freedom is enhanced greatly.

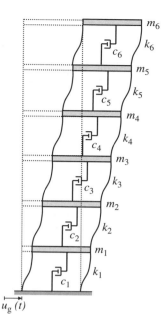

Figure 3.4   Six-story damped shear building model.

3.6 Numerical Examples

ptConsider a six-story damped shear building model, as shown in Figure 3.4, subjected to a stationary band-limited white noise $\ddot{u}_g(t)$ with zero mean whose PSD function is given by

$$S_g(\omega) = 0.01 \text{ m}^2/\text{s}^3 \quad (-2\pi \times 20 \leq \omega \leq -2\pi \times 0.2, \ 2\pi \times 0.2 \leq \omega \leq 2\pi \times 20)$$

$$S_g(\omega) = 0 \quad \text{otherwise}$$

For a simple presentation of the proposed design method, a rather simple excitation is dealt with. Application of the present method to more general excitations, such as one with a Kanai–Tajimi power spectrum or one with a Clough–Penzien power spectrum (Clough and Penzien, 1975), is straightforward. Only numerical integration including PSD functions has to be modified. The floor masses are prescribed as $m_i = 32.0 \times 10^3$ kg $(i = 1, \cdots, 6)$. The following three examples are considered to demonstrate the practical applicability of the present design method.

Example 3.1 Deformation Minimization   Consider Problem 3.1 first. Only stage (i) is considered in this example. The initial sums of stiffnesses and damping coefficients are $\overline{W}_{K0} = 3.38 \times 10^8$ N/m and $\overline{W}_{C0} = 7.50 \times 10^6$ N s/m. The initial uniform story stiffnesses and uniform dashpot damping coefficients are $k_i = 5.64 \times 10^7$ N/m $(i = 1, \cdots, 6)$ and $c_i = 1.25 \times 10^6$ N s/m $(i = 1, \cdots, 6)$. The number of redesign

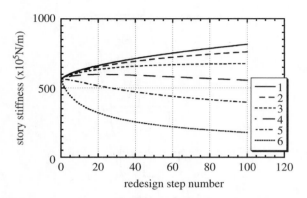

Figure 3.5   Plot of story stiffnesses with respect to redesign step number (Problem 3.1; only stage (i) is considered). (I. Takewaki, "Displacement-Acceleration Control via Stiffness-Damping Collaboration," *Earthquake Engineering and Structural Dynamics*, Vol.28, No.12. © 1999 John Wiley & Sons, Ltd).

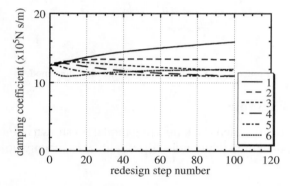

Figure 3.6   Variation of dashpot damping coefficients with respect to redesign step number. (I. Takewaki, "Displacement-Acceleration Control via Stiffness-Damping Collaboration," *Earthquake Engineering and Structural Dynamics*, Vol.28, No.12. © 1999 John Wiley & Sons, Ltd).

steps is $N_1 = 100$. Integration in Equations 3.8, 3.9, 3.21a, 3.21b, and 3.22a–3.22d has been made numerically.

Figure 3.5 shows the plots of story stiffnesses with respect to the redesign step number. On the other hand, Figure 3.6 illustrates the variations of dashpot damping coefficients with respect to the redesign step number. It can be observed from Figures 3.5 and 3.6 that story stiffnesses and dashpot damping coefficients exhibit strong nonlinearities. Figure 3.7 shows the variations of parameters $s_i = f^{,i+1}/f^{,1}$ and Figure 3.8 presents the variations of parameters $t_i = f_{,i+1}/f_{,1}$. It can be seen from Figures 3.7 and 3.8 that the optimality conditions $s_i = t_i = 1.0$ for all i are satisfied at the end of stage (i). Figure 3.9 illustrates the objective function f (the sum of mean-square interstory drifts) with respect to the redesign step number. It can be seen that the objective function indeed decreases as the redesign proceeds. On the other hand, Figure 3.10 shows the mean square of top-floor absolute acceleration with respect

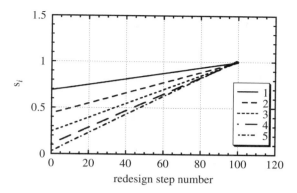

Figure 3.7 Variation of parameters $s_i = f^{,i+1}/f^{,1}$ with respect to redesign step number. (I. Takewaki, "Displacement-Acceleration Control via Stiffness-Damping Collaboration," *Earthquake Engineering and Structural Dynamics*, Vol.28, No.12. © 1999 John Wiley & Sons, Ltd).

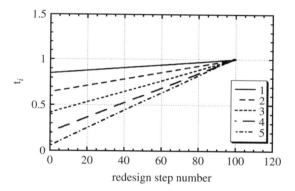

Figure 3.8 Variation of parameters $t_i = f_{,i+1}/f_{,1}$ with respect to redesign step number. (I. Takewaki, "Displacement-Acceleration Control via Stiffness-Damping Collaboration," *Earthquake Engineering and Structural Dynamics*, Vol.28, No.12. © 1999 John Wiley & Sons, Ltd).

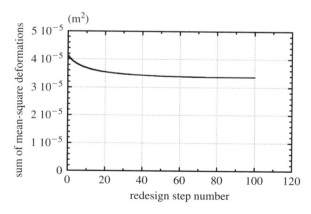

Figure 3.9 Objective function f with respect to redesign step number. (I. Takewaki, "Displacement-Acceleration Control via Stiffness-Damping Collaboration," *Earthquake Engineering and Structural Dynamics*, Vol.28, No.12. © 1999 John Wiley & Sons, Ltd).

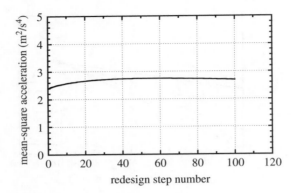

Figure 3.10 Mean-square top-floor absolute acceleration with respect to redesign step number. (I. Takewaki, "Displacement-Acceleration Control via Stiffness-Damping Collaboration," *Earthquake Engineering and Structural Dynamics*, Vol.28, No.12. © 1999 John Wiley & Sons, Ltd).

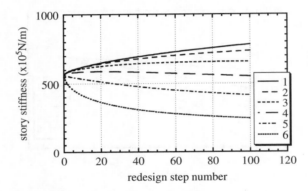

Figure 3.11 Plot of story stiffnesses with respect to redesign step number. (I. Takewaki, "Displacement-Acceleration Control via Stiffness-Damping Collaboration," *Earthquake Engineering and Structural Dynamics*, Vol.28, No.12. © 1999 John Wiley & Sons, Ltd).

to the redesign step number. While the sum of mean squares of interstory drifts decreases, the mean square of top-floor absolute acceleration increases. This means that the shear building model becomes stiffer during this redesign. The undamped fundamental natural period of the model has been shortened from 0.621 s (initial design) to 0.584 s. The computational errors in this numerical example are within 1.8 %.

Example 3.2 Acceleration Maximization   Consider Problem 3.2 next. Only stage (i) is considered in this example also. The initial sums of stiffnesses and damping coefficients are $\overline{W}_{K0} = 3.38 \times 10^8$ N/m and $\overline{W}_{C0} = 7.50 \times 10^6$ N s/m. The initial uniform story stiffnesses and uniform dashpot damping coefficients are the same as in Example 1. The number of redesign steps is $N_1 = 100$.

Figure 3.11 shows the plots of story stiffnesses with respect to the redesign step number. On the other hand, Figure 3.12 illustrates the variations of dashpot damping

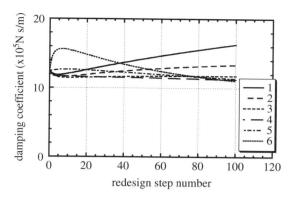

Figure 3.12 Variation of dashpot damping coefficients with respect to redesign step number. (I. Takewaki, "Displacement-Acceleration Control via Stiffness-Damping Collaboration," *Earthquake Engineering and Structural Dynamics*, Vol.28, No.12. © 1999 John Wiley & Sons, Ltd).

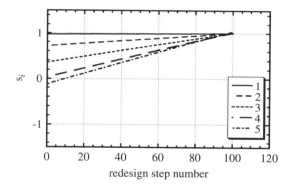

Figure 3.13 Variation of parameters $s_i = f^{,i+1}/f^{,1}$ with respect to redesign step number. (I. Takewaki, "Displacement-Acceleration Control via Stiffness-Damping Collaboration," *Earthquake Engineering and Structural Dynamics*, Vol.28, No.12. © 1999 John Wiley & Sons, Ltd).

coefficients with respect to the redesign step number. Figure 3.13 shows the variations of parameters $s_i = f^{,i+1}/f^{,1}$ and Figure 3.14 presents the variations of parameters $t_i = f_{,i+1}/f_{,1}$. It can be seen from Figures 3.13 and 3.14 that the optimality conditions $s_i = t_i = 1.0$ are satisfied at the end of stage (i). Figure 3.15 illustrates the objective function f (mean square of top-floor absolute acceleration) with respect to the redesign step number. It can be seen that the objective function indeed increases as the redesign proceeds. On the other hand, Figure 3.16 shows the sum of mean squares of interstory drifts with respect to the redesign step number. While the mean square of top-floor absolute acceleration increases, the sum of the mean squares of interstory drifts decreases. This means that the shear building model also becomes stiffer in this redesign. The undamped fundamental natural period of the model has been shortened from 0.621 s to 0.583 s. The computational errors in this numerical example are within 3.9 %.

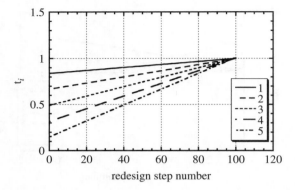

Figure 3.14 Variation of parameters $t_i = f_{,i+1}/f_{,1}$ with respect to redesign step number. (I. Takewaki, "Displacement-Acceleration Control via Stiffness-Damping Collaboration," *Earthquake Engineering and Structural Dynamics*, Vol.28, No.12. © 1999 John Wiley & Sons, Ltd).

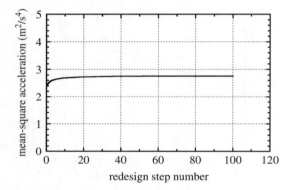

Figure 3.15 Objective function f with respect to redesign step number. (I. Takewaki, "Displacement-Acceleration Control via Stiffness-Damping Collaboration," *Earthquake Engineering and Structural Dynamics*, Vol.28, No.12. © 1999 John Wiley & Sons, Ltd).

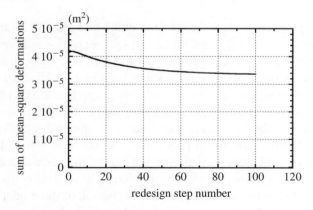

Figure 3.16 Sum of mean squares of interstory drifts with respect to redesign step number. (I. Takewaki, "Displacement-Acceleration Control via Stiffness-Damping Collaboration," *Earthquake Engineering and Structural Dynamics*, Vol.28, No.12. © 1999 John Wiley & Sons, Ltd).

Figure 3.17 shows the stiffness distributions for Examples 3.1 and 3.2. Figure 3.18 illustrates the damper damping coefficient distributions for Examples 3.1 and 3.2. It can be observed that deformation minimization and acceleration maximization are almost equivalent.

Example 3.3 Two-step Design (Deformation–Acceleration Control)   Consider Problem 3.1 again. An example for the two-step design method is presented. The sum of mean squares of interstory drifts has been adopted as the objective function throughout the redesign. Case (1) in Figure 3.2 has been employed here to reduce the acceleration in stage (ii). The initial sums of stiffnesses and damping coefficients in the stage (i) are $\overline{W}_{K0} = 3.38 \times 10^8$ N/m and $\overline{W}_{C0} = 7.50 \times 10^6$ N s/m. The initial uniform story stiffnesses and uniform dashpot damping coefficients are the same as in Examples 3.1 and 3.2. The number of redesign steps in stage (i) is $N_1 = 100$. The final sums of stiffnesses and damping coefficients in stage (ii) are $\overline{W}_{KF} = 3.38 \times 10^8$ N/m and $\overline{W}_{CF} = 8.40 \times 10^6$ N s/m. The number of redesign steps in stage (ii) is $N_2 = 100$.

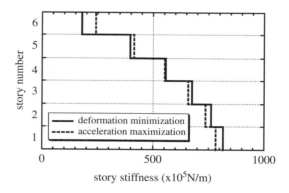

Figure 3.17   Stiffness distributions for Examples 3.1 and 3.2. (I. Takewaki, "Displacement-Acceleration Control via Stiffness-Damping Collaboration," *Earthquake Engineering and Structural Dynamics*, Vol.28, No.12. © 1999 John Wiley & Sons, Ltd).

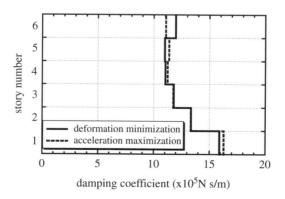

Figure 3.18   Damper damping coefficient distributions for Examples 3.1 and 3.2. (I. Takewaki, "Displacement-Acceleration Control via Stiffness-Damping Collaboration," *Earthquake Engineering and Structural Dynamics*, Vol.28, No.12. © 1999 John Wiley & Sons, Ltd).

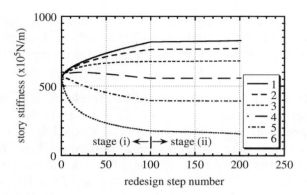

Figure 3.19 Plot of story stiffnesses with respect to redesign step number. (I. Takewaki, "Displacement-Acceleration Control via Stiffness-Damping Collaboration," *Earthquake Engineering and Structural Dynamics*, Vol.28, No.12. © 1999 John Wiley & Sons, Ltd).

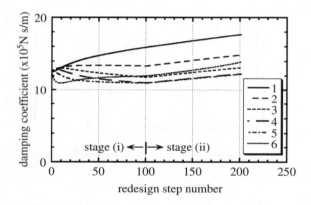

Figure 3.20 Variation of dashpot damping coefficients with respect to redesign step number. (I. Takewaki, "Displacement-Acceleration Control via Stiffness-Damping Collaboration," *Earthquake Engineering and Structural Dynamics*, Vol.28, No.12. © 1999 John Wiley & Sons, Ltd).

Figure 3.19 shows the plots of story stiffnesses with respect to the redesign step number. On the other hand, Figure 3.20 illustrates the variations of dashpot damping coefficients with respect to the redesign step number. It can be observed from Figures 3.19 and 3.20 that story stiffnesses and dashpot damping coefficients exhibit strong nonlinearities in stage (i). Figure 3.21 shows the variations of parameters $s_i = f^{,i+1}/f^{,1}$ and Figure 3.22 presents the variations of parameters $t_i = f_{,i+1}/f_{,1}$. It can be seen from Figures 3.21 and 3.22 that the optimality conditions $s_i = t_i = 1.0$ are satisfied at the end of stage (i) and those conditions *continue to be satisfied* during stage (ii). Figure 3.23 illustrates the objective function f (the sum of mean squares of interstory drifts) with respect to the redesign step number. On the other hand, Figure 3.24 shows the mean-square top-floor absolute acceleration with respect to the redesign step number. It can be seen that the objective function (the sum of mean squares of interstory drifts)

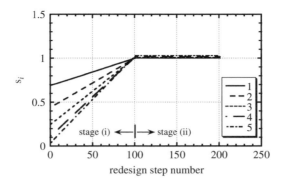

Figure 3.21 Variation of parameters $s_i = f^{,i+1}/f^{,1}$ with respect to redesign step number. (I. Takewaki, "Displacement-Acceleration Control via Stiffness-Damping Collaboration," *Earthquake Engineering and Structural Dynamics*, Vol.28, No.12. © 1999 John Wiley & Sons, Ltd).

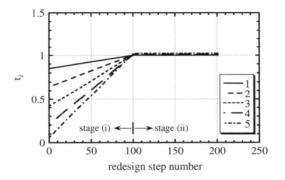

Figure 3.22 Variation of parameters $t_i = f_{,i+1}/f_{,1}$ with respect to redesign step number. (I. Takewaki, "Displacement-Acceleration Control via Stiffness-Damping Collaboration," *Earthquake Engineering and Structural Dynamics*, Vol.28, No.12. © 1999 John Wiley & Sons, Ltd).

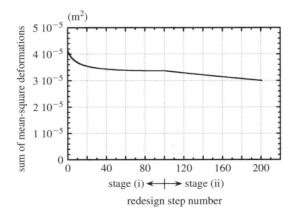

Figure 3.23 Objective function f with respect to redesign step number. (I. Takewaki, "Displacement-Acceleration Control via Stiffness-Damping Collaboration," *Earthquake Engineering and Structural Dynamics*, Vol.28, No.12. © 1999 John Wiley & Sons, Ltd).

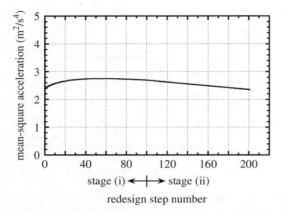

Figure 3.24 Mean-square top-floor absolute acceleration with respect to redesign step number. (I. Takewaki, "Displacement-Acceleration Control via Stiffness-Damping Collaboration," *Earthquake Engineering and Structural Dynamics*, Vol.28, No.12. © 1999 John Wiley & Sons, Ltd).

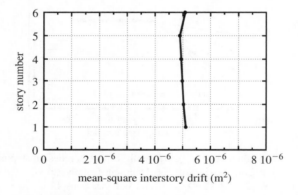

Figure 3.25 Story-wise distribution of mean-square interstory drifts of the model at the end of stage (ii). (I. Takewaki, "Displacement-Acceleration Control via Stiffness-Damping Collaboration," *Earthquake Engineering and Structural Dynamics*, Vol.28, No.12. © 1999 John Wiley & Sons, Ltd).

indeed decreases as the redesign proceeds and that the mean-square top-floor absolute acceleration decreases in stage (ii) via an increase of total damper capacity. From Figure 3.24, designers can find the target value of total damper capacity satisfying the acceleration constraint. The story-wise distribution of mean-square interstory drifts of the model at the end of stage (ii) is shown in Figure 3.25. It is found that the model at the end of stage (ii) (also in stage (ii)) has an almost uniform distribution of mean-square interstory drifts. The computational errors in this numerical example are within 1.8 % in stage (i) and within 1.1 % in stage (ii).

If the interstory drift constraints cannot be satisfied only by an increase of total damper capacity, then Cases (2) or (3) in Figure 3.2 will have to be utilized as another design path.

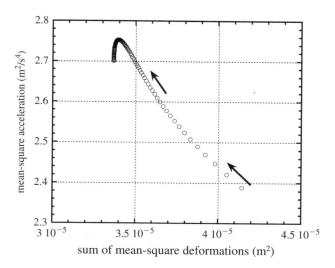

Figure 3.26 Multicriteria plot with respect to sum of mean-square deformations and mean-square acceleration which is derived from Figures 3.9 and 3.10.

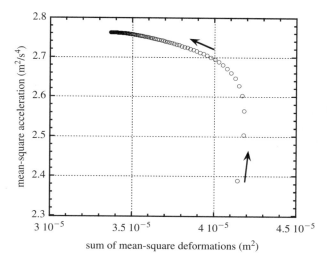

Figure 3.27 Multicriteria plot with respect to sum of mean-square deformations and mean-square acceleration which is obtained from Figures 3.15 and 3.16.

3.6.1 Multicriteria Plot

It may be useful to show the multicriteria plot with respect to the sum of mean-square deformations and mean-square acceleration. Figure 3.26 presents that plot for Example 3.1, which is derived from Figures 3.9 and 3.10. It can be observed that, as deformation decreases, acceleration increases up to a certain stage but starts to decrease from that point. Figure 3.27 illustrates that plot for Example 3.2, which is obtained

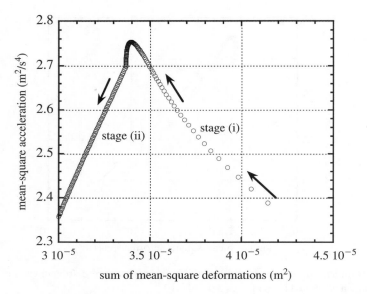

Figure 3.28 Multicriteria plot with respect to sum of mean-square deformations and mean-square acceleration which is derived from Figures 3.23 and 3.24.

from Figures 3.15 and 3.16. It is found that, as acceleration increases, deformation does not change at first but starts to decrease from a certain point. Figure 3.28 shows that plot for Example 3.3, which is derived from Figures 3.23 and 3.24. In stage (i), as deformation decreases, acceleration increases, as in Figure 3.26. Then, in stage (ii), as the total quantity of damper damping coefficients is increased, both deformation and acceleration are reduced simultaneously.

It may be concluded that the multicriteria plot with respect to the sum of mean-square deformations and mean-square acceleration provides structural designers with useful information on trade-off of deformation and acceleration caused by seismic excitations.

3.7 Summary

The results are summarized as follows.

1. An efficient and systematic design method has been explained for simultaneous optimization of story stiffness and damping distributions in a shear building model. Owing to the great dependence of the optimal damping distributions on stiffness distributions, such simultaneous optimization often causes numerical problems. This problem can be overcome by introducing a design problem to minimize the sum of mean-square deformations to stationary random excitations subject to a constraint on total stiffness capacity and that on total damping capacity. Another

problem has also been posed to maximize the mean-square response of top-floor absolute acceleration subject to the same constraints.

2. The optimal stiffness and damping distributions can be determined based upon the optimality criteria derived here through the concept of incremental inverse problems. It has been shown that deformation minimization and acceleration maximization are almost equivalent and the top-floor absolute acceleration can be reduced effectively by increasing the total damper capacity. The efficiency and reliability of the present two-step design procedure have been demonstrated through examples.

3. A multicriteria plot with respect to the sum of mean-square deformations and mean-square acceleration provides structural designers with useful information on the trade-off of deformation and acceleration caused by seismic excitations.

References

Casciati, F. (ed) (2002) *Proceedings of 3rd World Conference on Structural Control*, John Wiley & Sons, Ltd, Chichester.

Clough, R.W. and Penzien, J. (1975) *Dynamics of Structures*, McGraw-Hill, New York.

Housner, G.W., Masri, S.F., and Chassiakos, A.G. (eds) (1994) *Proceedings of 1st World Conference on Structural Control*, IASC, Los Angeles, CA.

Housner, G., Bergmann, L.A., and Caughey, T.A. (1997) Structural control: past, present, and future (special issue). *Journal of Engineering Mechanics*, **123** (9), 897–971.

Johnson, E. and Smyth, A. (eds) (2006) Proceedings of 4th World Conference on Structural Control and Monitoring (4WCSCM), San Diego, CA.

Kobori, T. (1996) Structural control for large earthquakes. Proceedings of IUTAM Symposium, Kyoto, Japan, pp. 3–28.

Kobori, T., Inoue, Y., Seto, K. *et al.* (eds) (1998) *Proceedings of the 2nd World Conference on Structural Control*, John Wiley & Sons, Ltd, Chichester.

Soong, T.T. and Dargush, G.F. (1997) *Passive Energy Dissipation Systems in Structural Engineering*, John Wiley & Sons, Ltd, Chichester.

Takewaki, I. (1997a) Optimal damper placement for minimum transfer functions. *Earthquake Engineering & Structural Dynamics*, **26** (11), 1113–1124.

Takewaki, I. (1997b) Efficient redesign of damped structural systems for target transfer functions. *Computer Methods in Applied Mechanics and Engineering*, **147** (3–4), 275–286.

Takewaki, I. (1998) Optimal damper positioning in beams for minimum dynamic compliance. *Computer Methods in Applied Mechanics and Engineering*, **156** (1–4), 363–373.

Takewaki, I. and Nakamura, T. (1995) Hybrid inverse mode problems for FEM-shear models. *Journal of Engineering Mechanics*, **121** (8), 873–880.

4

Optimal Sensitivity-based Design of Dampers in Moment-resisting Frames

4.1 Introduction

In Chapters 2 and 3, a planar shear building model has been treated. The problem considered in this chapter is to find the optimal damper positioning so as to minimize the dynamic compliance of a planar moment-resisting frame. In contrast to the shear building model treated in Chapters 2 and 3, the frame deformation influences the performance of dampers. The dynamic compliance is defined as the sum of the transfer function amplitudes of the interstory drifts evaluated at the undamped fundamental natural frequency of the building frame as in Chapter 2. Such an objective function is minimized with respect to various damper placements and sizing subject to a constraint on the sum of the viscous damping coefficients of dampers.

A systematic algorithm is explained for the optimal damper positioning first for frames with supplemental dampers modeled by a viscous damping system. The features of the present formulation are (i) possible treatment of any damping system (e.g., viscous-type or Maxwell-type, proportional or nonproportional), (ii) possible treatment of any structural system so far as it can be modeled with FE systems, and (iii) realization of a systematic algorithm without any indefinite iterative operation. While a different algorithm was devised and a variation from a uniform storywise distribution of added dampers was considered in Chapter 2 (Takewaki, 1997a), a variation from the null state is treated here (Takewaki, 2000). This treatment helps designers to understand simultaneously which position would be the best and what capacity of dampers would be required to attain a series of desired response performance levels.

In the latter part of this chapter, the corresponding algorithm is explained for the optimal damper positioning for frames with supplemental dampers modeled by a

Maxwell-type damping system. The supporting spring of the supplemental damper represents the stiffness of the damper itself or the stiffness of surrounding members. The influences of these support-member stiffnesses on the response suppression level and on the optimal damper positioning are also disclosed numerically.

4.2 Viscous-type Modeling of Damper Systems

The simplest model of a viscous damper system (e.g., an oil damper) is the viscous-type model. The damper force is related to the relative velocity between the two ends with a constant coefficient. In the practical structural design with the use of oil dampers, a relief mechanism is often employed which changes the constant viscous damping coefficient to a smaller value and the maximum damping force is limited. However, this relief mechanism is not taken into account in this book. Furthermore, it is well known that the stiffness of the damper system itself and the surrounding subassemblage affects the performance of the dampers. This influence will be considered later in this chapter through modeling into the Maxwell-type.

4.3 Problem of Optimal Damper Placement and Optimality Criteria (Viscous-type Modeling)

Supplemental dampers described by the viscous-type model discussed in the previous section are to be used in the building frame. Consider an n-story s-span planar building frame with brace-type supplemental dampers, as shown in Figure 4.1(a). Let $\mathbf{u}(t)$ denote a set of generalized displacements in the system coordinate system. The nodal mass and the mass moment of inertia around the node are taken into account at every node. $\mathbf{M}$ indicates the system mass matrix of the frame. All the member cross-sections of the frame and Young's moduli of the members are given and the system stiffness matrix is described by $\mathbf{K}$. For realistic modeling, the structural damping of the frame is taken into account. The system damping matrix due to this structural damping is assumed to be given and denoted by $\mathbf{C}$.

Consider X-brace-type supplemental dampers, as shown in Figure 4.1(b). Each damper is modeled by the viscous-type damper model. Let c_{Vi} denote the damping coefficient of the added supplemental damper (one element) in the ith story. The set $\mathbf{c}_V = \{c_{Vi}\}$ are the design variables in this case. The undamped fundamental natural circular frequency of the main frame is denoted by ω_1.

When this planar frame without added dampers is subjected to a horizontal acceleration $\ddot{u}_g(t)$ at the fixed base, the equations of motion for this model may be described as

$$\mathbf{K}\mathbf{u}(t) + \mathbf{C}\dot{\mathbf{u}}(t) + \mathbf{M}\ddot{\mathbf{u}}(t) = -\mathbf{M}\mathbf{r}\ddot{u}_g(t) \tag{4.1}$$

where $\mathbf{r}$ is the influence coefficient vector. The influence coefficient vector is the vector such that unity exists in the components corresponding to horizontal degrees of freedom and zero is allocated to the other components.

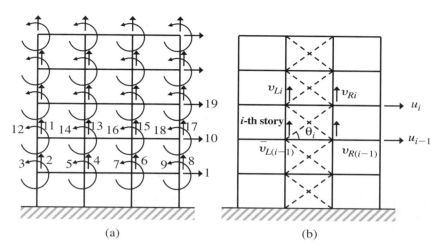

Figure 4.1 An n-story s-span planar building frame: (a) DOF in frame; (b) DOF in damper. (Reproduced with permission from *Structural and Multidisciplinary Optimization*, Vol. 20, No. 4, pp 280–287, 2000, I. Takewaki, Optimal Damper Placement for Planar Building Frames Using Transfer Functions, with permission from Springer.).

As in Chapters 2 and 3, the frequency-domain formulation is useful and effective. Let $U(\omega)$ and $\ddot{U}_g(\omega)$ respectively denote the Fourier transforms of the nodal displacements $u(t)$ and input acceleration $\ddot{u}_g(t)$ at the base. Fourier transformation of Equation 4.1 may be reduced to the following form:

$$(\mathbf{K} + i\omega\mathbf{C} - \omega^2\mathbf{M})\mathbf{U}(\omega) = -\mathbf{M}\mathbf{r}\ddot{U}_g(\omega) \qquad (4.2)$$

where i is the imaginary unit.

Consider that the added supplemental dampers are included in the main building frame. Then, Equation 4.2 may be modified to the following form:

$$\{\mathbf{K} + i\omega(\mathbf{C} + \mathbf{C}_V) - \omega^2\mathbf{M}\}\mathbf{U}_V(\omega) = -\mathbf{M}\mathbf{r}\ddot{U}_g(\omega) \qquad (4.3)$$

where $\mathbf{C}_V$ is the damping matrix due to the added dampers and $\mathbf{U}_V(\omega)$ are the Fourier transforms of the generalized displacements $u_V(t)$ of the model with the added dampers. The procedure of constructing $\mathbf{C}_V$ is shown in Appendix 4.A.

New complex-value quantities $\hat{\mathbf{U}}$ will be defined by

$$\hat{\mathbf{U}} \equiv \frac{\mathbf{U}_V(\omega_1)}{\ddot{U}_g(\omega_1)} \qquad (4.4)$$

In Equation 4.4, $\hat{U}_i$ indicates the value such that ω_1 is substituted in the frequency-response function obtained as $U_{Vi}(\omega)$ after substituting $\ddot{U}_g(\omega) = 1$ in Equation 4.3. A quantity similar to this quantity has been introduced and utilized in Chapter 2. Because

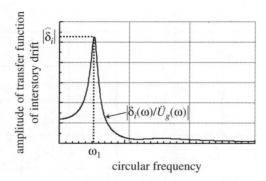

Figure 4.2 Amplitude of transfer function $\hat{\delta}_i$ at $\omega = \omega_1$ (fundamental natural circular frequency of frame) of interstory drift in ith story. (Reproduced with permission from *Structural and Multidisciplinary Optimization*, Vol. 20, No. 4, pp 280–287, 2000, I. Takewaki, Optimal Damper Placement for Planar Building Frames Using Transfer Functions, with permission from Springer.).

M and **K** are completely prescribed, ω_1 is regarded as a given value. By combining Equation 4.3 after substituting $\omega = \omega_1$ with Equation 4.4, the following relation for $\hat{\mathbf{U}}$ may be obtained:

$$\mathbf{A}\hat{\mathbf{U}} = -\mathbf{Mr} \tag{4.5}$$

In Equation 4.5, the coefficient matrix **A** indicates the following matrix:

$$\mathbf{A} = \mathbf{K} + i\omega_1(\mathbf{C} + \mathbf{C_V}) - \omega_1^2\mathbf{M} \tag{4.6}$$

Let $\hat{\delta}_i$ denote the transfer function at $\omega = \omega_1$ of the interstory drift in the ith story (see Figure 4.2). $\hat{\boldsymbol{\delta}} = \{\hat{\delta}_i\}$ can be derived from $\hat{\mathbf{U}}$ by using the relation $\hat{\boldsymbol{\delta}} = \mathbf{T}\hat{\mathbf{U}}$, where **T** is a constant transformation matrix from the nodal displacements into the interstory drifts.

It may be meaningful to note that the transfer function amplitude squared can be transformed into the mean squares of a response after multiplication with the PSD function of an external disturbance and integration in the frequency range. Since the transfer function amplitude of a nodal displacement evaluated at the undamped fundamental natural circular frequency can be associated with the level of the mean square of the response, this transfer function amplitude is treated in this book as a controlled quantity.

Let us now consider the problem of optimal damper positioning for a frame (PODPF) model.

Problem 4.1 PODPF Find the damping coefficients $\mathbf{c_V} = \{c_{Vi}\}$ of added supplemental viscous dampers so as to minimize the following sum of the transfer function amplitudes of the interstory drifts evaluated at the undamped fundamental natural circular frequency ω_1

$$f(\mathbf{c}_V) = \sum_{i=1}^{n} |\hat{\delta}_i(\mathbf{c}_V)| \tag{4.7}$$

subject to a constraint on the sum of the damping coefficients of added supplemental viscous dampers

$$\sum_{i=1}^{n} c_{Vi} = \overline{W} \quad (\overline{W}: \text{specified value}) \tag{4.8a}$$

and to constraints on the damping coefficients themselves of added supplemental viscous dampers

$$0 \le c_{Vi} \le \overline{c}_{Vi} \quad (i = 1, \cdots, n) \tag{4.8b}$$

where $\overline{c}_{Vi}$ is the upper bound of the damping coefficient of the added supplemental viscous damper in the ith story.

This problem can be said to be related to the H^∞ control in the sense that the amplitude of the transfer function is controlled.

4.3.1 Optimality Criteria

Since Problem 4.1 is an optimization problem, the Lagrange multiplier method can play an important role. The generalized Lagrangian L for Problem 4.1 may be expressed in terms of Lagrange multipliers λ, $\mathbf{\mu} = \{\mu_i\}$, and $\mathbf{v} = \{v_i\}$:

$$L(\mathbf{c}_V, \lambda, \mathbf{\mu}, \mathbf{v}) = f(\mathbf{c}_V) + \lambda \left(\sum_{i=1}^{n} c_{Vi} - \overline{W} \right) + \sum_{i=1}^{n} \mu_i(0 - c_{Vi}) + \sum_{i=1}^{n} v_i(c_{Vi} - \overline{c}_{Vi}) \tag{4.9}$$

For simplicity of expression, the argument $(\mathbf{c}_V)$ in the function $f(\mathbf{c}_V)$ will be omitted in the following formulation.

The principal (or major) optimality criteria for Problem 4.1 without active upper and lower bound constraints on damping coefficients of added supplemental viscous dampers may be derived from the stationarity conditions of the generalized Lagrangian $L(\mathbf{\mu} = \mathbf{0}, \mathbf{v} = \mathbf{0})$ with respect to the design variables $\mathbf{c}_V$ and the Lagrange multiplier λ.

$$f_{,j} + \lambda = 0 \quad \text{for } 0 < c_{vj} < \overline{c}_{vj} \quad (j = 1, \cdots, n) \tag{4.10}$$

$$\sum_{i=1}^{n} c_{Vi} - \overline{W} = 0 \tag{4.11}$$

Here, and in the following, the symbol $(\cdot)_{,j}$ denotes the partial differentiation with respect to a design variable $c_{\mathrm{V}j}$. In the case where the constraints (Equation 4.8b) are active, Equation 4.10 has to be modified into the following forms:

$$f_{,j} + \lambda \geq 0 \quad \text{for } c_{vj} = 0 \tag{4.12a}$$

$$f_{,j} + \lambda \leq 0 \quad \text{for } c_{vj} = \bar{c}_{vj} \tag{4.12b}$$

4.4 Solution Algorithm (Viscous-type Modeling)

A solution algorithm for Problem 4.1 is explained in this section. In the solution procedure, the model without added supplemental viscous dampers, namely $c_{\mathrm{V}j} = 0$ $(j = 1, \cdots, n)$, is used as the initial model. This treatment is well suited to the situation such that a structural designer is just starting the allocation and placement of added supplemental viscous dampers at desired positions. The damping coefficients of added dampers are *increased gradually* based on the optimality criteria stated above. This algorithm will be called a steepest direction search algorithm.

Let $\Delta c_{\mathrm{V}i}$ and ΔW denote the increment of the damping coefficient of the ith-story added damper in one cycle and the increment of the sum of the damping coefficients of added dampers in one cycle, respectively. Once ΔW is given, the problem is to determine simultaneously the effective position and amount $\{\Delta c_{\mathrm{V}i}\}$ of the increments of the damper damping coefficients. In order to develop this algorithm, the first and second-order sensitivities of the objective function with respect to a design variable are derived and explained in the following.

Differentiation of the principal equation, Equation 4.5, with respect to a design variable $c_{\mathrm{V}j}$ provides

$$\mathbf{A}_{,j}\hat{\mathbf{U}} + \mathbf{A}\hat{\mathbf{U}}_{,j} = \mathbf{0} \tag{4.13}$$

It is not difficult to show that the coefficient matrix $\mathbf{A}$ in Equation 4.5 is regular because the transfer function exists at $\omega = \omega_1$. Then the first-order sensitivities of the nodal displacements $\hat{\mathbf{U}}$ are derived from Equation 4.13 as

$$\hat{\mathbf{U}}_{,j} = -\mathbf{A}^{-1}\mathbf{A}_{,j}\hat{\mathbf{U}} \tag{4.14}$$

Once the nodal displacements $\hat{\mathbf{U}}$ and their derivatives $\hat{\mathbf{U}}_{,j}$ are calculated and obtained, the interstory drifts $\hat{\boldsymbol{\delta}} = \mathbf{T}\hat{\mathbf{U}}$ (see the procedure below Equation 4.6) and their derivatives $\hat{\boldsymbol{\delta}}_{,j}$ can be derived through simple manipulation.

The quantity $\hat{\delta}_i$, the ith component in $\hat{\boldsymbol{\delta}}$, may be expressed as

$$\hat{\delta}_i = \mathrm{Re}[\hat{\delta}_i] + \mathrm{i}\,\mathrm{Im}[\hat{\delta}_i] \tag{4.15}$$

where the symbols Re[ ] and Im[ ] indicate the real and imaginary parts respectively of a complex number. The first-order sensitivity of $\hat{\delta}_i$ with respect to the jth design variable may be formally expressed as

$$\hat{\delta}_{i,j} = (\mathrm{Re}[\hat{\delta}_i])_{,j} + \mathrm{i}(\mathrm{Im}[\hat{\delta}_i])_{,j} \tag{4.16}$$

It can be understood from Equation 4.16 that $\mathrm{Re}[\hat{\delta}_{i,j}] = (\mathrm{Re}[\hat{\delta}_i])_{,j}$ and $\mathrm{Im}[\hat{\delta}_{i,j}] = (\mathrm{Im}[\hat{\delta}_i])_{,j}$ hold. With the aid of mathematical manipulation, the first-order sensitivity of the absolute value $|\hat{\delta}_i| = \sqrt{(\mathrm{Re}[\hat{\delta}_i])^2 + (\mathrm{Im}[\hat{\delta}_i])^2}$ of $\hat{\delta}_i$ may then be expressed as

$$|\hat{\delta}_i|_{,j} = \frac{1}{|\hat{\delta}_i|}\{\mathrm{Re}[\hat{\delta}_i](\mathrm{Re}[\hat{\delta}_i])_{,j} + \mathrm{Im}[\hat{\delta}_i](\mathrm{Im}[\hat{\delta}_i])_{,j}\} \tag{4.17}$$

where $(\mathrm{Re}[\hat{\delta}_i])_{,j}$ and $(\mathrm{Im}[\hat{\delta}_i])_{,j}$ are calculated from Equations 4.14 and 4.16 and the relation $\hat{\boldsymbol{\delta}}_{,j} = \mathbf{T}\hat{\mathbf{U}}_{,j}$.

A general expression of $|\hat{\delta}_i|_{,j\ell}$ is derived and explained in this section for developing a solution procedure for Problem 4.1. Differentiation of Equation 4.17 with respect to a design variable $c_{V\ell}$ provides

$$|\hat{\delta}_i|_{,j\ell} = \frac{1}{|\hat{\delta}_i|^2}(|\hat{\delta}_i|\{(\mathrm{Re}[\hat{\delta}_i])_{,\ell}(\mathrm{Re}[\hat{\delta}_i])_{,j} + \mathrm{Re}[\hat{\delta}_i](\mathrm{Re}[\hat{\delta}_i])_{,j\ell} + (\mathrm{Im}[\hat{\delta}_i])_{,\ell}(\mathrm{Im}[\hat{\delta}_i])_{,j}$$
$$+ \mathrm{Im}[\hat{\delta}_i](\mathrm{Im}[\hat{\delta}_i])_{,j\ell}\} - |\hat{\delta}_i|_{,\ell}\{\mathrm{Re}[\hat{\delta}_i](\mathrm{Re}[\hat{\delta}_i])_{,j} + \mathrm{Im}[\hat{\delta}_i](\mathrm{Im}[\hat{\delta}_i])_{,j}\}) \tag{4.18}$$

$(\mathrm{Re}[\hat{\delta}_i])_{,j\ell}$ and $(\mathrm{Im}[\hat{\delta}_i])_{,j\ell}$ in Equation 4.18 are found from the second derivatives of $\hat{\mathbf{U}}$ and the relation $\hat{\boldsymbol{\delta}}_{,j\ell} = \mathbf{T}\hat{\mathbf{U}}_{,j\ell}$.

$$\hat{\mathbf{U}}_{,j\ell} = \mathbf{A}^{-1}\mathbf{A}_{,\ell}\mathbf{A}^{-1}\mathbf{A}_{,j}\hat{\mathbf{U}} - \mathbf{A}^{-1}\mathbf{A}_{,j}\hat{\mathbf{U}}_{,\ell} \tag{4.19}$$

which is derived by differentiating Equation 4.14 with respect to a design variable $c_{V\ell}$ and using the relation $\mathbf{A}_{,\ell}^{-1} = -\mathbf{A}^{-1}\mathbf{A}_{,\ell}\mathbf{A}^{-1}$ derived from the differentiation of $\mathbf{A}\mathbf{A}^{-1} = \mathbf{I}$. It is important to note that, since the components in the matrix $\mathbf{A}$ are linear functions of the design variables $\mathbf{c}_V$, $\mathbf{A}_{,j\ell}$ becomes a null matrix for all j and ℓ. Substitution of Equation 4.14 into Equation 4.19 leads to the following compact form:

$$\hat{\mathbf{U}}_{,j\ell} = \mathbf{A}^{-1}\mathbf{A}_{,\ell}\mathbf{A}^{-1}\mathbf{A}_{,j}\hat{\mathbf{U}} + \mathbf{A}^{-1}\mathbf{A}_{,j}\mathbf{A}^{-1}\mathbf{A}_{,\ell}\hat{\mathbf{U}}$$
$$= \mathbf{A}^{-1}(\mathbf{A}_{,\ell}\mathbf{A}^{-1}\mathbf{A}_{,j} + \mathbf{A}_{,j}\mathbf{A}^{-1}\mathbf{A}_{,\ell})\hat{\mathbf{U}} \tag{4.20}$$

The second-order derivatives $|\hat{\delta}_i|_{,j\ell}$ are derived from Equation 4.18. $(\mathrm{Re}[\hat{\delta}_i])_{,j}$ and $(\mathrm{Im}[\hat{\delta}_i])_{,j}$ in Equation 4.18 are calculated from Equation 4.14, and $(\mathrm{Re}[\hat{\delta}_i])_{,j\ell}$ and $(\mathrm{Im}[\hat{\delta}_i])_{,j\ell}$ in Equation 4.18 are found from Equation 4.20 and the relation $\hat{\boldsymbol{\delta}}_{,j\ell} = \mathbf{T}\hat{\mathbf{U}}_{,j\ell}$. It should be pointed out here that, although the model treated is different, the expressions in Equations 4.17 and 4.14 of first-order sensitivity have been derived in Chapter 2.

The solution algorithm in the case satisfying the conditions $c_{Vj} < \bar{c}_{Vj}$ for all j may be summarized as:

Step 0 Initialize all the added supplemental viscous dampers as $c_{Vj} = 0$ $(j = 1, \cdots, n)$. In the initial design stage, the structural damping alone exists in the frame. Assume the quantity ΔW.

Step 1 Compute the first-order derivative $f_{,i}$ of the objective function by Equation 4.17.

Step 2 Find the special index k such that

$$-f_{,k} = \max_{i}\{-f_{,i}\} \qquad (4.21)$$

Step 3 Update the objective function f by the linear approximation $f + f_{,k}\Delta c_{Vk}$, where $\Delta c_{Vk} = \Delta W$. This is because the supplemental damper is added only in the kth story in the initial design stage.

Step 4 Update the first-order sensitivity $f_{,i}$ of the objective function by the linear approximation $f_{,i} + f_{,ik}\Delta c_{Vk}$ using Equation 4.18.

Step 5 If, in Step 4, there exists a supplemental damper of an index j such that the condition

$$-f_{,k} = \max_{j, j \neq k}\{-f_{,j}\} \qquad (4.22)$$

is satisfied, then stop and compute the increment $\Delta \tilde{c}_{Vk}$ of the damping coefficient of the corresponding damper. At this stage, update the first-order sensitivity $f_{,i}$ of the objective function by the linear approximation $f_{,i} + f_{,ik}\Delta \tilde{c}_{Vk}$ by using Equation 4.18.

Step 6 Repeat the procedure from Step 2 to Step 5 until the constraint in Equation 4.8a (i.e., $\sum_{i=1}^{n} c_{Vi} = \overline{W}$) is satisfied.

The fundamental concept and algorithm of the present procedure are also summarized schematically in Figure 4.3. In Steps 2 and 3, the direction which decreases the objective function f most effectively under the condition $\sum_{i=1}^{n} \Delta c_{Vi} = \Delta W$ is searched and the design (the quantity of supplemental dampers) is updated in that direction. It is appropriate, therefore, to call the present algorithm explained above "the steepest direction search algorithm." A simple numerical example of damping sensitivity of the performance (sum of transfer function amplitudes of interstory drifts) in a two-story shear building model is presented in Appendix 2.A in Chapter 2. This sensitivity example just corresponds to the gradient direction of the performance function at the origin in the schematic diagram shown in Figure 4.4. This algorithm is similar to the conventional steepest descent method in the mathematical programming (see Figure 4.4). However, while the steepest descent method uses the gradient vector itself of the objective function as its redesign direction and does not utilize optimality criteria, the present algorithm takes advantage of the newly derived optimality criteria expressed by Equations 4.10, 4.12a, and 4.12b and does not adopt the gradient vector as its redesign direction. More specifically, the explained steepest direction search guarantees the automatic approximate satisfaction of the optimality criteria, although using first-order approximation. For example, if Δc_{Vk} is added to the kth added viscous damper in which Equation 4.21 is satisfied, then its damper ($c_{Vk} > 0$) satisfies the

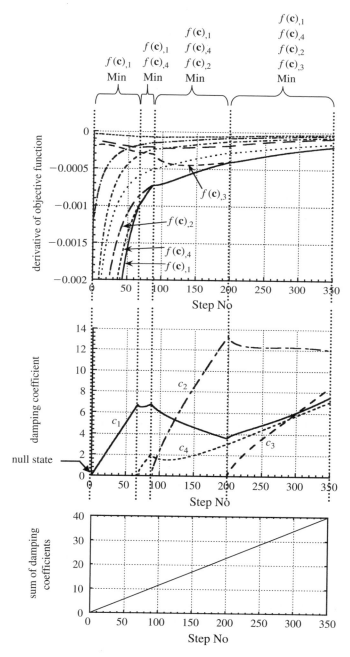

Figure 4.3 Fundamental algorithm of the present procedure. (Reproduced with permission from *Structural and Multidisciplinary Optimization*, Vol. 20, No. 4, pp 280–287, 2000, I. Takewaki, Optimal Damper Placement for Planar Building Frames Using Transfer Functions, with permission from Springer.).

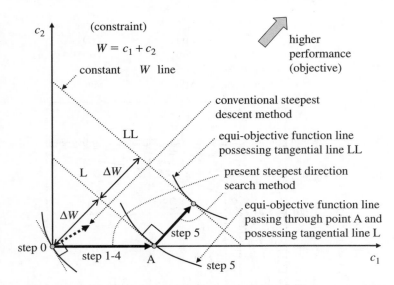

Figure 4.4 Comparison of the present steepest direction search method with the conventional steepest descent method (c_V is denoted here by c for simplicity of expression).

optimality condition (Equation 4.10) and the other dampers (($c_{Vj} = 0$, $j \neq k$), alternatively satisfy the optimality condition (Equation 4.12a). It is important to note that a series of subproblems is introduced here tentatively in which the total damper capacity level $\overline{W}$ is increased gradually by ΔW from null through the specified value.

It is necessary to investigate other possibilities. If multiple indices $k_1, \ldots, k_p$ exist in Step 2, then the objective function f and its derivative $f_{,j}$ have to be updated by the following rules:

$$f \rightarrow f + \sum_{i=k_1}^{k_p} f_{,i} \Delta c_{Vi} \qquad (4.23a)$$

$$f_{,j} \rightarrow f_{,j} + \sum_{\ell=k_1}^{k_p} f_{,j\ell} \Delta c_{V\ell} \qquad (4.23b)$$

Furthermore, the index k defined in Step 5 has to be replaced by the multiple indices $k_1, \ldots, k_p$. The ratios among the magnitudes Δc_{Vi} have to be determined so that the following relations are satisfied:

$$f_{,k_1} + \sum_{i=k_1}^{k_p} f_{,k_1 i} \Delta c_{Vi} = \cdots = f_{,k_p} + \sum_{i=k_1}^{k_p} f_{,k_p i} \Delta c_{Vi} \qquad (4.25)$$

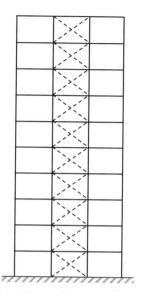

Figure 4.5 A 10-story three-span planar steel building frame. (Reproduced with permission from *Structural and Multidisciplinary Optimization*, Vol. 20, No. 4, pp 280–287, 2000, I. Takewaki, Optimal Damper Placement for Planar Building Frames Using Transfer Functions, with permission from Springer.).

Equation 4.24 requires that the optimality conditions (Equation 4.10) continue to be satisfied in the supplemental dampers with the indices $k_1, \cdots, k_p$.

It may be the case in realistic situations that the maximum quantity of supplemental dampers is limited by the requirements of building design and planning. In the case where the damping coefficients of some added supplemental dampers attain their upper bounds, such constraints must be incorporated in the aforementioned algorithm. In that case, the increment Δc_{Vk} of the supplemental damper is added subsequently to the damper in which $-f_{,k}$ attains the maximum among all the dampers except those attaining the upper bound.

4.5 Numerical Examples I (Viscous-type Modeling)

As a realistic example, consider the 10-story three-span planar steel building frame shown in Figure 4.5. All the bays have the same span length of 8 m and all the stories have the same story height of 4 m. The member properties, namely the cross-sectional areas and second moments of area, are shown in Table 4.1. The Young's modulus of the members is common throughout the frame and is specified as 2.06×10^{11} N/m^2. For simple and clear presentation of the present theory, shear deformation of the members is neglected. It is important to note that the present theory can be applied to the model including member shear deformation only by replacing the stiffness matrix of the frame in Equation 4.1 with that including shear deformation. While only bending

Table 4.1 Member property of an example ten-story, three-span steel frame. (Reproduced with permission from *Structural and Multidisciplinary Optimization*, Vol. 20, No. 4, pp 280–287, 2000, I. Takewaki, Optimal Damper Placement for Planar Building Frames Using Transfer Functions, with permission from Springer.).

Story	Cross-sectional area $(\times 10^{-4} \mathrm{m}^2)$	Second moment of area $(\times 10^{-8} \mathrm{m}^4)$
Column		
First–fourth	756	383 000
Fifth–seventh	683	353 000
Eighth–tenth	365	205 000
Beam		
First–fourth		383 000
Fifth–seventh		353 000
Eighth–tenth		205 000

deformation is considered in beams, both axial and bending deformations are taken into account in columns here.

A lumped mass of 51.2×10^3 kg is placed at every interior node and a lumped mass of 25.6×10^3 kg is placed at every exterior node. Every interior node is assumed to have a mass moment of inertia of 5.46×10^5 kg m^2 and every exterior node is assumed to possess a mass moment of inertia of 1.71×10^5 kg m^2. The undamped fundamental natural circular frequency of the main frame model is $\omega_1 = 4.75$ rad/s. The structural damping of the main frame is given by a critical damping ratio of 0.02 in the lowest eigenvibration.

In the present example model, a viscous damper is regarded as being effective with respect to a relative velocity between a pair of nodes at both ends of a member including the damper system. The model taking into account the flexibility in and around the viscous dampers will be treated in the following sections. It is also assumed that all the constraints on upper bounds of the damping coefficients are inactive; that is, $c_j < \bar{c}_{Vj}$ for all j. The final level of the sum of the damping coefficients of the added dampers is $\overline{W} = 2.94 \times 10^7$ N s/m. The increment of $\overline{W}$ is given by $\Delta W = \overline{W}/300$.

The distributions of the optimal viscous damping coefficients of supplemental dampers obtained via the present procedure are plotted in Figure 4.6. It can be observed that the supplemental dampers are added in the third story first and then in the second, fourth and fifth stories successively. This order of placement of supplemental dampers is based on the algorithm explained in the above section. Figure 4.7 shows the variation of the first-order derivatives (sensitivities) of the objective function with respect to the design variables; that is, the viscous damping coefficients of the supplemental dampers. It can be observed from Figure 4.7 that the optimality criteria are satisfied in all the stories. This process just indicates the automatic satisfaction of the optimality criteria in Equations 4.10 and 4.12a.

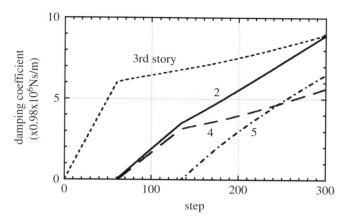

Figure 4.6 Distributions of optimal viscous damping coefficients. (Reproduced with permission from *Structural and Multidisciplinary Optimization*, Vol. 20, No. 4, pp 280–287, 2000, I. Takewaki, Optimal Damper Placement for Planar Building Frames Using Transfer Functions, with permission from Springer.).

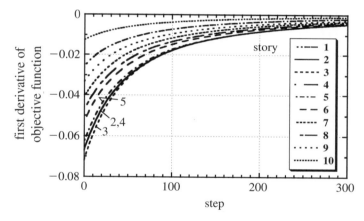

Figure 4.7 Variation of first-order derivatives of objective function with respect to design variables. (Reproduced with permission from *Structural and Multidisciplinary Optimization*, Vol. 20, No. 4, pp 280–287, 2000, I. Takewaki, Optimal Damper Placement for Planar Building Frames Using Transfer Functions, with permission from Springer.).

Figure 4.8 illustrates the comparison of the variations of the lowest-mode damping ratio for the optimal placement, the uniform placement and the restricted uniform placement to the second, third, fourth, and fifth stories. It can be seen from Figure 4.8 that the optimal damper placement actually increases the lowest-mode damping ratio effectively. Figure 4.9 shows the comparison of the variations of the objective function for the optimal placement, the uniform placement, and the restricted uniform placement to the second, third, fourth, and fifth stories. It can be observed that the optimal

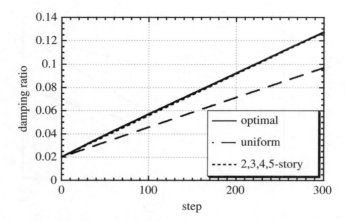

Figure 4.8 Variation of lowest mode damping ratio in optimal placement, uniform placement, and restricted uniform placement to the second, third, fourth, and fifth stories. (Reproduced with permission from *Structural and Multidisciplinary Optimization*, Vol. 20, No. 4, pp 280–287, 2000, I. Takewaki, Optimal Damper Placement for Planar Building Frames Using Transfer Functions, with permission from Springer.).

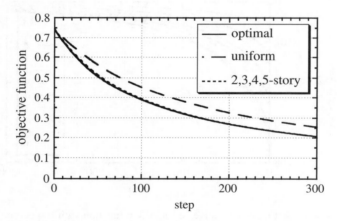

Figure 4.9 Variation of objective function in optimal placement, uniform placement, and restricted uniform placement to the second, third, fourth, and fifth stories. (Reproduced with permission from *Structural and Multidisciplinary Optimization*, Vol. 20, No. 4, pp 280–287, 2000, I. Takewaki, Optimal Damper Placement for Planar Building Frames Using Transfer Functions, with permission from Springer.).

damper placement obtained via the present procedure certainly reduces the objective function effectively compared with the corresponding uniform distribution. It may be said that, once optimal positions are found from the present procedure, the restricted uniform placement to these stories can reduce the objective function almost as effectively as the optimal placement.

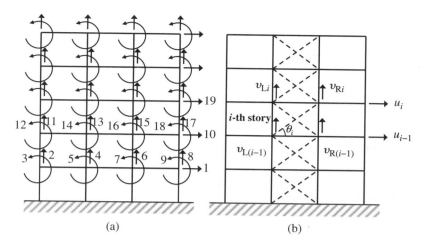

Figure 4.10   An n-story s-span planar building frame.

4.6 Maxwell-type Modeling of Damper Systems

4.6.1 Modeling of a Main Frame

Consider again an n-story s-span planar building frame, as shown in Figure 4.10(a). Let $\mathbf{u}(t)$ denote the generalized displacements in the system coordinate system (see Figure 4.10(b)). The nodal mass and moment of inertia are taken into account at every node. Let $\mathbf{M}$ denote the system mass matrix of the main frame. All the member stiffnesses of the main frame are assumed to be given and its system stiffness matrix is described by $\mathbf{K}$. The structural damping of the frame is considered and the system damping matrix due to this structural damping is given by $\mathbf{C}$. The undamped fundamental natural circular frequency of the main frame is denoted by ω_1.

4.6.2 Modeling of a Damper–Support-member System

Added viscous dampers installed in the stories are assumed to be supported by auxiliary members. The stiffnesses $\mathbf{k}_m = \{k_{mi}\}$ of the supporting members are given and the damping coefficients $\mathbf{c}_m = \{c_{mi}\}$ of the added viscous dampers are selected as the design variables. In this chapter, every damper–support-member system is to be modeled by a Maxwell-type damper–spring model, as shown in Figure 4.11 (Takewaki and Yoshitomi, 1998; Takewaki and Uetani, 1999). Let c_m and k_m denote the damping coefficient of the damper and the stiffness of the spring respectively in the representative story. It is well known that the complex axial member force $F(\omega)$ (ω: excitation frequency) of a Maxwell-type model can be related to the complex total elongation $\Delta(\omega)$ in terms of the following complex stiffness $K(\omega)$ (Flugge, 1967):

$$F(\omega) = K(\omega)\Delta(\omega) = \{K_R(\omega) + iK_I(\omega)\}\Delta(\omega) \qquad (4.26)$$

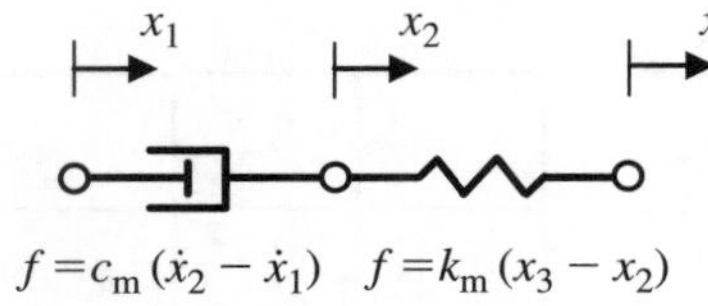

$$x_1 \quad\quad x_2 \quad\quad x_3$$

$$F = (K_R + iK_I)\Delta$$

$$\Delta = X_3 - X_1$$

$$f = c_m(\dot{x}_2 - \dot{x}_1) \quad f = k_m(x_3 - x_2)$$

Figure 4.11 Maxwell-type damper–spring model.

where i is the imaginary unit and $K_R(\omega)$ and $K_I(\omega)$ are as follows:

$$K_R(\omega) = \frac{k_m c_m^2 \omega^2}{k_m^2 + c_m^2 \omega^2} \tag{4.26a}$$

$$K_I(\omega) = \frac{k_m^2 c_m \omega}{k_m^2 + c_m^2 \omega^2} \tag{4.26b}$$

For comparison with a Kelvin–Voigt-type model, the following quantities are introduced:

$$k_M(\omega) = K_R(\omega) \tag{4.27a}$$

$$c_M(\omega) = \frac{K_I(\omega)}{\omega} \tag{4.27b}$$

Consider a forced steady-state vibration with the circular frequency ω_1 and define the following quantities for later convenience:

$$\hat{k}_M = K_R(\omega_1) \tag{4.28a}$$

$$\hat{c}_M = K_I(\omega_1)/\omega_1 \tag{4.28b}$$

It may be appropriate to define the following stiffness and damping matrices related to the added dampers:

$$\mathbf{K}_M = \sum_i \hat{k}_{Mi} \mathbf{B}_i \tag{4.29a}$$

$$\mathbf{C}_M = \sum_i \hat{c}_{Mi} \mathbf{B}_i \tag{4.29b}$$

In Equation 4.29, $\mathbf{B}_i$ is a square matrix of dimension $\{2 \times (s+1) + 1\} \times n$ which can be constructed from the following element matrix $\mathbf{B}_i^*$ in the ith story by allocating the member coordinates to the system coordinates:

$$\mathbf{B}_i^* = \begin{bmatrix} 2\cos^2\theta_i & \cos\theta_i\sin\theta_i & -\cos\theta_i\sin\theta_i & -2\cos^2\theta_i & -\cos\theta_i\sin\theta_i & \cos\theta_i\sin\theta_i \\ & \sin^2\theta_i & 0 & -\cos\theta_i\sin\theta_i & -\sin^2\theta_i & 0 \\ & & \sin^2\theta_i & \cos\theta_i\sin\theta_i & 0 & -\sin^2\theta_i \\ & & & 2\cos^2\theta_i & \cos\theta_i\sin\theta_i & -\cos\theta_i\sin\theta_i \\ & \text{sym.} & & & \sin^2\theta_i & 0 \\ & & & & & \sin^2\theta_i \end{bmatrix}$$

$$\tag{4.30}$$

where θ_i denotes the angle of the damper to the beam in the ith story. $\mathbf{B}_i^*$ is defined with respect to the displacements $\{u_{i-1} \; v_{L(i-1)} \; v_{R(i-1)} \; u_i \; v_{Li} \; v_{Ri}\}^T$ (see Figure 4.10(b)). It should be noted that the matrix $\mathbf{B}_i^*$ includes the effects of both dampers in the ith story.

4.6.3 Effects of Support-Member Stiffnesses on Performance of Dampers

To clarify the effect of support-member stiffness on performance of dampers, consider the five-story one-span planar steel building frame shown in Figure 4.12. Every column has the common square-tube-type cross-section (60 cm × 1.6 cm) and every beam has also the common wide-flange cross-section (65 cm × 20 cm × 1.2 cm × 2.8 cm). The Young's modulus of the members is $E = 2.1 \times 10^3$ tonf/cm^2. The nodal mass and moment of inertia at every node are assumed to be $m = 16.8$ ton and $I = 1.72 \times 10^5$ ton cm^2 respectively. The undamped fundamental natural circular frequency of the frame is $\omega_1 = 10.15$ rad/s. For several support-member stiffnesses, the damping coefficients of uniformly distributed added dampers have been increased gradually. Figure 4.13 shows the lowest mode damping ratio, due to complex eigenvalue analysis, with respect to the level of the damping coefficients. At the final state (400 steps), the sum of the damping coefficients of the added dampers is 10 tonf s/cm. In evaluating the lowest mode damping ratio, the stiffness matrix and the damping matrix have been fixed to $\mathbf{K} + \mathbf{K}_\mathrm{M}$ and $\mathbf{C} + \mathbf{C}_\mathrm{M}$ respectively. Figure 4.14 shows the objective function defined in Equation 4.13 with respect to the level of the damping coefficients. It can be observed from Figures 4.13 and 4.14 that, while the dampers work well with a rather stiff support member, they do not with a flexible one.

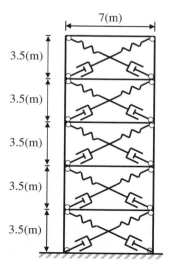

Figure 4.12 A five-story one-span planar steel building frame.

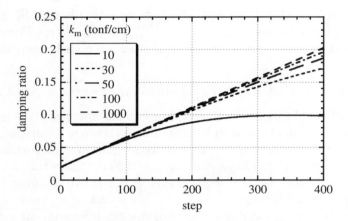

Figure 4.13 Lowest mode damping ratio, due to complex eigenvalue analysis, with respect to level of damping coefficients (uniform damper placement) for various cases of support-member stiffness (1 tonf/cm $= 0.98 \times 10^6$ N/m). (I. Takewaki and S. Yoshitomi, "Effects of Support Stiffnesses on Optimal Damper Placement for a Planar Building Frame," *Journal of the Structural Design of Tall Buildings*, Vol.7, No.4. © 1998 John Wiley & Sons, Ltd).

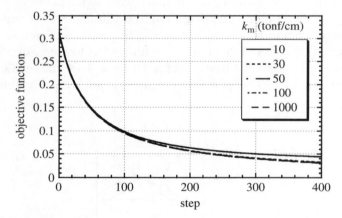

Figure 4.14 Objective function with respect to level of damping coefficients (uniform damper placement) for various cases of support-member stiffness (1 tonf/cm $= 0.98 \times 10^6$ N/m). (I. Takewaki and S. Yoshitomi, "Effects of Support Stiffnesses on Optimal Damper Placement for a Planar Building Frame," *Journal of the Structural Design of Tall Buildings*, Vol.7, No.4. © 1998 John Wiley & Sons, Ltd).

4.7 Problem of Optimal Damper Placement and Optimality Criteria (Maxwell-type Modeling)

When this planar frame without added dampers is subjected to a horizontal acceleration $\ddot{u}_g(t)$ at the fixed base, the equation of motion for this model may be written as

$$\mathbf{K}u(t) + \mathbf{C}\dot{u}(t) + \mathbf{M}\ddot{u}(t) = -\mathbf{M}r\ddot{u}_g(t) \tag{4.31}$$

where **r** is the influence coefficient vector.

Let $\mathbf{U}(\omega)$ and $\ddot{U}_g(\omega)$ denote the Fourier transforms of $\mathbf{u}(t)$ and $\ddot{u}_g(t)$ respectively. Fourier transformation of Equation 4.31 can be reduced to the following form:

$$(\mathbf{K} + i\omega\mathbf{C} - \omega^2\mathbf{M})\mathbf{U}(\omega) = -\mathbf{Mr}\ddot{U}_g(\omega) \qquad (4.32)$$

where i is the imaginary unit.

When the added dampers are incorporated, Equation 4.32 may be modified to the following form:

$$\{(\mathbf{K} + \mathbf{K}_M) + i\omega(\mathbf{C} + \mathbf{C}_M) - \omega^2\mathbf{M}\}\mathbf{U}_M(\omega) = -\mathbf{Mr}\ddot{U}_g(\omega) \qquad (4.33)$$

Let ω_1 denote the undamped fundamental natural circular frequency of the main frame (the frame without added viscous dampers) and let us define new quantities $\hat{\mathbf{U}}$ by

$$\hat{\mathbf{U}} \equiv \frac{\mathbf{U}_M(\omega_1)}{\ddot{U}_g(\omega_1)} \qquad (4.34)$$

$\hat{U}_i$ is equal to the value such that ω_1 is substituted in the frequency response function obtained as $U_{Mi}(\omega)$ after substituting $\ddot{U}_g(\omega) = 1$ in Equation 4.33. This quantity has been utilized by Takewaki (1997b) for structural redesign problems. It should be noted that, because $\mathbf{M}$ and $\mathbf{K}$ are prescribed, ω_1 is a given value. Owing to Equation 4.33 ($\omega = \omega_1$) and Equation 4.34, $\hat{\mathbf{U}}$ must satisfy

$$\mathbf{A}\hat{\mathbf{U}} = -\mathbf{Mr} \qquad (4.35)$$

where

$$\mathbf{A} = (\mathbf{K} + \mathbf{K}_M) + i\omega_1(\mathbf{C} + \mathbf{C}_M) - \omega_1^2\mathbf{M} \qquad (4.36)$$

Let $\hat{\delta}_i$ denote the transfer function at $\omega = \omega_1$ of the interstory drift in the ith story. $\hat{\boldsymbol{\delta}} = \{\hat{\delta}_i\}$ can be derived from $\hat{\mathbf{U}}$ by the transformation $\hat{\boldsymbol{\delta}} = \mathbf{T}\hat{\mathbf{U}}$, where $\mathbf{T}$ is a constant transformation matrix.

It should be remarked here that the squares of the transfer function amplitudes are useful from physical points of view because they can be transformed into response mean squares (statistical quantities) after multiplication with the PSD function of a disturbance and integration in the frequency range. Since the transfer function amplitude of a nodal displacement evaluated at the undamped fundamental natural circular frequency can be related to the level of this response mean square, these transfer function amplitudes are treated as controlled quantities here.

The problem of optimal damper positioning for a Maxwell (PODPM)-type model may be described as follows.

Problem 4.2 PODPM Find the damping coefficients $\mathbf{c}_m = \{c_{mi}\}$ of added dampers which minimize the sum of the transfer function amplitudes of the interstory drifts evaluated at the undamped fundamental natural circular frequency ω_1

$$V = \sum_{i=1}^{n} |\hat{\delta}_i(\mathbf{c}_m)| \qquad (4.37)$$

subject to a constraint on the sum of the damper damping coefficients

$$\sum_{i=1}^{n} c_{mi} = \overline{W} \quad (\overline{W}: \text{specified value}) \tag{4.38a}$$

and to constraints on the damping coefficients of added dampers

$$0 \le c_{mi} \le \overline{c}_{mi} \quad (i = 1, \cdots, n) \tag{4.38b}$$

where $\overline{c}_{mi}$ is the upper bound of the damping coefficient of the added damper in the ith story.

The generalized Lagrangian L for Problem 4.2 may be expressed in terms of Lagrange multipliers $\lambda, \boldsymbol{\mu} = \{\mu_i\}$, and $\boldsymbol{\nu} = \{\nu_i\}$:

$$L(\mathbf{c_m}, \lambda, \boldsymbol{\mu}, \boldsymbol{\nu}) = \sum_{i=1}^{n} |\hat{\delta}_i(\mathbf{c_m})| + \lambda \left(\sum_{i=1}^{n} c_{mi} - \overline{W} \right) + \sum_{i=1}^{n} \mu_i(0 - c_{mi}) + \sum_{i=1}^{n} \nu_i(c_{mi} - \overline{c}_{mi}) \tag{4.39}$$

For simplicity of expression, the argument $(\mathbf{c_m})$ will be omitted hereafter.

4.7.1 Optimality Criteria

The principal optimality criteria for Problem 4.2 without active upper and lower bound constraints on damping coefficients may be derived from the stationarity conditions of the generalized Lagrangian $L(\boldsymbol{\mu} = \mathbf{0}, \boldsymbol{\nu} = \mathbf{0})$ with respect to $\mathbf{c_m}$ and λ:

$$\left(\sum_{i=1}^{n} |\hat{\delta}_i| \right)_{,j} + \lambda = 0 \quad \text{for } 0 < c_{mj} < \overline{c}_{mj} \quad (j = 1, \cdots, n) \tag{4.40}$$

$$\sum_{i=1}^{n} c_{mi} - \overline{W} = 0 \tag{4.41}$$

Here, and in the following, $(\cdot)_{,j}$ denotes the partial differentiation with respect to c_{mj}. If the constraints of Equation 4.38b are active, then Equation 4.40 must be modified into the following forms:

$$\left(\sum_{i=1}^{n} |\hat{\delta}_i| \right)_{,j} + \lambda \ge 0 \quad \text{for } c_{mj} = 0 \tag{4.42a}$$

$$\left(\sum_{i=1}^{n} |\hat{\delta}_i| \right)_{,j} + \lambda \le 0 \quad \text{for } c_{mj} = \overline{c}_{mj} \tag{4.42b}$$

4.8 Solution Algorithm (Maxwell-type Modeling)

In the present procedure, the model without added dampers, namely $c_{mj} = 0$ ($j = 1, \cdots, n$), is employed as the initial model. The damping coefficients of added dampers are *increased gradually* in the present solution algorithm. Let Δc_{mi} and ΔW respectively denote the increment of the damping coefficient of the ith-story added damper and the increment of the sum of the damping coefficients of added dampers. Given ΔW, the problem is to determine the effective position and amount of the increments of the damping coefficients of added dampers. To develop this algorithm, the first- and second-order sensitivities of the objective function with respect to a design variable are derived in the following.

Differentiation of Equation 4.35 with respect to a design variable c_{mj} provides

$$\mathbf{A}_{,j}\hat{\mathbf{U}} + \mathbf{A}\hat{\mathbf{U}}_{,j} = \mathbf{0} \tag{4.43}$$

From Equation 4.36, $\mathbf{A}_{,j}$ may be expressed as follows:

$$\mathbf{A}_{,j} = \mathbf{K}_{M,j} + i\omega_1 \mathbf{C}_{M,j} \tag{4.44}$$

where

$$\mathbf{K}_{M,j} = \hat{k}_{Mj,j}\mathbf{B}_j \tag{4.45a}$$

$$\mathbf{C}_{M,j} = \hat{c}_{Mj,j}\mathbf{B}_j \tag{4.45b}$$

The quantities $\hat{k}_{Mj,j}$ and $\hat{c}_{Mj,j}$ are derived as

$$\hat{k}_{Mj,j} = \frac{2c_{mj}\omega_1^2}{k_{mj}^2 + c_{mj}^2\omega_1^2}(k_{mj} - \hat{k}_{Mj}) \tag{4.46a}$$

$$\hat{c}_{Mj,j} = \frac{k_{mj}}{k_{mj}^2 + c_{mj}^2\omega_1^2}(k_{mj} - 2\hat{k}_{Mj}) \tag{4.46b}$$

Since $\mathbf{A}$ is regular, the first-order sensitivities of $\hat{\mathbf{U}}$ are derived from Equation 4.43 as

$$\hat{\mathbf{U}}_{,j} = -\mathbf{A}^{-1}\mathbf{A}_{,j}\hat{\mathbf{U}} \tag{4.47a}$$

The first-order sensitivities of the interstory drift $\hat{\delta}$ are then expressed as

$$\hat{\delta}_{,j} = \mathbf{T}\hat{\mathbf{U}}_{,j} = -\mathbf{T}\mathbf{A}^{-1}\mathbf{A}_{,j}\hat{\mathbf{U}} \tag{4.47b}$$

The quantity $\hat{\delta}_i$ may be rewritten formally as

$$\hat{\delta}_i = \text{Re}[\hat{\delta}_i] + i\text{Im}[\hat{\delta}_i] \tag{4.48}$$

where Re[ ] and Im[ ] indicate the real and imaginary parts respectively of a complex number. The first-order sensitivity of $\hat{\delta}_i$ may be formally expressed as

$$\hat{\delta}_{i,j} = (\text{Re}[\hat{\delta}_i])_{,j} + i(\text{Im}[\hat{\delta}_i])_{,j} \tag{4.49}$$

The absolute value of $\hat{\delta}_i$ is defined by

$$|\hat{\delta}_i| = \sqrt{(\text{Re}[\hat{\delta}_i])^2 + (\text{Im}[\hat{\delta}_i])^2} \qquad (4.50)$$

The first-order sensitivity of $|\hat{\delta}_i|$ may then be expressed as

$$|\hat{\delta}_i|_{,j} = \frac{1}{|\hat{\delta}_i|}\{\text{Re}[\hat{\delta}_i](\text{Re}[\hat{\delta}_i])_{,j} + \text{Im}[\hat{\delta}_i](\text{Im}[\hat{\delta}_i])_{,j}\} \qquad (4.51)$$

where $(\text{Re}[\hat{\delta}_i])_{,j}$ and $(\text{Im}[\hat{\delta}_i])_{,j}$ are calculated from Equations 4.47b and 4.49.

A general expression $|\hat{\delta}_i|_{,j\ell}$ is derived here for the purpose of developing a solution procedure for Problem 4.2. Differentiation of Equation 4.51 with respect to $c_{m\ell}$ leads to

$$|\hat{\delta}_i|_{,j\ell} = \frac{1}{|\hat{\delta}_i|^2}(|\hat{\delta}_i|\{(\text{Re}[\hat{\delta}_i])_{,\ell}(\text{Re}[\hat{\delta}_i])_{,j} + \text{Re}[\hat{\delta}_i](\text{Re}[\hat{\delta}_i])_{,j\ell} + (\text{Im}[\hat{\delta}_i])_{,\ell}(\text{Im}[\hat{\delta}_i])_{,j}$$

$$+ \text{Im}[\hat{\delta}_i](\text{Im}[\hat{\delta}_i])_{,j\ell}\} - |\hat{\delta}_i|_{,\ell}\{\text{Re}[\hat{\delta}_i](\text{Re}[\hat{\delta}_i])_{,j} + \text{Im}[\hat{\delta}_i](\text{Im}[\hat{\delta}_i])_{,j}\}) \qquad (4.52)$$

$(\text{Re}[\hat{\delta}_i])_{,j\ell}$ and $(\text{Im}[\hat{\delta}_i])_{,j\ell}$ in Equation 4.52 are found from

$$\hat{\delta}_{,j\ell} = \mathbf{TA}^{-1}\mathbf{A}_{,\ell}\mathbf{A}^{-1}\mathbf{A}_{,j}\hat{\mathbf{U}} - \mathbf{TA}^{-1}\mathbf{A}_{,j\ell}\hat{\mathbf{U}} - \mathbf{TA}^{-1}\mathbf{A}_{,j}\hat{\mathbf{U}}_{,\ell} \qquad (4.53)$$

which is derived by differentiating Equation 4.47b with respect to $c_{m\ell}$ and using the relation $\mathbf{A}_{,\ell}^{-1} = -\mathbf{A}^{-1}\mathbf{A}_{,\ell}\mathbf{A}^{-1}$. It should be noted here that, while the components in the matrix $\mathbf{A}$ are linear functions of design variables and $\mathbf{A}_{,j\ell}$ becomes a null matrix for all j and ℓ in Takewaki (1997a), those are nonlinear functions of design variables $\mathbf{c}_m$ in the present problem for a Maxwell-type model. $\mathbf{A}_{,j\ell}$ may be written as

$$\mathbf{A}_{,j\ell} = \mathbf{K}_{\text{M},j\ell} + i\omega_1\mathbf{C}_{\text{M},j\ell} = (\hat{k}_{\text{M}j,j\ell} + i\omega_1\hat{c}_{\text{M}j,j\ell})\delta_{j\ell}\mathbf{B}_j \qquad (4.54)$$

where $\delta_{j\ell}$ is the Kronecker delta and $\hat{k}_{\text{M}j,jj}$ and $\hat{c}_{\text{M}j,jj}$ are as follows:

$$\hat{k}_{\text{M}j,jj} = \frac{2\omega_1^2}{k_{\text{m}j}^2}\{\hat{c}_{\text{M}j,j}(k_{\text{m}j} - \hat{k}_{\text{M}j}) - \hat{c}_{\text{M}j}\hat{k}_{\text{M}j,j}\} \qquad (4.55a)$$

$$\hat{c}_{\text{M}j,jj} = \frac{2\omega_1^2}{k_{\text{m}j}^2 + c_{\text{m}j}^2\omega_1^2}\left(\frac{2}{k_{\text{m}j}}\hat{k}_{\text{M}j}\hat{c}_{\text{M}j} - \hat{c}_{\text{M}j} - \frac{k_{\text{m}j}}{\omega_1^2}\hat{k}_{\text{M}j,j}\right) \qquad (4.55b)$$

Substitution of Equation 4.47a into Equation 4.53 leads to the following form:

$$\hat{\delta}_{,j\ell} = \mathbf{TA}^{-1}\mathbf{A}_{,\ell}\mathbf{A}^{-1}\mathbf{A}_{,j}\hat{\mathbf{U}} - \mathbf{TA}^{-1}\mathbf{A}_{,j\ell}\hat{\mathbf{U}} + \mathbf{TA}^{-1}\mathbf{A}_{,j}\mathbf{A}^{-1}\mathbf{A}_{,\ell}\hat{\mathbf{U}}$$

$$= \mathbf{TA}^{-1}(\mathbf{A}_{,\ell}\mathbf{A}^{-1}\mathbf{A}_{,j} - \mathbf{A}_{,j\ell} + \mathbf{A}_{,j}\mathbf{A}^{-1}\mathbf{A}_{,\ell})\hat{\mathbf{U}} \qquad (4.56)$$

The derivatives $|\hat{\delta}_i|_{,j\ell}$ are derived from Equation 4.52. $(\text{Re}[\hat{\delta}_i])_{,j}$ and $(\text{Im}[\hat{\delta}_i])_{,j}$ in Equation 4.52 are calculated from Equation 4.47b and $(\text{Re}[\hat{\delta}_i])_{,j\ell}$ and $(\text{Im}[\hat{\delta}_i])_{,j\ell}$ in Equation 4.52 are found from Equation 4.56.

The solution algorithm in the case of $c_{mj} < \bar{c}_{mj}$ for all j may be summarized as follows:

Step 0 Initialize all the added dampers as $c_{mj} = 0$ ($j = 1, \ldots, n$). In the initial state, the damping is the structural damping alone in the frame. Assume ΔW.

Step 1 Compute the first-order sensitivity $(\sum_j |\hat{\delta}_j|)_{,i}$ of the objective function by Equation 4.51.

Step 2 Find the index k such that

$$- \left(\sum_j |\hat{\delta}_j| \right)_{,k} = \max_i \left\{ - \left(\sum_j |\hat{\delta}_j| \right)_{,i} \right\} \tag{4.57}$$

Step 3 Update the objective function $\sum_j |\hat{\delta}_j|$ by the first-order approximation $\sum_j |\hat{\delta}_j| + (\sum_j |\hat{\delta}_j|)_{,k} \Delta c_{mk}$, where $\Delta c_{mk} = \Delta W$.

Step 4 Update the first-order sensitivity $(\sum_j |\hat{\delta}_j|)_{,i}$ of the objective function by the first-order approximation $(\sum_j |\hat{\delta}_j|)_{,i} + (\sum_j |\hat{\delta}_j|)_{,ik} \Delta c_{mk}$ using Equation 4.52.

Step 5 If, in Step 4, there exists a damper of an index j such that the condition

$$- \left(\sum_i |\hat{\delta}_i| \right)_{,k} = \max_{j, j \neq k} \left\{ - \left(\sum_i |\hat{\delta}_i| \right)_{,j} \right\} \tag{4.58}$$

is satisfied, then stop and compute the corresponding variation $\Delta \tilde{c}_{mk}$ of the damping coefficient of the supplemental damper. Update the first-order sensitivity $(\sum_j |\hat{\delta}_j|)_{,i}$ of the objective function by the first-order approximation $(\sum_j |\hat{\delta}_j|)_{,i} + (\sum_j |\hat{\delta}_j|)_{,ik} \Delta \tilde{c}_{mk}$ using Equation 4.52.

Step 6 Repeat from Step 2 through Step 5 until the constraint in Equation 4.38a (i.e., $\sum_{i=1}^{n} c_{mi} = \overline{W}$) is satisfied.

In Step 2 and Step 3, the direction which reduces the objective function most effectively under the condition $\sum_{i=1}^{n} \Delta c_{mi} = \Delta W$ is found and the design is updated in that direction. It is appropriate, therefore, to call the present algorithm "the steepest direction search algorithm." This algorithm may be similar to the well-known steepest descent method in the mathematical programming (see Figure 4.4 for viscous-type modeling). However, while the conventional steepest descent method uses the gradient vector itself of the objective function as the direction and does not utilize optimality criteria, the present algorithm takes full advantage of the newly derived optimality criteria in Equations 4.40, 4.42a, and 4.42b and does not adopt the gradient vector as the direction. In other words, the steepest direction search guarantees the satisfaction of the optimality criteria. For example, if Δc_{mk} is added to the kth added damper in which Equation 4.57 is satisfied, then its damper ($c_{mk} > 0$) satisfies the optimality condition in Equation 4.40 and the other dampers ($c_{mj} = 0$, $j \neq k$) satisfy the optimality condition in Equation 4.42a. It should be noted that a series of subproblems is introduced here tentatively in which the damper level $\overline{W}$ is increased gradually by ΔW from zero through the specified value.

If multiple indices $k_1, \cdots, k_p$ exist in Step 2, then $\sum_j |\hat{\delta}_j|$ and $(\sum_i |\hat{\delta}_i|)_{,j}$ have to be updated by the following rules:

$$\sum_j |\hat{\delta}_j| \rightarrow \sum_j |\hat{\delta}_j| + \sum_{i=k_1}^{k_p} \left(\sum_j |\hat{\delta}_j| \right)_{,i} \Delta c_{mi} \tag{4.59a}$$

$$\left(\sum_i |\hat{\delta}_i| \right)_{,j} \rightarrow \left(\sum_i |\hat{\delta}_i| \right)_{,j} + \sum_{\ell=k_1}^{k_p} \left(\sum_i |\hat{\delta}_i| \right)_{,j\ell} \Delta c_{m\ell} \tag{4.59b}$$

Furthermore, the index k in Step 5 has to be replaced by the indices $k_1, \cdots, k_p$. The ratios among the magnitudes Δc_{mi} must be determined so that the following relations are satisfied:

$$\left(\sum_j |\hat{\delta}_j| \right)_{,k_1} + \sum_{i=k_1}^{k_p} \left(\sum_j |\hat{\delta}_j| \right)_{,k_1 i} \Delta c_{mi} = \cdots = \left(\sum_j |\hat{\delta}_j| \right)_{,k_p}$$

$$+ \sum_{i=k_1}^{k_p} \left(\sum_j |\hat{\delta}_j| \right)_{,k_p i} \Delta c_{mi} \tag{4.60}$$

Equation 4.60 requires that the optimality condition Equation 4.40 continues to be satisfied in the dampers with the indices $k_1, \ldots, k_p$.

In the case where the damping coefficients of some added dampers attain their upper bounds, such constraints must be incorporated in the aforementioned algorithm. In that case, the increment Δc_{mk} is added subsequently to the damper in which $-(\sum_j |\hat{\delta}_j|)_{,k}$ attains the maximum among all the dampers, except those attaining the upper bound.

4.9 Numerical Examples II (Maxwell-type Modeling)

Consider the five-story one-span planar steel building frame shown in Figure 4.12. All the properties of the main frame are stated in Section 4.6. The undamped fundamental natural circular frequency of the model is $\omega_1 = 10.15$ rad/s. The structural damping of the main frame is given by a critical damping ratio of 0.02 in the lowest eigenvibration. Optimal damper positioning has been derived for three support-member stiffnesses, namely $k_m = 50$, 100, and 1000 tonf/cm (equal to (50, 100, and 1000) $\times 9.8 \times 10^5$ N/m). These values have been determined based upon identification on several existing buildings. It is assumed that all the constraints on upper bounds of the damping coefficients are inactive; that is, $c_j < \overline{c}_{mj}$ for all j. The final level of the sum of the damping coefficients of the added dampers is $\overline{W} = 10$ tonf s/cm. The increment of $\overline{W}$ is given by $\Delta W = \overline{W}/400$.

The distributions of the optimal damping coefficients obtained via the present procedure are plotted in Figure 4.15 for $k_m = 50$, 100, and 1000 tonf/cm. It can be observed that the dampers are added in the second story first and then in the third story. The ratio of the damping coefficients in these two stories is strongly dependent on the support-member stiffness. In Figure 4.15(a), the dampers in the second and third stories exhibit constant values after 360 steps. This phenomenon results from the fact that the damping coefficient in one of the other stories becomes negative and the procedure stops. Figure 4.16 shows the variation of the first-order derivatives of the objective function with respect to the design variables. It can be observed from Figure 4.16 that the

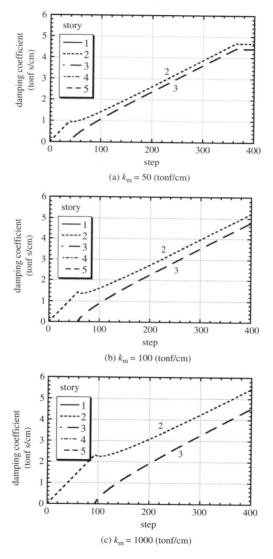

Figure 4.15 Distribution of optimal damping coefficients obtained via the present procedure for various cases of support-member stiffness $k_m = 50$, 100, and 1000 tonf/cm (1 tonf/cm $= 0.98 \times 10^6$ N/m). (I. Takewaki and S. Yoshitomi, "Effects of Support Stiffnesses on Optimal Damper Placement for a Planar Building Frame," *Journal of the Structural Design of Tall Buildings*, Vol.7, No.4. © 1998 John Wiley & Sons, Ltd).

optimality criteria are satisfied in all the stories. In Figure 4.16(c), for example, the
first-order derivative of the objective function with respect to the design variable in
the second story exhibits the maximum absolute value until 90 steps where the damping coefficient of the damper in the third story begins to increase. After 90 steps, the
first-order derivatives of the objective function with respect to the design variables in

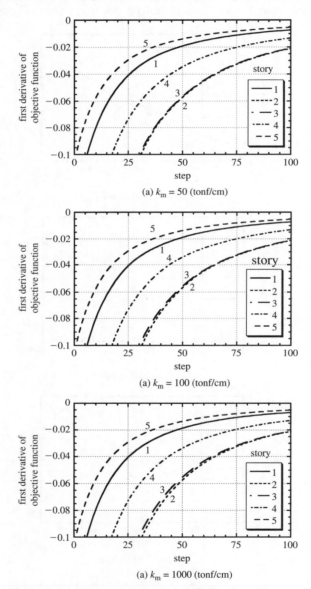

Figure 4.16 Variation of first-order derivatives of objective function with respect to the design vari-
ables for various cases of support-member stiffness (1 tonf/cm = 0.98 × 10^6 N/m). (I. Takewaki and
S. Yoshitomi, "Effects of Support Stiffnesses on Optimal Damper Placement for a Planar Building Frame,"
Journal of the Structural Design of Tall Buildings, Vol.7, No.4. © 1998 John Wiley & Sons, Ltd).

the second and third stories exhibit the same maximum absolute value. This process just indicates the satisfaction of the optimality criteria in Equations 4.49 and 4.42a. Figure 4.17 illustrates the variation of the lowest-mode damping ratio and Figure 4.18 shows the variation of the objective function. It can be observed from these figures that, while the lowest mode damping ratio is affected significantly by the support-member stiffness in this range of support-member stiffnesses, the objective function is affected slightly by the support-member stiffness. However, in case of rather flexible support-member stiffnesses, as shown in Figures 4.17 and 4.18, these influences may become serious and should be taken into account in the design practice.

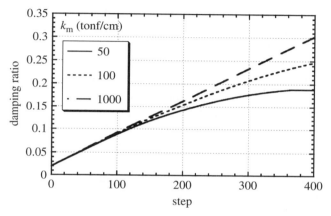

Figure 4.17 Variation of lowest mode damping ratio in optimal damper placement with respect to level of damping coefficients for various cases of support-member stiffness (1 tonf/cm = 0.98 × 10⁶ N/m). (I. Takewaki and S. Yoshitomi, "Effects of Support Stiffnesses on Optimal Damper Placement for a Planar Building Frame," *Journal of the Structural Design of Tall Buildings*, Vol.7, No.4. © 1998 John Wiley & Sons, Ltd).

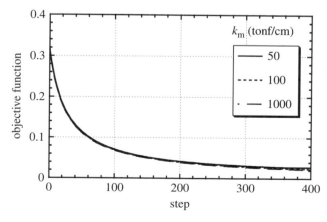

Figure 4.18 Variation of objective function in optimal damper placement with respect to level of damping coefficients for various cases of support-member stiffness (1 tonf/cm = 0.98 × 10⁶ N/m). (I. Takewaki and S. Yoshitomi, "Effects of Support Stiffnesses on Optimal Damper Placement for a Planar Building Frame," *Journal of the Structural Design of Tall Buildings*, Vol.7, No.4. © 1998 John Wiley & Sons, Ltd).

4.10 Nonmonotonic Sensitivity Case

When a negative damping coefficient appears in the fundamental algorithm explained in the previous section, an augmented algorithm has to be devised. In this section, a new augmented algorithm is explained of updating the damping coefficients to $\mathbf{c}_m - \Delta\mathbf{c}_m$ instead of $\mathbf{c}_m + \Delta\mathbf{c}_m$. Since the fundamental algorithm requires the satisfaction of $f(\mathbf{c}_m)_{,k_1} = \cdots = f(\mathbf{c}_m)_{,k_p}$ and Equation 4.60, the following optimality conditions continue to be satisfied automatically within a first-order approximation:

$$f(\mathbf{c}_m - \Delta\mathbf{c}_m)_{,k_1} = \cdots = f(\mathbf{c}_m - \Delta\mathbf{c}_m)_{,k_p} \tag{4.61}$$

Actually, the damping coefficient of the damper with a negative damping coefficient in the fundamental algorithm is taken as a design parameter tentatively. Then, the sum of the damping coefficients is taken again as a design parameter after the value of the sum of the damping coefficients returns to the original value at which a negative damping coefficient starts to appear in the fundamental algorithm (see Figure 4.19).

Consider again the five-story one-span planar steel building frame shown in Figure 4.12. All the properties of the main frame are stated in Section 4.6. The structural damping of the main frame is given by a critical damping ratio of 0.02 in the lowest eigenvibration. Optimal damper positioning has been derived for the support-member stiffness $k_m = 50$ tonf/cm. This value has been determined based upon identification on several existing buildings. It is assumed that all the constraints on upper bounds of the damping coefficients are inactive; that is, $c_j < \bar{c}_{mj}$ for all j. The final level of the sum of the damping coefficients of the added dampers is $\overline{W} = 10$ tonf s/cm. The increment of $\overline{W}$ is given by $\Delta W = \overline{W}/400$.

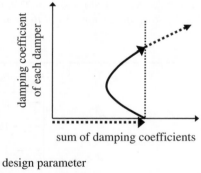

design parameter

Figure 4.19 Augmented algorithm. (I. Takewaki and K. Uetani, "Optimal Damper Placement in Shear-Flexural Building Models," *Proceedings of the 2nd World Conference on Structural Control*, Vol.2. © 1999 John Wiley & Sons, Ltd).

The distributions of the optimal damping coefficients obtained via the fundamental algorithm are plotted in Figure 4.20 for $k_m = 50$ (tonf/cm). It can be observed that the dampers are added in the second story first and then in the third story. The ratio of the damping coefficients in these two stories has been found to be strongly dependent on the support-member stiffness. The damping coefficient in the fourth story exhibits a negative value after 370 steps. To avoid this negative value, the proposed augmented algorithm has been applied to this model. Figure 4.21 illustrates the distributions of the optimal damping coefficients obtained via the augmented algorithm. Figure 4.22 shows the variation of the first-order derivatives of the objective function with respect to the design variables. It can be observed that the optimality criteria are satisfied in all the stories. The first-order derivative of the objective function with respect to

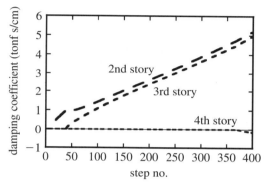

Figure 4.20 Distributions of optimal damping coefficients obtained via the fundamental algorithm for $k_m = 50$ (tonf/cm) (1 tonf/cm $= 0.98 \times 10^6$ N/m). (I. Takewaki and K. Uetani, "Optimal Damper Placement in Shear-Flexural Building Models," *Proceedings of the 2nd World Conference on Structural Control*, Vol.2. © 1999 John Wiley & Sons, Ltd).

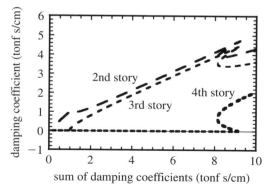

Figure 4.21 Distribution of optimal damping coefficients obtained via the augmented algorithm (1 tonf/cm $= 0.98 \times 10^6$ N/m). (I. Takewaki and K. Uetani, "Optimal Damper Placement in Shear-Flexural Building Models," *Proceedings of the 2nd World Conference on Structural Control*, Vol.2. © 1999 John Wiley & Sons, Ltd).

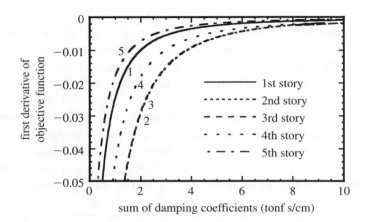

Figure 4.22 Variation of first-order derivatives of objective function with respect to design variables (1 tonf/cm = 0.98×10^6 N/m). (I. Takewaki and K. Uetani, "Optimal Damper Placement in Shear-Flexural Building Models," *Proceedings of the 2nd World Conference on Structural Control*, Vol.2. © 1999 John Wiley & Sons, Ltd).

the design variable in the second story exhibits the maximum absolute value until 40 steps, where the damping coefficient of the damper in the third story begins to increase. After 40 steps, the first-order derivatives of the objective function with respect to the design variables in the second and third stories exhibit the same maximum absolute value. This process just indicates the satisfaction of the optimality criteria in Equations 4.40 and 4.42a. Figure 4.23 illustrates the variation of the lowest mode damping ratio and Figure 4.24 shows the variation of the objective function. Figure 4.25 shows the variation of the sum of seismic-response interstory drifts for a Newmark–Hall design spectrum (maximum ground velocity of 25 cm/s) (Newmark and Hall, 1982). It is observed that the model designed by the proposed method exhibits a smaller seismic response compared with the model with uniform damper capacity.

4.11 Summary

The results are summarized as follows.

1. A systematic procedure called a steepest direction search algorithm has been explained for finding the optimal damper positioning in a planar moment-resisting frame. This problem is aimed at minimizing the sum of the transfer function amplitudes of the interstory drifts evaluated at the undamped fundamental natural frequency subject to a constraint on the sum of the damping coefficients of dampers. The optimal damper positioning is determined based upon the optimality criteria. The features of the present formulation are to be able to deal with any damping system (e.g., viscous-type or Maxwell-type, proportional or nonproportional) to be able to treat any structural system so far as it can be modeled with FE systems

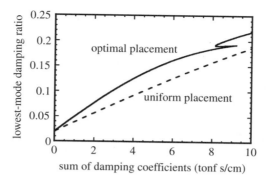

Figure 4.23 Variation of lowest mode damping ratio in optimal placement and uniform placement (1 tonf/cm = 0.98 × 10⁶ N/m). (I. Takewaki and K. Uetani, "Optimal Damper Placement in Shear-Flexural Building Models," *Proceedings of the 2nd World Conference on Structural Control*, Vol.2. © 1999 John Wiley & Sons, Ltd).

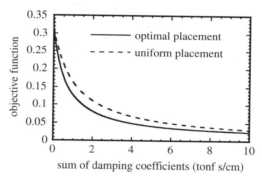

Figure 4.24 Variation of objective function in optimal placement and uniform placement (1 tonf/cm = 0.98 × 10⁶ N/m). (I. Takewaki and K. Uetani, "Optimal Damper Placement in Shear-Flexural Building Models," *Proceedings of the 2nd World Conference on Structural Control*, Vol.2. © 1999 John Wiley & Sons, Ltd).

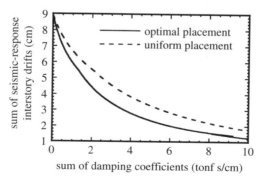

Figure 4.25 Variation of sum of seismic-response interstory drifts for a Newmark–Hall design spectrum in optimal placement and uniform placement (1 tonf/cm = 0.98 × 10⁶ N/m). (I. Takewaki and K. Uetani, "Optimal Damper Placement in Shear-Flexural Building Models," *Proceedings of the 2nd World Conference on Structural Control*, Vol.2. © 1999 John Wiley & Sons, Ltd).

and to consist of a systematic algorithm without any indefinite iterative operation. Efficiency and reliability of the present procedure have been demonstrated through an example.

2. The proposed technique is general and is expected to be applicable to other structural systems. In the case where seismic responses are treated as the objective functions directly, higher mode effects should be included adequately. Even in such a case, the present procedure will play a principal role and a slightly modified algorithm could be developed without much difficulty (see Chapters 6–9). It is well known that the stiffness of members supporting added viscous dampers influences the performance of the dampers and that the seismic response of a building structure is greatly influenced by dynamic soil–structure interaction effects (Takewaki *et al.*, 1998; Takewaki, 1998). A problem of optimal damper placement in a soil–structure interaction model including damper–support-member systems would be of interest in the future research.

3. The support-member stiffness affects greatly the optimal damper positioning and the response suppression level due to added viscous dampers. This stiffness should be taken into account in the design of sizing and positioning of the added dampers. The Maxwell-type damper–spring model, a proper model for a damper–support member system, can be treated adequately by the present transfer function formulation.

Appendix 4.A: Construction of C_V

The damping matrix C_V of the added viscous dampers may be expressed as

$$C_V = \sum_i c_{Vi} B_i \tag{A4.1}$$

B_i is a square matrix of dimension $= \{2 \times (s+1)+1\} \times n$ which can be constructed from the following element matrix B_i^* in the ith story by allocating the member coordinates to the system coordinates:

$$B_i^* = \begin{bmatrix} 2\cos^2\theta_i & \cos\theta_i\sin\theta_i & -\cos\theta_i\sin\theta_i & -2\cos^2\theta_i & -\cos\theta_i\sin\theta_i & \cos\theta_i\sin\theta_i \\ & \sin^2\theta_i & 0 & -\cos\theta_i\sin\theta_i & -\sin^2\theta_i & 0 \\ & & \sin^2\theta_i & \cos\theta_i\sin\theta_i & 0 & -\sin^2\theta_i \\ & & & 2\cos^2\theta_i & \cos\theta_i\sin\theta_i & -\cos\theta_i\sin\theta_i \\ & \text{sym.} & & & \sin^2\theta_i & 0 \\ & & & & & \sin^2\theta_i \end{bmatrix} \tag{A4.2}$$

where θ_i denotes the angle of the damper to the beam in the ith story. B_i^* is defined with respect to the displacements $\{u_{i-1} \, \upsilon_{L(i-1)} \, \upsilon_{R(i-1)} \, u_i \, \upsilon_{Li} \, \upsilon_{Ri}\}^T$ (see Figure 4.10(b)). It should be noted that the matrix B_i^* includes the effects of both dampers in the ith story.

References

Flugge, W. (1967) *Viscoelasticity*, Blaisdell Publishing Company.

Newmark, N.M. and Hall, W.J. (1982) *Earthquake Spectrum and Design*, EERI, Oakland, CA.

Takewaki, I. (1997a) Optimal damper placement for minimum transfer functions. *Earthquake Engineering & Structural Dynamics*, **26** (11), 1113–1124.

Takewaki, I. (1997b) Efficient redesign of damped structural systems for target transfer functions. *Computer Methods in Applied Mechanics and Engineering*, **147** (3–4), 275–286.

Takewaki, I. (1998) Remarkable response amplification of building frames due to resonance with the surface ground. *Soil Dynamics and Earthquake Engineering*, **17** (4), 211–218.

Takewaki, I. (2000) Optimal damper placement for planar building frames using transfer functions. *Structural and Multidisciplinary Optimization*, **20** (4), 280–287.

Takewaki, I. and Uetani, K. (1999) Optimal damper placement in shear-flexural building models. *Proceedings of the Second World Conference on Structural Control*, vol. **2**, John Wiley & Sons, Ltd, Chichester, pp. 1273–1282.

Takewaki, I. and Yoshitomi, S. (1998) Effects of support stiffnesses on optimal damper placement for a planar building frame. *Journal of the Structural Design of Tall Buildings*, **7** (4), 323–336.

Takewaki, I., Nakamura, T., and Hirayama, K. (1998) Seismic frame design via inverse mode design of frame-ground systems. *Soil Dynamics and Earthquake Engineering*, **17** (3), 153–163.

5

Optimal Sensitivity-based Design of Dampers in Three-dimensional Buildings

5.1 Introduction

The problem treated in this chapter is to find the optimal damper placement in a three-dimensional (3-D) shear building model so as to minimize the dynamic compliance of the 3-D shear building model. The dynamic compliance is defined in terms of the sum of the transfer function amplitudes of local interstory drifts computed at the undamped fundamental natural frequency of the 3-D shear building model. The transfer function amplitude indicates the absolute value of a transfer function of a local interstory drift at the undamped fundamental natural frequency. That objective function is minimized subject to a constraint on the sum of damper capacities (damper damping coefficients). In contrast to two-dimensional building structures, torsional effects arising from the difference of the center of mass and the center of rigidity will play an important role in 3-D structures. It is well known that eccentricity can occur even in a structure with uniform mass and stiffness distributions. This is because of irregularities in the building plan or additionally mounted masses after construction being possible sources of the eccentricity. It is expected that optimal damper placement can resolve the issues arising from this torsional effect and improve the seismic performance of 3-D structures.

A set of optimality criteria is derived and a systematic algorithm based on the optimality criteria is introduced first for the optimal damper placement. The features of the present formulation are (i) possible treatment of any damping system, (ii) possible treatment of any structural system so far as it can be modeled with FE systems, and (iii) realization of a systematic algorithm without any indefinite iterative operation. While a different algorithm was devised and a variation from a *uniform* storywise distribution of added dampers was considered in Chapter 2, a variation from the null state is treated in this chapter as in Chapter 4. This procedure helps designers to understand simultaneously which position would be the best and what capacity of dampers would be required to attain a series of desired response performance levels. In the fundamental algorithm explained in the first part of this chapter, it sometimes happens that negative damping coefficients are required to satisfy the optimality conditions. To avoid this unrealistic situation, an updated and augmented algorithm via *parameter switching* is explained.

5.2 Problem of Optimal Damper Placement

5.2.1 Modeling of Structure

Consider the 3-D n-story shear building model shown in Figure 5.1. Let m_i and I_i denote the mass of the ith floor and the mass moment of inertia of the ith floor respectively. The radius of gyration in the ith floor is denoted by $r_i = \sqrt{I_i/m_i}$. For simplicity of presentation, this model is assumed to have a mono-eccentricity and the lateral-torsional vibration only in the x-direction is considered. Furthermore, an excitation in the x-direction is considered. This model consists of m planar frames. It is assumed that added viscous dampers can be installed in all the stories in every frame. The numbering of added dampers is made sequentially. The translational displacement of the center of mass G in the x-direction and the angle of rotation of the floor around a vertical axis through G are the displacements to be considered. Let u_i denote the translational displacement of the center of mass G at ith floor in the x-direction and θ_i denote the angle of rotation of the ith floor around a vertical axis through G. In this chapter, a new displacement $w_i = r_i\theta_i$ is adopted in place of θ_i and the displacement in this direction is called the z-directional displacement. Let $\mathbf{u}(t) = \{u_1 \ w_1 \cdots u_i \ w_i \cdots u_n \ w_n\}^{\mathrm{T}}$ denote the generalized displacements in the system coordinate system.

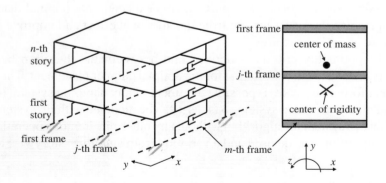

Figure 5.1 A 3-D n-story shear building model.

5.2.2 Mass, Stiffness, and Damping Matrices

The system mass matrix $\mathbf{M}$ of the model can be expressed by

$$\mathbf{M} = \text{diag}(m_1 \ m_1 \cdots m_i \ m_i \cdots m_n \ m_n) \tag{5.1}$$

The symbol diag( ) denotes a diagonal matrix including the terms in the parentheses as the diagonal terms.

Let $K(i, j)$ denote the stiffness element of frame j in the ith story and let $k_{(i,j)}$ denote the translational stiffness of $K(i, j)$. All the member stiffnesses of the model are given and its system stiffness matrix is described by $\mathbf{K}$. The system stiffness matrix may be derived as follows.

5.2.3 Relation of Element-end Displacements with Displacements at Center of Mass

The relations of element-end displacements with displacements at the center of mass may be expressed as

$$\left\{ \begin{array}{c} u^L_{K(i,j)} \\ w^L_{K(i,j)} \\ u^U_{K(i,j)} \\ w^U_{K(i,j)} \end{array} \right\} = \left[\begin{array}{cccc} 1 & -\bar{y}^L_{K(i,j)} & 0 & 0 \\ 0 & 1 & 0 & 0 \\ 0 & 0 & 1 & -\bar{y}^U_{K(i,j)} \\ 0 & 0 & 0 & 1 \end{array} \right] \left\{ \begin{array}{c} u_{i-1} \\ w_{i-1} \\ u_i \\ w_i \end{array} \right\} \tag{5.2}$$

These relations can be rewritten compactly as

$$\mathbf{u}_{K(i,j)} = \mathbf{t}_{K(i,j)} \left\{ \begin{array}{c} \mathbf{u}_{i-1} \\ \mathbf{u}_i \end{array} \right\} \tag{5.3}$$

where

$$\bar{y}^U_{K(i,j)} = \frac{y_{K(i,j)} - y_i}{r_i}$$

$$\bar{y}^L_{K(i,j)} = \frac{y_{K(i,j)} - y_{i-1}}{r_{i-1}}$$

and where $y_{K(i,j)}$ is the y-coordinate of the stiffness element $K(i, j)$, y_i is the y-coordinate of the center of mass of the ith floor, $u^L_{K(i,j)}$ is the x-directional displacement of the lower-end of stiffness element $K(i, j)$, $w^L_{K(i,j)}$ is the z-directional displacement of the lower-end of stiffness element $K(i, j)$, $u^U_{K(i,j)}$ is the x-directional displacement of the upper-end of stiffness element $K(i, j)$, and $w^U_{K(i,j)}$ is the z-directional displacement of the upper-end of stiffness element $K(i, j)$. The origin of the coordinate system is placed at an arbitrary point (see Figure 5.1).

5.2.4 Relation of Forces at Center of Mass due to Stiffness Element K(i, j) with Element-end Forces

The relations of forces $\mathbf{F}_{K(i,j)}$ at the center of mass due to $K(i,j)$ with element-end forces $\mathbf{f}_{K(i,j)}$ may be expressed as

$$\mathbf{F}_{K(i,j)} = \mathbf{t}_{K(i,j)}{}^{\mathrm{T}}\mathbf{f}_{K(i,j)} \tag{5.4}$$

where

$$\mathbf{F}_{K(i,j)} = \{F_{xK(i,j)}^{\mathrm{L}} \ F_{zK(i,j)}^{\mathrm{L}} \ F_{xK(i,j)}^{\mathrm{U}} \ F_{zK(i,j)}^{\mathrm{U}}\}^{\mathrm{T}}$$

$$\mathbf{f}_{K(i,j)} = \{f_{xK(i,j)}^{\mathrm{L}} \ f_{zK(i,j)}^{\mathrm{L}} \ f_{xK(i,j)}^{\mathrm{U}} \ f_{zK(i,j)}^{\mathrm{U}}\}^{\mathrm{T}}$$

and where $F_{xK(i,j)}^{\mathrm{L}}$ is the x-directional element-end force at the lower-end of stiffness element $K(i,j)$ defined at the center of mass, $F_{zK(i,j)}^{\mathrm{L}}$ is the z-directional element-end force at the lower-end of stiffness element $K(i,j)$ defined at the center of mass, $F_{xK(i,j)}^{\mathrm{U}}$ is the x-directional element-end force at the upper-end of stiffness element $K(i,j)$ defined at the center of mass, $F_{zK(i,j)}^{\mathrm{U}}$ is the z-directional element-end force at the upper-end of stiffness element $K(i,j)$ defined at the center of mass, $f_{xK(i,j)}^{\mathrm{L}}$ is the x-directional element-end force at the lower- end of stiffness element $K(i,j)$, $f_{zK(i,j)}^{\mathrm{L}}$ is the z-directional element-end force at the lower-end of stiffness element $K(i,j)$, $f_{xK(i,j)}^{\mathrm{U}}$ is the x-directional element-end force at the upper-end of stiffness element $K(i,j)$, and $f_{zK(i,j)}^{\mathrm{U}}$ is the z-directional element-end force at the upper-end of stiffness element $K(i,j)$.

5.2.5 Relation of Element-end Forces with Element-end Displacements

The relations of element-end forces $\mathbf{f}_{K(i,j)}$ with element-end displacements $\mathbf{u}_{K(i,j)}$ may be expressed as

$$\begin{Bmatrix} f_{xK(i,j)}^{\mathrm{L}} \\ f_{zK(i,j)}^{\mathrm{L}} \\ f_{xK(i,j)}^{\mathrm{U}} \\ f_{zK(i,j)}^{\mathrm{U}} \end{Bmatrix} = \begin{bmatrix} k_{(i,j)} & 0 & -k_{(i,j)} & 0 \\ 0 & 0 & 0 & 0 \\ -k_{(i,j)} & 0 & k_{(i,j)} & 0 \\ 0 & 0 & 0 & 0 \end{bmatrix} \begin{Bmatrix} u_{K(i,j)}^{\mathrm{L}} \\ w_{K(i,j)}^{\mathrm{L}} \\ u_{K(i,j)}^{\mathrm{U}} \\ w_{K(i,j)}^{\mathrm{U}} \end{Bmatrix} \tag{5.5}$$

These relations can be rewritten compactly as

$$\mathbf{f}_{K(i,j)} = \mathbf{k}_{K(i,j)}\mathbf{u}_{K(i,j)} \tag{5.6}$$

5.2.6 Relation of Forces at Center of Mass due to Stiffness Element K(i, j) with Displacements at Center of Mass

The relations of forces $\mathbf{F}_{K(i,j)}$ at the center of mass due to $K(i,j)$ with displacements $\mathbf{u}_i$ at the center of mass may be expressed as

$$\mathbf{F}_{K(i,j)} = \mathbf{K}_{K(i,j)} \begin{Bmatrix} \mathbf{u}_{i-1} \\ \mathbf{u}_i \end{Bmatrix} \tag{5.7}$$

where

$$\mathbf{K}_{K(i,j)} = \mathbf{t}_{K(i,j)}{}^{\mathrm{T}} \mathbf{k}_{K(i,j)} \mathbf{t}_{K(i,j)}$$

$$= \begin{bmatrix} k_{(i,j)} & -k_{(i,j)}\overline{y}^{\mathrm{L}}_{K(i,j)} & -k_{(i,j)} & k_{(i,j)}\overline{y}^{\mathrm{U}}_{K(i,j)} \\ -k_{(i,j)}\overline{y}^{\mathrm{L}}_{K(i,j)} & k_{(i,j)}\overline{y}^{\mathrm{L}}_{K(i,j)}\overline{y}^{\mathrm{L}}_{K(i,j)} & k_{(i,j)}\overline{y}^{\mathrm{L}}_{K(i,j)} & -k_{(i,j)}\overline{y}^{\mathrm{L}}_{K(i,j)}\overline{y}^{\mathrm{U}}_{K(i,j)} \\ -k_{(i,j)} & k_{(i,j)}\overline{y}^{\mathrm{L}}_{K(i,j)} & k_{(i,j)} & -k_{(i,j)}\overline{y}^{\mathrm{U}}_{K(i,j)} \\ k_{(i,j)}\overline{y}^{\mathrm{U}}_{K(i,j)} & -k_{(i,j)}\overline{y}^{\mathrm{L}}_{K(i,j)}\overline{y}^{\mathrm{U}}_{K(i,j)} & -k_{(i,j)}\overline{y}^{\mathrm{U}}_{K(i,j)} & k_{(i,j)}\overline{y}^{\mathrm{U}}_{K(i,j)}\overline{y}^{\mathrm{U}}_{K(i,j)} \end{bmatrix} \tag{5.8}$$

The stiffness matrix $\mathbf{K}_{K(i)}$ for the ith story may then be expressed as

$$\mathbf{K}_{K(i)} = \sum_j \mathbf{K}_{K(i,j)} = \begin{bmatrix} K_{x(i)} & -K^{\mathrm{L}}_{xz(i)} & -K_{x(i)} & K^{\mathrm{U}}_{xz(i)} \\ -K^{L}_{xz(i)} & K^{\mathrm{LL}}_{xz(i)} & K^{\mathrm{L}}_{xz(i)} & -K^{\mathrm{LU}}_{xz(i)} \\ -K_{x(i)} & K^{\mathrm{L}}_{xz(i)} & K_{x(i)} & -K^{\mathrm{U}}_{xz(i)} \\ K^{\mathrm{U}}_{xz(i)} & -K^{\mathrm{LU}}_{xz(i)} & -K^{\mathrm{U}}_{xz(i)} & K^{\mathrm{UU}}_{xz(i)} \end{bmatrix} \tag{5.9}$$

where

$$K_{x(i)} = \sum_j k_{(i,j)}$$

$$K^{\mathrm{L}}_{xz(i)} = \sum_j k_{(i,j)}\overline{y}^{\mathrm{L}}_{K(i,j)}$$

$$K^{\mathrm{U}}_{xz(i)} = \sum_j k_{(i,j)}\overline{y}^{\mathrm{U}}_{K(i,j)}$$

$$K^{\mathrm{LL}}_{xz(i)} = \sum_j k_{(i,j)}\overline{y}^{\mathrm{L}}_{K(i,j)}\overline{y}^{\mathrm{L}}_{K(i,j)}$$

$$K^{\mathrm{LU}}_{xz(i)} = \sum_j k_{(i,j)}\overline{y}^{\mathrm{L}}_{K(i,j)}\overline{y}^{\mathrm{U}}_{K(i,j)}$$

$$K^{\mathrm{UU}}_{xz(i)} = \sum_j k_{(i,j)}\overline{y}^{\mathrm{U}}_{K(i,j)}\overline{y}^{\mathrm{U}}_{K(i,j)} \tag{5.10a–f}$$

The system stiffness matrix $\mathbf{K}$ can finally be obtained by superposing $\mathbf{K}_{K(i)}$ through all the stories.

The structural damping of the model is considered and the system damping matrix due to this structural damping is given by $\mathbf{C}$. Let $c_{(i,j)}$ denote the damping coefficient at frame j in the ith story. $\mathbf{C}$ can be derived by replacing $k_{(i,j)}$ in the stiffness matrix by $c_{(i,j)}$.

See Chopra (1995) for more information on modeling.

5.2.7 Equations of Motion and Transfer Function Amplitude

When this 3-D shear building model without added dampers is subjected to a horizontal base acceleration $\ddot{u}_g(t)$ in the x-direction, the equations of motion for this model can be written as

$$\mathbf{K}\mathbf{u}(t) + \mathbf{C}\dot{\mathbf{u}}(t) + \mathbf{M}\ddot{\mathbf{u}}(t) = -\mathbf{M}\mathbf{r}\ddot{u}_g(t) \tag{5.11}$$

where $\mathbf{r}$ is the influence coefficient vector defined by $\mathbf{r} = \{1\ 0 \cdots 1\ 0 \cdots 1\ 0\}^{\mathrm{T}}$.

Let $\mathbf{U}(\omega)$ and $\ddot{U}_g(\omega)$ denote the Fourier transforms of $\mathbf{u}(t)$ and $\ddot{u}_g(t)$ respectively. Fourier transformation of Equation 5.11 may be reduced to the following form:

$$(\mathbf{K} + \mathrm{i}\omega\mathbf{C} - \omega^2\mathbf{M})\mathbf{U}(\omega) = -\mathbf{M}\mathbf{r}\ddot{U}_g(\omega) \tag{5.12}$$

where i is the imaginary unit.

Let $\mathbf{C}_V$ denote the damping matrix due to the added viscous dampers. When the added dampers are included, Equation 5.12 may be modified to the following form:

$$\{\mathbf{K} + \mathrm{i}\omega(\mathbf{C} + \mathbf{C}_V) - \omega^2\mathbf{M}\}\mathbf{U}_V(\omega) = -\mathbf{M}\mathbf{r}\ddot{U}_g(\omega) \tag{5.13}$$

In this chapter it is assumed that the masses of the dampers are negligible in comparison with the floor masses. It is interesting to note that, since the present formulation is developed in the frequency domain, it is possible to deal with various other damping systems in terms of complex stiffnesses (see Chapter 1).

Let c_i denote the damping coefficient of the ith added damper. The total number of added dampers is denoted by $N = n \times m$. The undamped fundamental natural circular frequency of the model in the x-direction (lateral-torsional vibration) is denoted by ω_1. Let us define new quantities $\hat{\mathbf{U}}$ by

$$\hat{\mathbf{U}} \equiv \frac{\mathbf{U}_V(\omega_1)}{\ddot{U}_g(\omega_1)} \tag{5.14}$$

$\hat{U}_i$ is equal to the value such that ω_1 is substituted in the frequency response function obtained as $U_{Vi}(\omega)$ after substituting $\ddot{U}_g(\omega) = 1$ in Equation 5.13 (see Figure 5.2). This quantity has been utilized first by Takewaki (1997a, 1997b) for structural redesign problems. While a steady-state resonant response at the fundamental natural frequency

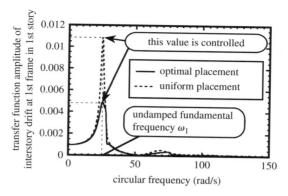

Figure 5.2 Transfer function amplitude of interstory drift at the first frame in the first story. (I. Takewaki, S. Yoshitomi, K. Uetani and M. Tsuji, "Non-Monotonic Optimal Damper Placement via Steepest Direction Search," *Earthquake Engineering and Structural Dynamics*, Vol.28, No.6. © 1999 John Wiley & Sons, Ltd).

is treated here to explain the fundamental feature of the present optimization algorithm, optimization for random earthquake inputs with wide-band frequency contents may be possible based on the present theory as given in Chapters 6–9. It should be noted that, because $\mathbf{M}$ and $\mathbf{K}$ are prescribed, ω_1 is a given value. Owing to Equation 5.13 ($\omega = \omega_1$) and Equation 5.14, $\hat{\mathbf{U}}$ must satisfy

$$\mathbf{A}\hat{\mathbf{U}} = -\mathbf{Mr} \tag{5.15}$$

where

$$\mathbf{A} = \mathbf{K} + i\omega_1(\mathbf{C} + \mathbf{C_V}) - \omega_1^2\mathbf{M} \tag{5.16}$$

Let $\hat{\delta}_i$ denote the transfer function at $\omega = \omega_1$ of the local interstory drift at the location where the ith damper exists. $\hat{\boldsymbol{\delta}} = \{\hat{\delta}_i\}$ can be derived from $\hat{\mathbf{U}}$ by the transformation $\hat{\boldsymbol{\delta}} = \mathbf{T}\hat{\mathbf{U}}$, where $\mathbf{T}$ is a constant matrix including given geometrical parameters only. The local interstory drift will be simply called the interstory drift hereafter.

It should be remarked here that the squares of the transfer function amplitudes are useful from physical points of view because they can be transformed into response mean squares (statistical quantities) after multiplication with the PSD function of a disturbance and integration in the frequency range. Since the transfer function amplitude of a nodal displacement evaluated at the undamped fundamental natural circular frequency can be related to the level of this response mean square, these transfer function amplitudes are treated as controlled quantities in this chapter.

5.2.8 Problem of Optimal Damper Positioning

The problem of optimal damper positioning for a 3-D (PODPT) shear building model may be described as follows.

Problem 5.1 PODPT   Find the damping coefficients $\mathbf{c} = \{c_i\}$ of added dampers which minimize the sum of the transfer function amplitudes of the interstory drifts evaluated at the undamped fundamental natural circular frequency ω_1

$$f(\mathbf{c}) = \sum_{i=1}^{N} \left| \hat{\delta}_i(\mathbf{c}) \right| \tag{5.17}$$

subject to a constraint on the sum of the damping coefficients of added dampers

$$\sum_{i=1}^{N} c_i = \overline{W} \qquad (\overline{W}: \text{specified value}) \tag{5.18a}$$

and to constraints on the damping coefficients of added dampers

$$0 \le c_i \le \overline{c}_i \quad (i = 1, \cdots, N) \tag{5.18b}$$

where $\overline{c}_i$ is the upper bound of the damping coefficient of the ith added damper.

The generalized Lagrangian L for Problem 5.1 may be expressed in terms of Lagrange multipliers $\lambda, \boldsymbol{\mu} = \{\mu_i\}$, and $\boldsymbol{v} = \{v_i\}$:

$$L(\mathbf{c}, \lambda, \boldsymbol{\mu}, \boldsymbol{v}) = f(\mathbf{c}) + \lambda \left(\sum_{i=1}^{N} c_i - \overline{W} \right) + \sum_{i=1}^{N} \mu_i (0 - c_i) + \sum_{i=1}^{N} v_i (c_i - \overline{c}_i) \tag{5.19}$$

For simplicity of expression, the argument $(\mathbf{c})$ will be omitted hereafter.

5.3 Optimality Criteria and Solution Algorithm

The principal optimality criteria for Problem 5.1 without active upper and lower bound constraints on damping coefficients may be derived from the stationarity conditions of the generalized Lagrangian L ($\boldsymbol{\mu} = \mathbf{0}$, $\boldsymbol{v} = \mathbf{0}$) with respect to $\mathbf{c}$ and λ:

$$f_{,j} + \lambda = 0 \quad \text{for } 0 < c_j < \overline{c}_j \quad (j = 1, \cdots, N) \tag{5.20}$$

$$\sum_{i=1}^{N} c_i - \overline{W} = 0 \tag{5.21}$$

Here, and in the following, $(\,\cdot\,)_{,j}$ denotes the partial differentiation with respect to c_j. If the constraints (Equation 5.18b) are active, then Equation 5.20 must be modified into the following forms:

$$f_{,j} + \lambda \ge 0 \quad \text{for} \quad c_j = 0 \tag{5.22a}$$

$$f_{,j} + \lambda \le 0 \quad \text{for} \quad c_j = \overline{c}_j \tag{5.22b}$$

In the present procedure, the model without added dampers, namely $c_j = 0$ ($j = 1, \cdots, N$), is employed as the initial model. This treatment is well suited to the situation where a structural designer is just starting the allocation and placement of added supplemental viscous dampers at desired positions. The damping coefficients of added dampers are *increased gradually* based on the optimality criteria derived above. This algorithm will be called a steepest direction search algorithm, as in Chapter 4.

Let Δc_i and $\Delta \overline{W}$ denote the increment of the damping coefficient of the ith added damper and the increment of the sum of the damping coefficients of added dampers respectively. Given the increment $\Delta \overline{W}$, the problem is to determine the effective position and amount of the increments of the damping coefficients of supplemental dampers. To develop this algorithm, the first- and second-order sensitivities of the objective function with respect to a design variable are derived in the following.

Differentiation of Equation 5.15 with respect to a design variable c_j provides

$$\mathbf{A}_{,j}\hat{\mathbf{U}} + \mathbf{A}\hat{\mathbf{U}}_{,j} = \mathbf{0} \qquad (5.23)$$

From Equation 5.16, $\mathbf{A}_{,j}$ may be expressed as follows.

$$\mathbf{A}_{,j} = i\omega_1 \mathbf{C}_{\mathrm{V},j} \qquad (5.24)$$

Since the coefficient matrix $\mathbf{A}$ is regular, the first-order sensitivities of $\hat{\mathbf{U}}$ are derived from Equation 5.23 as

$$\hat{\mathbf{U}}_{,j} = -\mathbf{A}^{-1}\mathbf{A}_{,j}\hat{\mathbf{U}} \qquad (5.25a)$$

The first-order sensitivities of the interstory drift $\hat{\boldsymbol{\delta}}$ are then expressed as

$$\hat{\boldsymbol{\delta}}_{,j} = \mathbf{T}\hat{\mathbf{U}}_{,j} = -\mathbf{T}\mathbf{A}^{-1}\mathbf{A}_{,j}\hat{\mathbf{U}} \qquad (5.25b)$$

The quantity $\hat{\delta}_i$ may be rewritten formally as

$$\hat{\delta}_i = \mathrm{Re}[\hat{\delta}_i] + i\,\mathrm{Im}[\hat{\delta}_i] \qquad (5.26)$$

where $\mathrm{Re}[\ ]$ and $\mathrm{Im}[\ ]$ indicate the real and imaginary parts respectively of a complex number. Furthermore, the first-order sensitivity of $\hat{\delta}_i$ may be formally expressed as

$$\hat{\delta}_{i,j} = (\mathrm{Re}[\hat{\delta}_i])_{,j} + i(\mathrm{Im}[\hat{\delta}_i])_{,j} \qquad (5.27)$$

The absolute value of $\hat{\delta}_i$ is defined by

$$|\hat{\delta}_i| = \sqrt{(\mathrm{Re}[\hat{\delta}_i])^2 + (\mathrm{Im}[\hat{\delta}_i])^2} \qquad (5.28)$$

The first-order sensitivity of $|\hat{\delta}_i|$ may then be expressed as

$$|\hat{\delta}_i|_{,j} = \frac{1}{|\hat{\delta}_i|}\{\mathrm{Re}[\hat{\delta}_i](\mathrm{Re}[\hat{\delta}_i])_{,j} + \mathrm{Im}[\hat{\delta}_i](\mathrm{Im}[\hat{\delta}_i])_{,j}\} \qquad (5.29)$$

where $(\mathrm{Re}[\hat{\delta}_i])_{,j}$ and $(\mathrm{Im}[\hat{\delta}_i])_{,j}$ are calculated from Equations 5.25b and 5.27.

A general expression $|\hat{\delta}_i|_{,j\ell}$ is derived here for the purpose of developing a solution procedure for Problem 5.1. Differentiation of Equation 5.29 with respect to c_ℓ leads to

$$|\hat{\delta}_i|_{,j\ell} = \frac{1}{|\hat{\delta}_i|^2}(|\hat{\delta}_i|\{(\text{Re}[\hat{\delta}_i])_{,\ell}(\text{Re}[\hat{\delta}_i])_{,j} + \text{Re}[\hat{\delta}_i](\text{Re}[\hat{\delta}_i])_{,j\ell} + (\text{Im}[\hat{\delta}_i])_{,\ell}(\text{Im}[\hat{\delta}_i])_{,j}$$

$$+ \text{Im}[\hat{\delta}_i](\text{Im}[\hat{\delta}_i])_{,j\ell}\} - |\hat{\delta}_i|_{,\ell}\{\text{Re}[\hat{\delta}_i](\text{Re}[\hat{\delta}_i])_{,j} + \text{Im}[\hat{\delta}_i](\text{Im}[\hat{\delta}_i])_{,j}\}) \quad (5.30)$$

$(\text{Re}[\hat{\delta}_i])_{,j\ell}$ and $(\text{Im}[\hat{\delta}_i])_{,j\ell}$ in Equation 5.30 can be found from

$$\hat{\boldsymbol{\delta}}_{,j\ell} = \mathbf{TA}^{-1}(\mathbf{A}_{,\ell}\,\mathbf{A}^{-1}\mathbf{A}_{,j}\,\hat{\mathbf{U}} - \mathbf{A}_{,j}\,\hat{\mathbf{U}}_{,\ell}) \quad (5.31)$$

which is derived by differentiating Equation 5.25b with respect to c_ℓ and using the relation $\mathbf{A}_{,\ell}^{-1} = -\mathbf{A}^{-1}\mathbf{A}_{,\ell}\mathbf{A}^{-1}$. Substitution of Equation 5.25a into Equation 5.31 leads to the following form:

$$\hat{\boldsymbol{\delta}}_{,j\ell} = \mathbf{TA}^{-1}(\mathbf{A}_{,\ell}\,\mathbf{A}^{-1}\mathbf{A}_{,j} + \mathbf{A}_{,j}\,\mathbf{A}^{-1}\mathbf{A}_{,\ell})\hat{\mathbf{U}} \quad (5.32)$$

The second-order derivatives $|\hat{\delta}_i|_{,j\ell}$ are derived from Equation 5.30. $(\text{Re}[\hat{\delta}_i])_{,j}$ and $(\text{Im}[\hat{\delta}_i])_{,j}$ in Equation 5.30 are calculated from Equation 5.25b and $(\text{Re}[\hat{\delta}_i])_{,j\ell}$ and $(\text{Im}[\hat{\delta}_i])_{,j\ell}$ in Equation 5.30 are found from Equation 5.32.

The fundamental solution algorithm in the case of $c_j < \overline{c}_j$ for all j may be summarized as follows:

Step 0 Initialize all the added supplemental viscous dampers as $c_j = 0$ ($j = 1, \cdots, N$). In the initial state, the damping is the structural damping alone in the 3-D shear building model. Assume the quantity $\Delta\overline{W}$.

Step 1 Compute the first-order derivative $f_{,i}$ of the objective function by Equation 5.29.

Step 2 Find the index k satisfying the condition

$$-f_{,k} = \max_i\{-f_{,i}\} \quad (5.33)$$

Step 3 Update the objective function f by the linear approximation $f + f_{,k}\Delta c_k$, where $\Delta c_k = \Delta\overline{W}$. This is because the supplemental damper is added only in the kth damper in the initial design stage.

Step 4 Update the first-order sensitivity $f_{,i}$ of the objective function by the linear approximation $f_{,i} + f_{,ik}\Delta c_k$ using Equation 5.30.

Step 5 If, in Step 4, there exists a supplemental damper of an index j such that the condition

$$-f_{,k} = \max_{j, j\neq k}\{-f_{,j}\} \quad (5.34)$$

is satisfied, then stop and compute the increment $\Delta\tilde{c}_k$ of the damping coefficient of the corresponding damper. At this stage, update the first-order sensitivity

$f_{,i}$ of the objective function by the linear approximation $f_{,i} + f_{,ik}\Delta\tilde{c}_k$ using Equation 5.30.

Step 6 Repeat Step 2 through Step 5 until the constraint in Equation 5.18a, $\sum_{i=1}^{N} c_i = \overline{W}$, is satisfied.

The schematic diagram for this fundamental solution algorithm is shown in Figure 5.3. The relation between the first-order derivatives of the objective function and the damping coefficients is explained there in detail. Once the first-order derivative of the objective function starts to attain the maximum absolute value, the corresponding damper begins possessing a nonzero value. The dampers begin possessing nonzero damping coefficients in the order c_1, c_4, c_2, c_3 in Figure 5.3.

In Step 2 and Step 3 in the aforementioned algorithm, the direction which decreases the objective function most effectively under the condition $\sum_{i=1}^{N} \Delta c_i = \Delta\overline{W}$ is found and the design is updated in that direction. It is appropriate, therefore, to call the present algorithm "the steepest direction search algorithm," as stated before. This algorithm is similar to the well-known steepest descent method in mathematical programming (see Figure 4.4 in Chapter 4 to understand the concept). However, while the conventional steepest descent method uses the gradient vector itself of the objective function as the direction and does not utilize optimality criteria, the present algorithm takes full advantage of the newly derived optimality criteria in Equations 5.20, 5.22a, and 5.22b and does not adopt the gradient vector as the direction. In other words, the steepest direction search guarantees the automatic satisfaction of the optimality criteria. For example, if Δc_k is added to the kth added damper in which Equation 5.33 is satisfied, then its damper ($c_k > 0$) satisfies the optimality condition (Equation 5.20) and the other dampers ($c_j = 0$, $j \neq k$) satisfy the optimality condition (Equation 5.22a). It should be noted that a series of subproblems is introduced here tentatively in which the damper level $\overline{W}$ is increased gradually by $\Delta\overline{W}$ from zero through the specified value.

If multiple indices $k_1, \cdots, k_p$ exist in Step 2, then f and $f_{,j}$ have to be updated by

$$f \rightarrow f + \sum_{i=k_1}^{k_p} f_{,i}\Delta c_i \tag{5.35a}$$

$$f_{,j} \rightarrow f_{,j} + \sum_{\ell=k_1}^{k_p} f_{,j\ell}\Delta c_\ell \tag{5.35b}$$

Furthermore, the index k in Step 5 has to be replaced by the indices $k_1, \cdots, k_p$. The ratios among the magnitudes Δc_i must be determined so that the following relations are satisfied:

$$f_{,k_1} + \sum_{i=k_1}^{k_p} f_{,k_1 i}\Delta c_i = \cdots = f_{,k_p} + \sum_{i=k_1}^{k_p} f_{,k_p i}\Delta c_i \tag{5.36}$$

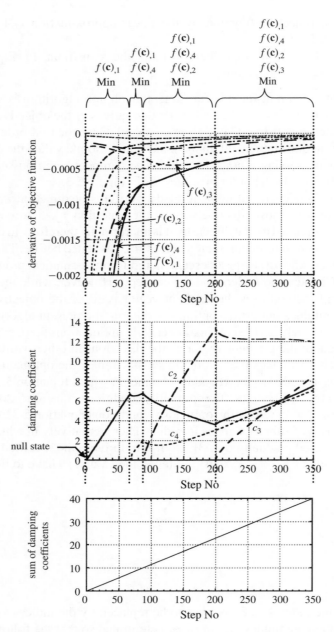

Figure 5.3 Schematic diagram of the fundamental solution algorithm. (Reproduced with permission from *Structural and Multidisciplinary Optimization*, Vol. 20, No. 4, pp 280–287, 2000, I. Takewaki, Optimal Damper Placement for Planar Building Frames Using Transfer Functions, with permission from Springer.).

Equation 5.36 requires that the optimality condition (Equation 5.20) continues to be satisfied in the dampers with the indices $k_1, \cdots, k_p$.

In the case where the damping coefficients of some added dampers attain their upper bounds, such constraints must be incorporated in the aforementioned algorithm. In that case, the increment Δc_k is added subsequently to the damper in which $-f_{,k}$ attains the maximum among all the dampers, except those attaining the upper bound.

5.4 Nonmonotonic Path with Respect to Damper Level

In most cases, an increase of member stiffnesses reduces the displacement (or deformation) response of a structure under disturbances with a wide-band frequency content. On the other hand, an increase of the damping coefficient of added dampers does not necessarily lead to a reduction of the response because of complicated damping characteristics. An example can be found in the increase of acceleration arising from the increase of damping in the base-isolation story. Nonproportional damping may cause further complicated phenomena. It has been found through numerical experiments that the algorithm mentioned above can lead to the appearance of negative damping coefficients. In the real world, negative damping coefficients are difficult to understand, and this situation should be excluded with the constraints in Equation 5.18b. In this section a new method is explained for preventing the appearance of negative damping coefficients and finding a path continuously within such complex design regions.

Figure 5.4(a) shows a situation in which a damping coefficient of a specific added damper c_2 starts to attain a negative value. At this stage, the proposed augmented algorithm recommends changing the signs of the damping coefficients of the dampers satisfying the optimality condition in Equation 5.20. In the following step, the design parameter is switched from the total damper capacity level (sum of the damping coefficients) to the damping coefficient of that damper c_2. In this range, the sum of the damping coefficients begins decreasing and afterwards increases. It is noted that, while the increment of the sum of the damping coefficients has been specified in the range without the emergence of negative damping coefficients, the optimal damper placement is obtained for a specified increment of the damping coefficient of that damper c_2. The validity of this treatment is shown in the following.

The optimality condition, Equation 5.20, is satisfied in the following form:

$$f(\mathbf{c})_{,k_1} = \cdots = f(\mathbf{c})_{,k_p} \tag{5.37}$$

Let $\Delta\mathbf{c}$ denote the set of increments of the damping coefficients including a negative damping coefficient. In order for these optimality conditions to continue to be satisfied for the increments $\Delta\mathbf{c}$, the following relations must be satisfied:

$$f(\mathbf{c} + \Delta\mathbf{c})_{,k_1} = \cdots = f(\mathbf{c} + \Delta\mathbf{c})_{,k_p} \tag{5.38}$$

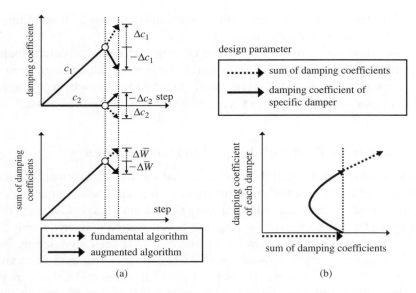

Figure 5.4 (a) Situation at which a damping coefficient of a specific added damper c_2 starts to attain a negative value; (b) switch of the main parameter. (I. Takewaki, S. Yoshitomi, K. Uetani and M. Tsuji, "Non-Monotonic Optimal Damper Placement via Steepest Direction Search," *Earthquake Engineering and Structural Dynamics*, Vol.28, No.6. © 1999 John Wiley & Sons, Ltd).

The first-order Taylor series expansion of the first term may be written as

$$f(\mathbf{c} + \Delta\mathbf{c})_{,k_1} = f(\mathbf{c})_{,k_1} + \sum_{i=1}^{p} f(\mathbf{c})_{,k_1 k_i} \Delta\tilde{c}_{k_i} \qquad (5.39)$$

From Equation 5.39, the following conditions must be satisfied in order that Equations 5.37 and 5.38 hold:

$$\sum_{i=1}^{p} f(\mathbf{c})_{,k_1 k_i} \Delta\tilde{c}_{k_i} = \cdots = \sum_{i=1}^{p} f(\mathbf{c})_{,k_j k_i} \Delta\tilde{c}_{k_i} = \cdots = \sum_{i=1}^{p} f(\mathbf{c})_{,k_p k_i} \Delta\tilde{c}_{k_i} \qquad (5.40)$$

It should be noted that (except for dampers with negative damping coefficients) if the damping coefficients of the dampers in the fundamental algorithm are relatively large, then the new set $\mathbf{c} - \Delta\mathbf{c}$ does not include a negative damping coefficient. In order that the optimality conditions are satisfied for the new design variables $\mathbf{c} - \Delta\mathbf{c}$, the following relation must hold:

$$f(\mathbf{c} - \Delta\mathbf{c})_{,k_1} = \cdots = f(\mathbf{c} - \Delta\mathbf{c})_{,k_p} \qquad (5.41)$$

The first-order Taylor series expansion of the first term may be written as

$$f(\mathbf{c} - \Delta\mathbf{c})_{,k_1} = f(\mathbf{c})_{,k_1} - \sum_{i=1}^{p} f(\mathbf{c})_{,k_1 k_i} \Delta\tilde{c}_{k_i} \qquad (5.42)$$

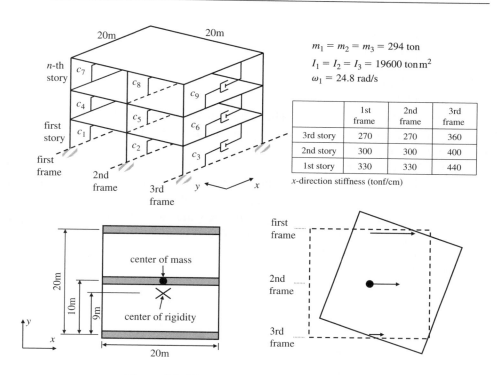

$m_1 = m_2 = m_3 = 294$ ton

$I_1 = I_2 = I_3 = 19600$ ton m^2

$\omega_1 = 24.8$ rad/s

	1st frame	2nd frame	3rd frame
3rd story	270	270	360
2nd story	300	300	400
1st story	330	330	440

x-direction stiffness (tonf/cm)

Figure 5.5   Three-story shear building model.

From Equations 5.37, 5.40, and 5.42, it is apparent that Equation 5.41 holds. This guarantees the validity of the new algorithm for avoiding negative damping coefficients.

After the sum of the damping coefficients returns to a value from which the damping coefficient of a specific damper starts to attain a negative value, the design parameter is re-switched from the damping coefficient of that damper to the sum of the damping coefficients (see Figure 5.4(b)).

5.5 Numerical Examples

Consider the three-story shear building model shown in Figure 5.5. The shear building model consists of three planar frames with different story stiffnesses. The center of mass and center of rigidity in every floor coincide. Every floor mass and mass moment of inertia of the floor around a vertical axis are assumed to be $m = 294$ ton and $I = 1.96 \times 10^4$ ton cm^2 respectively. The element story stiffnesses in the three frames are shown in Figure 5.5. The stiffness of the third frame is strongest and the center of rigidity is located slightly near to the third frame. The undamped fundamental natural circular frequency of the model (lateral-torsional vibration) is

$\omega_1 = 24.8$ rad/s. The structural damping of the shear building model is given by a critical damping ratio of 0.02 in the lowest eigenvibration. The numbering of the added dampers is shown in Figure 5.5. It is assumed that all the constraints on upper bounds of the damping coefficients are inactive; that is, $c_j < \overline{c}_j$ for all j. The final level of the sum of the damping coefficients of the added dampers is $\overline{W} = 40$ tonf s/cm (1 tonf s/cm = 0.98×10^6 N/m). The increment of $\overline{W}$ is given by $\Delta\overline{W} = \overline{W}/400$. In the nonmonotonic range, 25 steps are required for the sum of the damping coefficients to return to the original value.

The distributions of the optimal damping coefficients obtained via the present fundamental procedure are plotted in Figure 5.6. It can be observed that the dampers are added in order c_1, c_4, c_2 and the damper c_3 begins to attain a negative damping coefficient around $\overline{W} = 27$ tonf s/cm. This indicates that the optimal damper placement requires that added dampers be placed in order to suppress the torsional vibration component. Figure 5.7 shows the distributions of the optimal damping coefficients obtained via the augmented procedure described in Section 5.4. The damping

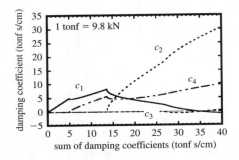

Figure 5.6 Distributions of the optimal damping coefficients obtained via the present fundamental procedure (1 tonf s/cm = 0.98×10^6 N/m). (I. Takewaki, S. Yoshitomi, K. Uetani and M. Tsuji, "Non-Monotonic Optimal Damper Placement via Steepest Direction Search," *Earthquake Engineering and Structural Dynamics*, Vol.28, No.6. © 1999 John Wiley & Sons, Ltd).

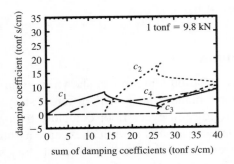

Figure 5.7 Distributions of the optimal damping coefficients obtained via the augmented procedure (1 tonf s/cm = 0.98×10^6 N/m). (I. Takewaki, S. Yoshitomi, K. Uetani and M. Tsuji, "Non-Monotonic Optimal Damper Placement via Steepest Direction Search," *Earthquake Engineering and Structural Dynamics*, Vol.28, No.6. © 1999 John Wiley & Sons, Ltd).

coefficient of damper c_3 is taken as a design parameter in the nonmonotonic range. It is found that the modification by c_3 greatly influences the optimal damping distributions of dampers c_1 and c_2. It is also found that multiple stationary solutions exist in the nonmonotonic range. It is possible to obtain the optimal solution by comparing the objective functions of the stationary solutions. Figure 5.8 shows the variation of the first-order derivatives of the objective function. In addition, Figure 5.9 illustrates the satisfaction level of the optimality conditions (Equation 5.20). The ordinate indicates the value of the left-hand side in Equation 5.20 divided by λ. It can be observed from Figure 5.9 that the optimality criteria, Equation 5.20, and the inequality in Equation 5.22a are satisfied in all the dampers for every $\overline{W}$.

Figure 5.10 shows the variations of the lowest-mode damping ratio for the optimal placement and the uniform placement. The uniform placement means the case where the same increment $\Delta\overline{W}/9$ of the damping coefficient is added to every damper. It

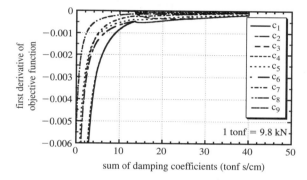

Figure 5.8 Variation of the first-order derivatives of the objective function (1 tonf s/cm $= 0.98 \times$ 10^6 N/m). (I. Takewaki, S. Yoshitomi, K. Uetani and M. Tsuji, "Non-Monotonic Optimal Damper Placement via Steepest Direction Search," *Earthquake Engineering and Structural Dynamics*, Vol.28, No.6. © 1999 John Wiley & Sons, Ltd).

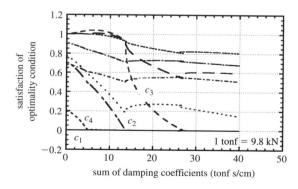

Figure 5.9 Satisfaction level of the optimality conditions (1 tonf s/cm $= 0.98 \times 10^6$ N/m). (I. Takewaki, S. Yoshitomi, K. Uetani and M. Tsuji, "Non-Monotonic Optimal Damper Placement via Steepest Direction Search," *Earthquake Engineering and Structural Dynamics*, Vol.28, No.6. © 1999 John Wiley & Sons, Ltd).

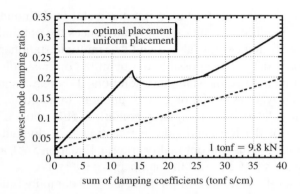

Figure 5.10 Variations of the lowest mode damping ratio for the optimal placement and the uniform placement (1 tonf s/cm = 0.98×10^6 N/m). (I. Takewaki, S. Yoshitomi, K. Uetani and M. Tsuji, "Non-Monotonic Optimal Damper Placement via Steepest Direction Search," *Earthquake Engineering and Structural Dynamics*, Vol.28, No.6. © 1999 John Wiley & Sons, Ltd).

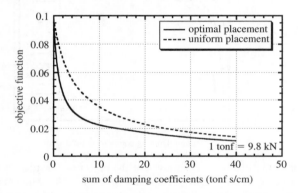

Figure 5.11 Objective function with respect to the sum of the damping coefficients for the optimal placement and the uniform placement (1 tonf s/cm = 0.98×10^6 N/m). (I. Takewaki, S. Yoshitomi, K. Uetani and M. Tsuji, "Non-Monotonic Optimal Damper Placement via Steepest Direction Search," *Earthquake Engineering and Structural Dynamics*, Vol.28, No.6. © 1999 John Wiley & Sons, Ltd).

can be seen that optimal placement increases the lowest mode damping ratio more effectively than uniform placement does. Figure 5.11 illustrates the objective function with respect to the sum of the damping coefficients for the optimal placement and the uniform placement. The optimal placement actually reduces the objective function more rapidly than the uniform placement, especially at the lower damping capacity level. Figure 5.12 shows the sum of the seismic-response interstory drifts corresponding to the objective function. Each transfer function amplitude of the local interstory drift has been replaced by the mean peak local interstory drift to the design earthquakes represented by the response spectrum due to Newmark and Hall (1982). The response spectrum method by Yang *et al.* (1990) for a nonproportional damping has been employed to estimate the mean peak local interstory drift. It can be found

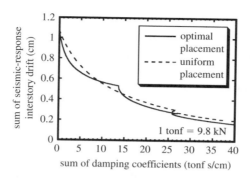

Figure 5.12 Sum of seismic-response interstory drifts corresponding to objective function (1 tonf s/cm $= 0.98 \times 10^6$ N/m). (I. Takewaki, S. Yoshitomi, K. Uetani and M. Tsuji, "Non-Monotonic Optimal Damper Placement via Steepest Direction Search," *Earthquake Engineering and Structural Dynamics*, Vol.28, No.6. © 1999 John Wiley & Sons, Ltd).

from Figure 5.12 that, in a certain narrow range, the seismic response of the model designed optimally for the steady-state resonant response becomes larger than that for the uniform placement. Higher mode effects on seismic responses to wide-band inputs and nonproportional damping effects may be a cause of this phenomenon. Direct introduction of the seismic response into the objective function (Equation 5.17) and application of the random vibration theories in the seismic response evaluation may lead to more realistic optimal damper placement (see Chapters 3 and 6–9). Numerical integration in the frequency range and sensitivity expressions of frequency-response functions will be required in the computation. However, this treatment may require unrealistically high computational resources.

5.6 Summary

The results are summarized as follows.

1. An efficient and systematic procedure called a steepest direction search algorithm has been explained for finding the optimal damper positioning in a 3-D shear building model. This problem is aimed at minimizing the sum of the transfer function amplitudes of local interstory drifts evaluated at the undamped fundamental natural frequency subject to a constraint on the sum of the damping coefficients of supplemental dampers. The transfer function amplitude indicates the absolute value of a transfer function of a local interstory drift at the undamped fundamental natural frequency. The optimal damper positioning is determined based upon the optimality criteria derived here. The features of the present formulation are to be able to deal with any damping system, to be able to treat any structural system so far as it can be modeled with FE systems, and to consist of a systematic algorithm without any indefinite iterative operation.

2. An algorithm to avoid the emergence of negative damping coefficients has been developed by introducing a devised procedure of parameter switching. Efficiency and reliability of the present procedure have been demonstrated through an example.
3. Optimal damper placement requires that supplemental dampers be placed in order to suppress the torsional vibration component.

References

Chopra, A.K. (1995) *Dynamics of Structures: Theory and Applications to Earthquake Engineering,* Prentice-Hall Inc.

Newmark, N.M. and Hall, W.J. (1982) *Earthquake Spectrum and Design,* EERI, Oakland, CA.

Takewaki, I. (1997a) Efficient redesign of damped structural systems for target transfer functions. *Computer Methods in Applied Mechanics and Engineering,* **147** (3–4), 275–286.

Takewaki, I. (1997b) Optimal damper placement for minimum transfer functions. *Earthquake Engineering & Structural Dynamics,* **26** (11), 1113–1124.

Yang, J.N., Sarkani, S., and Long, F.X. (1990). A response spectrum approach for seismic analysis of nonclassically damped structure. *Engineering Structures,* **12** (3), 173–184.

6

Optimal Sensitivity-based Design of Dampers in Shear Buildings on Surface Ground under Earthquake Loading

6.1 Introduction

The soil or ground under a structure greatly influences the structural vibration properties. This fact has been demonstrated in past earthquakes. It is important, therefore, to develop a seismic-resistant design method for such an interaction model. While a design philosophy depending largely on stiffness and strength of building structures has been employed for a long time, passive control devices have recently been used as effective substitutes of stiffness and strength-type resisting elements for suppressing the structural responses and upgrading the structural performances. It is not difficult to imagine that a building structure including such control systems is also affected by the soil–structure interaction.

The purpose of this chapter is to explain a systematic method for optimal placement of supplemental viscous dampers in building structures including response amplification due to the surface ground. The supplemental viscous dampers are installed in the stories of a shear building model. Nonlinear amplification of ground motion within the surface ground is described by an equivalent linear model. Both hysteretic damping of the surface ground and radiation damping into the semi-infinite ground are taken into account in the model.

While several useful investigations have been reported on active control of building structures including soil–structure interaction effects (Wong and Luco, 1991; Alam and Baba, 1993; Smith *et al.*, 1994; Smith and Wu, 1997), it should be remarked that there are a very few on optimal damper placement in a building structure taking into account

Building Control with Passive Dampers   Izuru Takewaki
© 2009 John Wiley & Sons (Asia) Pte Ltd

interaction with the surface ground. A method called the steepest direction search algorithm (Takewaki, 1998a; Takewaki and Yoshitomi, 1998) has been explained in Chapters 4 and 5 for fixed-base planar frames or 3-D structures. In this chapter, it is shown that it can be extended to a structure–surface ground interaction model. While resonant steady-state responses of structures with fixed bases were treated in Chapters 2–5 (Takewaki, 1998a; Takewaki and Yoshitomi, 1998), earthquake responses to random ground motions are treated in this chapter as controlled parameters. It is shown that closed-form expressions of the inverse of the tri-diagonal coefficient matrix in the governing equations lead to a drastic reduction of computational time for mean-square responses of the building structure to a random earthquake input and their derivatives with respect to the design variables (damping coefficients of added dampers). Optimal placement and quantity of the supplemental dampers are found simultaneously and automatically via the steepest direction search algorithm, which is based on successive approximate satisfaction of the optimality conditions. Several examples for various surface ground properties are presented to demonstrate the effectiveness and validity of the present method and to examine the effects of surface ground characteristics on optimal damper placement.

6.2 Building and Ground Model

In this chapter, a building structure is assumed to exist on a surface ground. For simple and clear presentation, the building structure is described by a shear building model and the surface ground is represented by a shear beam model (see Figure 6.1). It is also assumed here that many identical structure–ground systems are arranged in the two horizontal directions and one of the systems is taken into account as a representative model. It should be kept in mind that, while the response of the building is affected by the surface ground and the response of the surface ground is not affected by the building in a model adopting a wave propagation theory for the surface ground (the case where the mass of surface ground is much larger than that of the building), both the building and surface ground are affected by each other in the present model. Furthermore, the local interaction between the building foundation and the surface ground is not considered here. This effect will be considered in Chapters 7 and 8.

Let n and N denote the number of stories of the shear building model and the number of surface soil layers respectively. A random horizontal input is defined at the level just below the engineering bedrock on which the layered surface ground lies.

As in the case where the surface ground is modeled by a FE model, so-called radiation damping into the semi-infinite visco-elastic ground is taken into account here by a viscous boundary (Lysmer and Kuhlemeyer, 1969). Let $c_b = \rho_0 V_S A = \sqrt{\rho_0 G_0} A$ denote the total damping coefficient of the viscous boundary where ρ_0, V_S, and G_0 are respectively the mass density, shear wave velocity, and shear modulus of the semi-infinite visco-elastic ground and A is the horizontal governing area of the surface ground. The dynamic performance of this viscous boundary was investigated and

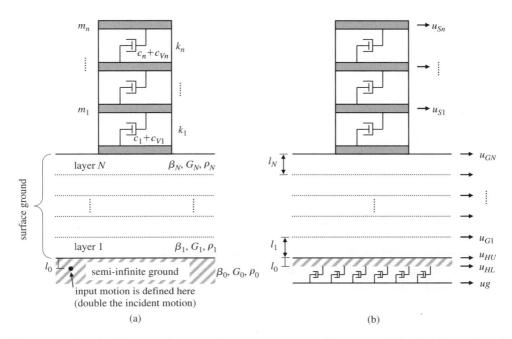

Figure 6.1   Shear building model supported by a surface ground. (Originally published in I. Takewaki and K. Uetani, "Optimal damper placement for building structures including surface ground amplification," *Soil Dynamics and Earthquake Engineering*, **18**, no. 5, 363–371, 1999, Elsevier B.V.).

tested by Lysmer and Kuhlemeyer (1969), and it has been demonstrated that there is no accuracy problem in treating one-dimensional models.

Let l_i, ρ_i, G_i, and β_i denote the thickness, mass density, shear modulus, and hysteretic damping ratio respectively in layer i. A linear displacement function is used in evaluating the stiffness and consistent mass matrices in the FE model of the surface ground. Nonlinear amplification of ground motion within the surface ground is taken into account through an equivalent linear model (Schnabel *et al.*, 1972). The equivalent linear model used for the surface ground is almost equivalent to that used in the SHAKE program, as shown by Takewaki *et al.* (2002a, 2002b). This model can be regarded as a deterministic equivalent linear model and its accuracy has been demonstrated by many researchers (e.g., Takewaki *et al.*, 2002a, 2002b). The method explained here uses a peak factor (Der Kiureghian, 1980) multiplied by the standard deviation of the shear strain under random vibration. The validity of this treatment has been demonstrated by Der Kiureghian (1980). Once the surface ground is modeled by the equivalent linear model, the whole system consisting of a structure and the surface ground is a linear system. The random vibration theory for linear models is well established and its accuracy has been checked by many investigators.

On the other hand, the parameters in the building may be specified as follows. Let m_i and k_i denote the floor mass and the story stiffness respectively in the ith story of the building. In addition, let c_i and c_{Vi} denote the structural damping coefficient in the ith story of the building and the damping coefficient of the supplemental damper in the ith story respectively. The set $\{c_{Vi}\}$ of damping coefficients of supplemental dampers is the design variable set here.

6.3 Seismic Response

The method of dynamic response evaluation in this interaction model will be explained in this section.

Assume that this shear building–surface ground model is subjected to a horizontal acceleration $\ddot{u}_g(t)$ at a level l_0 below the bedrock surface. $\mathbf{M}$, $\mathbf{C}$, $\mathbf{D}$, $\mathbf{K}$, and $\mathbf{r}$ are respectively the system mass matrix (combination of a lumped mass matrix and a consistent mass matrix), the system viscous damping matrix for the building and the viscous boundary, the system hysteretic damping matrix for the soil layers, the system stiffness matrix, and the influence coefficient vector. An example of these matrices $\mathbf{M}$, $\mathbf{C}$, $\mathbf{D}$, and $\mathbf{K}$ is shown in Appendix 6.A for a two-story shear building model supported by two soil layers. It is noted that the system viscous damping matrix for the building includes the damping coefficients of the supplemental viscous dampers and that the damping coefficient in the ith story may be expressed by $C_i = c_i + c_{Vi}$ in terms of the structural damping coefficient c_i in the ith story of the building and the damping coefficient c_{Vi} of the supplemental damper in the ith story.

Equations of motion of this building–ground interaction system in the frequency domain can be expressed by

$$(-\omega^2 \mathbf{M} + i\omega \mathbf{C} + i\mathbf{D} + \mathbf{K})\mathbf{U}(\omega) = -\mathbf{M}\mathbf{r}\ddot{U}_g(\omega) \tag{6.1}$$

In Equation 6.1, $\mathbf{U}(\omega)$ and $\ddot{U}_g(\omega)$ are the Fourier transforms of the horizontal displacements of the nodes (floors of the building and layer boundaries of the surface ground) and the Fourier transform of the horizontal input acceleration $\ddot{u}_g(t)$ defined at the level l_0 below the bedrock. It should be noted that, because the present model includes a hysteretic damping, the equations of motion in the time domain for $\ddot{u}_g(t)$ cannot be expressed and only those in the frequency-domain can be derived. The horizontal input acceleration $\ddot{u}_g(t)$ is assumed to be a stationary Gaussian random process with zero mean.

Equation 6.1 can then be reduced to the following form:

$$\mathbf{A}\mathbf{U}(\omega) = \mathbf{B}\ddot{U}_g(\omega) \tag{6.2}$$

where $\mathbf{A}$ and $\mathbf{B}$ are the coefficient matrix and the vector and are given by

$$\mathbf{A} = (-\omega^2 \mathbf{M} + i\omega \mathbf{C} + i\mathbf{D} + \mathbf{K}) \tag{6.3a}$$

$$\mathbf{B} = -\mathbf{M}\mathbf{r} \tag{6.3b}$$

A is a tri-diagonal matrix and its example for a two-story shear building model supported by two soil layers can be found in Appendix 6.A.

Because the interstory drift in a building can be a good indicator of the overall stiffness of the building, its control plays an important role in the stiffness design of the building. For this reason, the interstory drift is treated here as the controlled parameter.

Let us define the time-domain interstory drifts $\mathbf{d}(t) = \{d_i(t)\}$ of the shear building model and consider their Fourier transforms $\mathbf{\Delta}(\omega) = \{\Delta_i(\omega)\}$. The Fourier transforms $\mathbf{\Delta}(\omega)$ of interstory drifts are related to the Fourier transforms $\mathbf{U}(\omega)$ of horizontal nodal displacements with the use of the transformation matrix $\mathbf{T}$:

$$\mathbf{\Delta}(\omega) = \mathbf{T}\mathbf{U}(\omega) \tag{6.4}$$

The transformation matrix $\mathbf{T}$ consists of 1, -1, and 0 and is a rectangular matrix. Substitution of a modified expression of Equation 6.2 into Equation 6.4 provides

$$\mathbf{\Delta}(\omega) = \mathbf{T}\mathbf{A}^{-1}\mathbf{B}\ddot{U}_g(\omega) \tag{6.5}$$

Equation 6.5 is simply expressed as

$$\mathbf{\Delta}(\omega) = \mathbf{H}_\Delta(\omega)\ddot{U}_g(\omega) \tag{6.6}$$

where $\mathbf{H}_\Delta(\omega) = \{H_{\Delta_i}(\omega)\}$ are the transfer functions of the interstory drifts with respect to the input acceleration $\ddot{U}_g(\omega)$ and are described by

$$\mathbf{H}_\Delta(\omega) = \mathbf{T}\mathbf{A}^{-1}\mathbf{B} \tag{6.7}$$

As pointed out before, because **A** is a tri-diagonal matrix, its inverse can be obtained in closed form (see Appendix 6.B). This property enables one to compute the transfer functions of interstory drifts very efficiently. In particular, as the number of soil layers and building stories increases, this advantage will be accelerated and remarkable.

The statistical characteristics of stationary random signals can be described by its PSD function. Let $S_g(\omega)$ denote the PSD function of the horizontal random input $\ddot{u}_g(t)$ at the bedrock. The PSD function of the interstory drift can be expressed as the product of the transfer function squared $|H_{\Delta_i}(\omega)|^2$ and $S_g(\omega)$. Using the random vibration theory, the mean-square response of the ith interstory drift can then be computed from

$$\sigma_{\Delta_i}^2 = \int_{-\infty}^{\infty} |H_{\Delta_i}(\omega)|^2 S_g(\omega)\mathrm{d}\omega = \int_{-\infty}^{\infty} H_{\Delta_i}(\omega)H_{\Delta_i}^*(\omega)S_g(\omega)\mathrm{d}\omega \tag{6.8}$$

where $(\ )^*$ denotes the complex conjugate.

6.4 Problem of Optimal Damper Placement and Optimality Criteria

In this section, the problem of optimal damper placement in a super-building is formulated. The problem of optimal damper placement for the building structure–ground interaction (PODP-BGI) models explained above may be described as follows.

Problem 6.1 PODP-BGI   Minimize the sum of the mean squares of the interstory drifts as a flexibility measure

$$f = \sum_{i=1}^{n} \sigma_{\Delta i}^2 \tag{6.9}$$

subject to the constraint on total quantity of supplemental dampers

$$\sum_{i=1}^{n} c_{Vi} = \overline{W} \quad (\overline{W}: \text{specified total quantity of supplemental dampers}) \tag{6.10}$$

and to the constraints on each damper quantity

$$0 \le c_{Vi} \le \overline{c}_{Vi} \quad (i = 1, \cdots, n) \quad (\overline{c}_{Vi}: \text{upper bound of damping coefficient}) \tag{6.11}$$

6.4.1 Optimality Conditions

It is straightforward to use the Lagrange multiplier method for solving constrained optimization problems. The constrained optimization problem stated above can be formulated mathematically by the generalized Lagrangian formulation. The generalized Lagrangian for Problem 6.1 may be defined as

$$L(\mathbf{c}_V, \lambda, \boldsymbol{\mu}, \boldsymbol{v}) = \sum_{i=1}^{n} \sigma_{\Delta i}^2 + \lambda \left(\sum_{i=1}^{n} c_{Vi} - \overline{W} \right) + \sum_{i=1}^{n} \mu_i (0 - c_{Vi}) + \sum_{i=1}^{n} v_i (c_{Vi} - \overline{c}_{Vi}) \tag{6.12}$$

In Equation 6.12, $\boldsymbol{\mu} = \{\mu_i\}$ and $\boldsymbol{v} = \{v_i\}$ are the Lagrangian multipliers together with λ. The principal (or major) optimality criteria for Problem 6.1 without active upper and lower bound constraints on damping coefficients of supplemental dampers may be derived from the stationarity conditions of the generalized Lagrangian $L(\boldsymbol{\mu} = \mathbf{0}, \boldsymbol{v} = \mathbf{0})$ with respect to each component of $\mathbf{c}_V$ and λ:

$$f_{,j} + \lambda = 0 \quad \text{for } 0 < c_{Vj} < \overline{c}_{Vj} \ (j = 1, \cdots, n) \tag{6.13}$$

$$\sum_{i=1}^{n} c_{Vi} - \overline{W} = 0 \tag{6.14}$$

Equation 6.13 has been derived by the differentiation of Equation 6.12 with respect to c_{Vj} and Equation 6.14 has been obtained by the differentiation of Equation 6.12 with respect to λ.

Here, and in the following, the mathematical symbol $(\cdot)_{,j}$ indicates the partial differentiation with respect to the damping coefficient c_{Vj} of the supplemental damper in the jth story. When the lower or upper bound of the constraint on damping coefficients of supplemental dampers is active, the optimality condition should be modified to the following forms:

$$f_{,j} + \lambda \geq 0 \quad \text{for } c_{Vj} = 0 \tag{6.15}$$

$$f_{,j} + \lambda \leq 0 \quad \text{for } c_{Vj} = \overline{c}_{Vj} \tag{6.16}$$

6.5 Solution Algorithm

A solution algorithm for the problem of optimal damper placement stated above is explained in this section. In the solution procedure, the model without supplemental viscous dampers, namely $c_{Vj} = 0$ ($j = 1, \cdots, n$), is used as the initial model. This treatment is well suited to the situation where a structural designer is just starting the allocation and placement of added supplemental viscous dampers at desired positions. The damping coefficients of added supplemental dampers are *increased gradually* based on the optimality criteria stated above. This algorithm is called the steepest direction search algorithm as in Chapters 4 and 5, while the transfer function amplitudes were treated in Chapters 4 and 5 in place of seismic responses.

Let Δc_{Vi} and ΔW denote the increment of the damping coefficient of the ith added supplemental damper (damper in the ith story) and the increment of the sum of the damping coefficients of added supplemental dampers respectively. Once ΔW is given, the problem is to determine simultaneously the effective position and the amount of the increments of the damping coefficients of added supplemental dampers. In order to develop this algorithm, the first- and second-order sensitivities of the objective function with respect to a design variable are needed. Those quantities can be derived by differentiating successively Equation 6.8 by the design variables.

First-order derivative of mean-square interstory drift:

$$(\sigma_{\Delta_i}^2)_{,j} = \int_{-\infty}^{\infty} \{H_{\Delta_i}(\omega)\}_{,j} H_{\Delta_i}^*(\omega) S_g(\omega)\, d\omega + \int_{-\infty}^{\infty} H_{\Delta_i}(\omega) \{H_{\Delta_i}^*(\omega)\}_{,j} S_g(\omega) d\omega \tag{6.17}$$

Second-order derivative of mean-square interstory drift:

$$(\sigma_{\Delta_i}^2)_{,jl} = \int_{-\infty}^{\infty} \{H_{\Delta_i}(\omega)\}_{,j} \{H_{\Delta_i}^*(\omega)\}_{,l} S_g(\omega) d\omega + \int_{-\infty}^{\infty} \{H_{\Delta_i}(\omega)\}_{,l} \{H_{\Delta_i}^*(\omega)\}_{,j} S_g(\omega) d\omega$$

$$+ \int_{-\infty}^{\infty} \{H_{\Delta_i}(\omega)\}_{,jl} H_{\Delta_i}^*(\omega) S_g(\omega) d\omega + \int_{-\infty}^{\infty} H_{\Delta_i}(\omega) \{H_{\Delta_i}^*(\omega)\}_{,jl} S_g(\omega) d\omega \tag{6.18}$$

In Equations 6.17 and 6.18, the first- and second-order derivatives of transfer functions and their complex conjugates may be expressed as

$$\{H_{\Delta_i}(\omega)\}_{,j} = \mathbf{T}_i \mathbf{A}_{,j}^{-1} \mathbf{B} \tag{6.19}$$

$$\{H_{\Delta_i}(\omega)\}_{,jl} = \mathbf{T}_i \mathbf{A}_{,jl}^{-1} \mathbf{B} \tag{6.20}$$

$$\{H_{\Delta_i}^*(\omega)\}_{,j} = \mathbf{T}_i \mathbf{A}_{,j}^{-1*} \mathbf{B} \tag{6.21}$$

$$\{H_{\Delta_i}^*(\omega)\}_{,jl} = \mathbf{T}_i \mathbf{A}_{,jl}^{-1*} \mathbf{B} \tag{6.22}$$

In Equations 6.19–6.22, $\mathbf{T}_i$ is the ith row vector in the rectangular transformation matrix $\mathbf{T}$.

The first-order derivative of the inverse $\mathbf{A}^{-1}$ of the coefficient matrix is computed by $(\mathbf{A}^{-1})_{,j} = -\mathbf{A}^{-1}\mathbf{A}_{,j}\mathbf{A}^{-1}$, which is obtained by differentiating the identity $\mathbf{A}\mathbf{A}^{-1} = \mathbf{I}$. Because the components in the coefficient matrix $\mathbf{A}$ are linear functions of design variables, the expression $\mathbf{A}_{,jl} = \mathbf{0}$ is derived. Then, the second-order derivative of the inverse $\mathbf{A}^{-1}$ is obtained from

$$(\mathbf{A}^{-1})_{,jl} = \mathbf{A}^{-1}(\mathbf{A}_{,l}\mathbf{A}^{-1}\mathbf{A}_{,j} + \mathbf{A}_{,j}\mathbf{A}^{-1}\mathbf{A}_{,l})\mathbf{A}^{-1} \tag{6.23}$$

The first-order derivative of the complex conjugate $\mathbf{A}^{-1*}$ of the inverse can be computed as $\{\mathbf{A}_{,j}^{-1}\}^*$ and the second-order derivative of $\mathbf{A}^{-1*}$ can be found as $\mathbf{A}_{,jl}^{-1*} = \{\mathbf{A}_{,jl}^{-1}\}^*$.

The solution algorithm in the case satisfying the conditions $c_{\mathrm{V}j} < \bar{c}_{\mathrm{V}j}$ for all j may be summarized as follows:

Step 0 Initialize all the added supplemental viscous dampers as $c_{\mathrm{V}j} = 0$ ($j = 1, \cdots, n$). In the initial design stage, the structural damping alone exists in the shear building model. Assume the quantity ΔW.

Step 1 Compute the first-order derivative $f_{,i}$ of the objective function by Equation 6.17.

Step 2 Find the index p satisfying the condition

$$-f_{,p} = \max_i (-f_{,i}) \tag{6.24}$$

Step 3 Update the objective function f by the linear approximation $f + f_{,p}\Delta c_{\mathrm{V}p}$, where $\Delta c_{\mathrm{V}p} = \Delta W$. This is because the supplemental damper is added only in the pth story in the initial design stage.

Step 4 Update the first-order sensitivity $f_{,i}$ of the objective function by the linear approximation $f_{,i} + f_{,ip}\Delta c_{\mathrm{V}p}$ using Equation 6.18.

Step 5 If, in Step 4, there exists a supplemental damper of an index j such that the condition

$$-f_{,p} = \max_{j, j \neq p} (-f_{,j}) \tag{6.25}$$

is satisfied, then stop and compute the increment $\Delta \tilde{c}_{Vp}$ of the damping coefficient of the corresponding damper. At this stage, update the first-order sensitivity $f_{,i}$ by the linear approximation $f_{,i} + f_{,ip}\Delta \tilde{c}_{Vp}$ using Equation 6.18.

Step 6 Repeat the procedure from Step 2 to Step 5 until the constraint in Equation 6.10 (i.e., $\sum_{i=1}^{n} c_{Vi} = \overline{W}$) is satisfied.

In Steps 2 and 3, the direction which decreases the objective function most effectively under the condition $\sum_{i=1}^{n} \Delta c_{Vi} = \Delta W$ is found and the design (the quantity of supplemental dampers) is updated in that direction. It is suitable, therefore, to call the present algorithm "the steepest direction search algorithm," as in Chapters 4 and 5. As explained before, this algorithm is similar to the conventional steepest descent method in the mathematical programming (see Figure 4.4 in Chapter 4 to understand the concept). However, while the conventional steepest descent method uses the gradient vector itself of the objective function as the direction and does not utilize optimality criteria, the present algorithm takes advantage of the newly derived optimality criteria expressed by Equations 6.13, 6.15, and 6.16 and does not adopt the gradient vector as its direction. More specifically, the explained steepest direction search guarantees the successive and approximate satisfaction of the optimality criteria by a linear approximation. For example, if Δc_{Vp} is added to the pth added supplemental viscous damper in which Equation 6.24 is satisfied, then its damper ($c_p > 0$) satisfies the optimality condition in Equation 6.13 and the other dampers ($c_j = 0$, $j \neq p$) alternatively satisfy the optimality condition in Equation 6.15. It is important to note that a series of subproblems is introduced here tentatively in which the total damper capacity level $\overline{W}$ is increased gradually by ΔW from zero through the specified value.

It is necessary to investigate other possibilities. If multiple indices $p_1, \cdots, p_m$ exist in Step 2, then the objective function f and its derivative $f_{,j}$ have to be updated by the following rules:

$$f \rightarrow f + \sum_{i=p_1}^{p_m} f_{,i}\Delta c_{Vi} \tag{6.26a}$$

$$f_{,j} \rightarrow f_{,j} + \sum_{i=p_1}^{p_m} f_{,ji}\Delta c_{Vi} \tag{6.26b}$$

Furthermore, the index p defined in Step 5 has to be replaced by the multiple indices $p_1, \cdots, p_m$. The ratios among the magnitudes Δc_{Vi} have to be determined so that the following relations are satisfied:

$$f_{,p_1} + \sum_{i=p_1}^{p_m} f_{,p_1 i}\Delta c_{Vi} = \cdots = f_{,p_m} + \sum_{i=p_1}^{p_m} f_{,p_m i}\Delta c_{Vi} \tag{6.27}$$

Equation 6.27 requires that the optimality condition in Equation 6.9 continues to be satisfied by a linear approximation in the supplemental dampers with the indices $p_1, \cdots, p_m$.

It may be the case in realistic situations that the maximum quantity of supplemental dampers is limited by the requirements of building design and planning. In the case where the damping coefficients of some added supplemental dampers attain their upper bounds, such constraints must be incorporated in the aforementioned algorithm. In that case, the increment Δc_{Vp} of the supplemental dampers is added subsequently to the damper in which $-f_{,p}$ attains the maximum value among all the dampers, except for those attaining the upper bound.

6.6 Numerical Examples

A band-limited white noise is often used as a test input of random disturbance. In this section, a band-limited white noise with the following PSD function is input to the building–ground system as the input acceleration $\ddot{u}_g(t)$:

$$S_g(\omega) = 0.01 \text{ m}^2/s^3 \quad (-2\pi \times 16 \le \omega \le -2\pi \times 0.2, 2\pi \times 0.2 \le \omega \le 2\pi \times 16)$$

$$S_g(\omega) = 0 \quad \text{otherwise}$$

As stated at the beginning of this chapter, the soil under a building structure influences greatly the seismic response of the building structure. In order to investigate in detail the influence of soil conditions on the seismic response and optimal distributions of supplemental dampers, three different soil conditions are considered. These three soil types are specified as very stiff ground, stiff ground, and soft ground. Every ground model consists of three soil layers of identical thickness $l_i = 6$ m and these layers rest on a semi-infinite visco-elastic ground (bedrock). The shear modulus, damping ratio, and mass density of the semi-infinite visco-elastic ground are given as 8.0×10^8 N/m^2, 0.01, and 2.0×10^3 kg/m^3 respectively.

The input acceleration is defined at $l_0 = 6$ m below the boundary (engineering bedrock surface) between the three soil layers and the semi-infinite visco-elastic ground. The mass densities of the three soil layers are assumed to be identical and take the same value of 1.8×10^3 kg/m^3 over the very stiff ground, stiff ground, and soft ground.

The shear moduli of the soil layers are also identical within each of the three layers and take the values 1.6×10^8 N/m^2 in the very stiff ground, 0.8×10^8 N/m^2 in the stiff ground, and 0.2×10^8 N/m^2 in the soft ground. Nonlinear amplification of ground motion within the surface ground is considered here. The dependence of the shear moduli and damping ratios on the strain level is taken into account through an equivalent linear model. Their relations used in this numerical example are shown in Figure 6.2 for clay, sand, and gravel. It is assumed that layer 1 consists of gravel, layer 2 of clay, and layer 3 of sand. The influence of the mean effective pressure has been taken into account. Linear interpolation has been used in evaluating the intermediate values between the selected points. The effective strain level (Schnabel

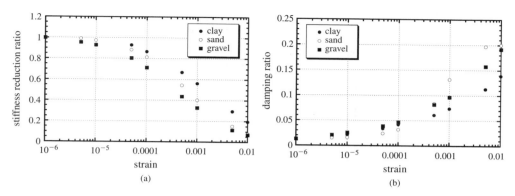

Figure 6.2 Dependence of shear moduli and damping ratios on strain level. (Originally published in I. Takewaki and K. Uetani, "Optimal damper placement for building structures including surface ground amplification," *Soil Dynamics and Earthquake Engineering*, **18**, no. 5, 363–371, 1999, Elsevier B.V.).

et al., 1972) has been evaluated from the relation $2.5\sigma_{\Delta_i} \times 0.65$, where the coefficient 2.5 represents the peak factor (Der Kiureghian, 1980; Takewaki *et al.* 2002a, 2002b) and 0.65 was introduced by Schnabel *et al.* (1972). The convergent characteristics of the stiffness reduction ratios of the shear moduli and damping ratios are shown in Figure 6.3 and Figure 6.4 respectively. This evaluation of equivalent values has been conducted for the structure–ground interaction system without added dampers.

While the fundamental natural periods of these three surface grounds at the small strain level are 0.239 s, 0.338 s, and 0.675 s for the very stiff ground, stiff ground, and soft ground respectively, those computed from the equivalent stiffnesses are 0.322 s, 0.617 s, and 1.48 s for the very stiff ground, stiff ground, and soft ground respectively.

As for super-buildings, we consider a three-story shear building model for simple and clear presentation of the theory developed in this chapter. The floor masses are specified as $m_0 = 96 \times 10^3$ kg and $m_i = 32 \times 10^3$ kg ($i = 1,2,3$) and the story stiffnesses are given by $k_i = 3.76 \times 10^7$ N/m ($i = 1,2,3$). The fundamental natural period of the super-structure with a fixed base is 0.412 s.

It is important to investigate the relation between the fundamental natural period of the structure and that of the surface ground in order to understand the effect of the soil property on the seismic response of the building structure. Figure 6.5 shows the relation between the fundamental natural period of the structure and that of the surface ground. In the present case, the fundamental natural period of the structure is close to that of the very stiff surface ground. The structural viscous damping matrix of the shear building model has been given so that it is proportional to the stiffness matrix of the shear building model, and the lowest mode damping ratio attains the value of 0.02.

Figure 6.6(a) shows the variation of the viscous damping coefficients of added supplemental dampers in the optimal placement for the very stiff ground. It is observed

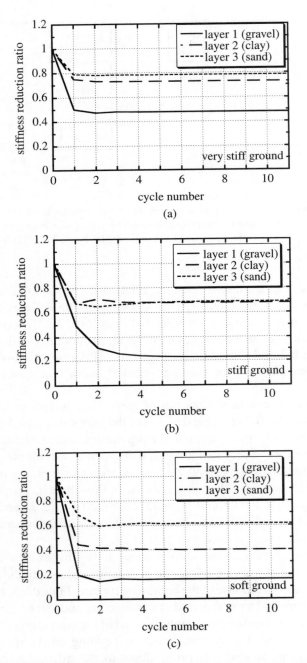

Figure 6.3 Convergence processes of stiffness reduction ratios of shear moduli. (Originally published in I. Takewaki and K. Uetani, "Optimal damper placement for building structures including surface ground amplification," *Soil Dynamics and Earthquake Engineering*, **18**, no. 5, 363–371, 1999, Elsevier B.V.).

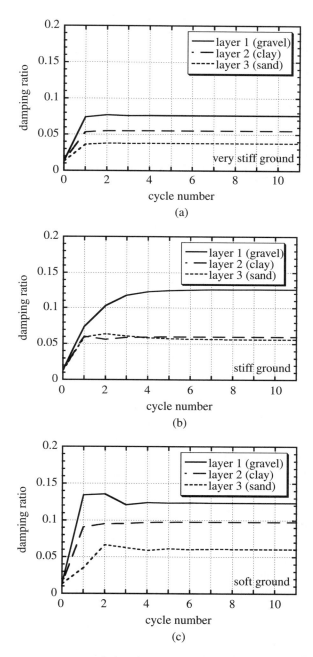

Figure 6.4 Convergence processes of damping ratio. (Originally published in I. Takewaki and K. Uetani, "Optimal damper placement for building structures including surface ground amplification," *Soil Dynamics and Earthquake Engineering*, **18**, no. 5, 363–371, 1999, Elsevier B.V.).

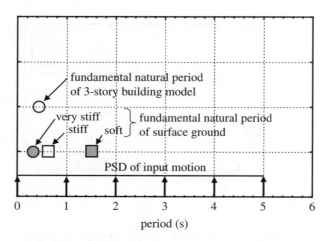

Figure 6.5 Relation between the fundamental natural period of the structure and that of the surface ground. (Originally published in I. Takewaki and K. Uetani, "Optimal damper placement for building structures including surface ground amplification," *Soil Dynamics and Earthquake Engineering*, **18**, no. 5, 363–371, 1999, Elsevier B.V.).

that the viscous damping coefficient of the supplemental damper is first added in the first story and then in the second story. Figure 6.6(b) and (c) illustrates the variations of viscous damping coefficients of supplemental dampers in the optimal placement for the stiff ground and the soft ground. It can be understood from Figure 6.6(a)–(c) that the optimal supplemental damper placement is strongly dependent on the surface ground properties. It is well known that resonance of the fundamental natural frequency of the building structure with the predominant frequency of the surface ground can sometimes cause amplified responses (Takewaki, 1998b). It is noted that, because the soil exhibits a high nonlinearity under earthquake loading, the evaluation of the equivalent stiffness of the ground needs careful treatment. It may be concluded, therefore, that the relation of the fundamental natural frequency of the building structure with the predominant frequency of the surface ground (equivalent stiffness) is a key parameter for characterizing the optimal damper placement.

Figure 6.7 presents the comparison of the objective functions between the model of the optimal placement of supplemental dampers and that of uniform placement. In the same step number, both placements have the same total damper capacity. It can be seen that the optimal placement can reduce the objective function effectively.

For further investigation, consider two nine-story shear building models with different distributions of story stiffnesses. The floor masses are specified as $m_0 = 96 \times 10^3$ kg and $m_i = 32 \times 10^3$ kg ($i = 1, \cdots, 9$), and the story stiffnesses are specified as $k_i = 5.64 \times 10^7$ N/m ($i = 1, \cdots, 9$) for model A and $k_i = 7.53 \times 10^7$ N/m ($i = 1, 2, 3$), $k_i = 5.64 \times 10^7$ N/m ($i = 4, 5, 6$), and $k_i = 3.76 \times 10^7$ N/m ($i = 7, 8, 9$) for model B. The story stiffness distributions of models A and B are shown in Figure 6.8.

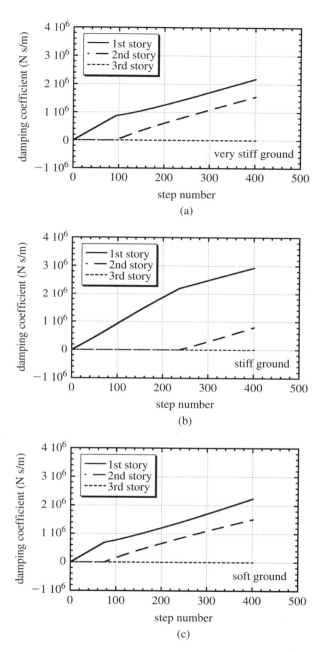

Figure 6.6 Variation of damping coefficients in optimal damper placement. (Originally published in I. Takewaki and K. Uetani, "Optimal damper placement for building structures including surface ground amplification," *Soil Dynamics and Earthquake Engineering*, **18**, no. 5, 363–371, 1999, Elsevier B.V.).

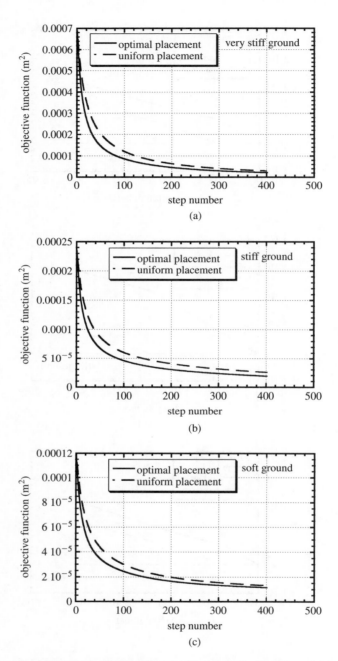

Figure 6.7 Comparison of the objective functions between the optimal placement and uniform placement. (Originally published in I. Takewaki and K. Uetani, "Optimal damper placement for building structures including surface ground amplification," *Soil Dynamics and Earthquake Engineering*, **18**, no. 5, 363–371, 1999, Elsevier B.V.).

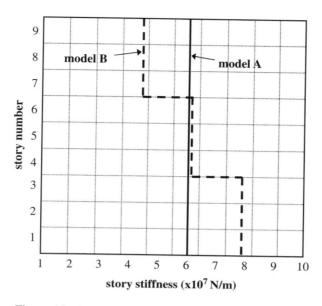

Figure 6.8   Story stiffness distributions of models A and B.

The fundamental natural period of the structure with the fixed base of model A is 0.906 s and that of model B is 0.858 s. The lowest mode structural damping ratio of the building is 0.02.

Figure 6.9 illustrates the distributions of the optimal viscous damping coefficients of supplemental dampers with respect to step number in very stiff ground, stiff ground, and soft ground for these two nine-story buildings (uniform stiffness and varied stiffness). It is observed that the optimal supplemental damper placement is also strongly dependent on the surface ground properties and the story stiffness distributions. It is again expected that the relation of the fundamental natural frequency of the building structure with that of the surface ground is a key parameter for the optimal supplemental damper placement.

6.7 Summary

The results are summarized as follows.

1. Viscous damping of supplemental dampers in a building structure, hysteretic damping in a surface ground, and viscous damping at the ground viscous boundary can be taken into account exactly in the present unified formulation based on a frequency-domain approach. A so-called steepest direction search algorithm has been developed and explained in detail.

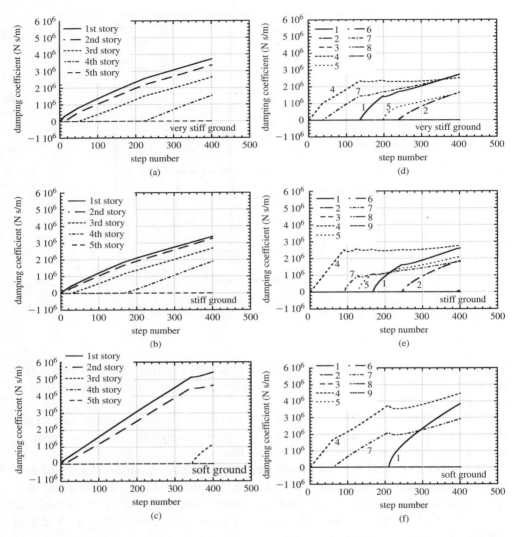

Figure 6.9 Distributions of optimal damping coefficients with respect to step number in very stiff ground, stiff ground, and soft ground for nine-story buildings: (a)–(c) uniform stiffness; (d)–(f): varied stiffness.

2. Because a transfer function with respect to the input acceleration at the engineering bedrock can be obtained in a closed form for MDOF systems with nonproportional damping owing to the tri-diagonal property of the coefficient matrix, the mean-square interstory drifts of the building structure to the random earthquake input and their first- and second-order derivatives with respect to the design variables (viscous damping coefficients of supplemental dampers) can be computed very efficiently.

3. It has been demonstrated numerically that the optimal placement of supplemental dampers obtained from the present formulation can actually reduce the objective function effectively compared with uniform damper placement.
4. The effects of surface ground properties on optimal supplemental damper placement in building structures have been examined numerically by considering three soil conditions. It has been clarified that the relation of the fundamental natural frequency of the building structure with the predominant frequency of the surface ground (equivalent stiffness) is a key parameter for characterizing the optimal supplemental damper placement. The story stiffness distributions also influence the optimal supplemental damper placement.

Appendix 6.A: System Mass, Damping, and Stiffness Matrices for a Two-story Shear Building Model Supported by Two Soil Layers

The system mass, damping, and stiffness matrices for a two-story shear building model supported by two soil layers may be expressed by

$$
\mathbf{M} =
\begin{bmatrix}
m_2 & & & & & \mathbf{0} \\
& m_1 & & & & \\
& & m_0 + \mu_2/3 & \mu_2/6 & & \\
& & \mu_2/6 & (\mu_2+\mu_1)/3 & \mu_1/6 & \\
& & & \mu_1/6 & (\mu_1+\mu_0)/3 & \mu_0/6 \\
\mathbf{0} & & & & \mu_0/6 & \mu_0/3
\end{bmatrix}
\tag{A6.1}
$$

$$
\mathbf{C} =
\begin{bmatrix}
C_2 & -C_2 & & & & \mathbf{0} \\
-C_2 & C_2+C_1 & -C_1 & & & \\
& -C_1 & C_1 & & & \\
& & & 0 & & \\
& & & & 0 & \\
\mathbf{0} & & & & & c_b
\end{bmatrix}
\tag{A6.2}
$$

$$
\mathbf{iD} + \mathbf{K} =
\begin{bmatrix}
k_2 & -k_2 & & & & \mathbf{0} \\
-k_2 & k_2+k_1 & -k_1 & & & \\
& -k_1 & k_1+K_2 & -K_2 & & \\
& & -K_2 & K_2+K_1 & -K_1 & \\
& & & -K_1 & K_1+K_0 & -K_0 \\
\mathbf{0} & & & & -K_0 & K_0
\end{bmatrix}
\tag{A6.3}
$$

where $\mu_i = \rho_i A l_i$, $C_i = c_i + c_{Vi}$, and $K_i = (G_i A / l_i)(1 + i2\beta_i)$.

Appendix 6.B: Closed-form Expression of the Inverse of a Tri-diagonal Matrix

Consider the following symmetric tri-diagonal matrix of $M \times M$:

$$
\mathbf{A} = \begin{bmatrix}
d_M & -e_M & & & & 0 \\
-e_M & \ddots & & \ddots & & \\
& & \ddots & & \ddots & \\
& & & \ddots & \ddots & -e_3 \\
& & & -e_3 & d_2 & -e_2 \\
0 & & & & -e_2 & d_1
\end{bmatrix} \tag{B6.1}
$$

Let us define the following principal minors of the matrix $\mathbf{A}$:

$$
P_0 = 1, \quad P_1 = d_1, \quad P_2 = \begin{vmatrix} d_2 & -e_2 \\ -e_2 & d_1 \end{vmatrix}, \dots, P_M = \det \mathbf{A} \tag{B6.2}
$$

$$
P_{R0} = 1, \quad P_{R1} = d_M, \quad P_{R2} = \begin{vmatrix} d_M & -e_M \\ -e_M & d_{M-1} \end{vmatrix}, \dots, P_{RM} = \det \mathbf{A} \tag{B6.3}
$$

The principal minors satisfy the following recurrence formula:

$$
P_{j-1} = d_{j-1} P_{j-2} - e_{j-1}^2 P_{j-3} \ (j = 3, \dots, M) \tag{B6.4}
$$

The jth column of $\mathbf{A}^{-1}$ may be expressed as

$$
\frac{1}{\det \mathbf{A}} \left\{ \left(\prod_{i=M-j+2}^{M} e_i \right) P_{M-j} P_{R0} \quad \left(\prod_{i=M-j+2}^{M-1} e_i \right) P_{M-j} P_{R1} \quad \cdots \right.
$$

$$
\left(\prod_{i=M-j+2}^{M-j+2} e_i \right) P_{M-j} P_{R(j-2)} \quad P_{M-j} P_{R(j-1)} \quad \left(\prod_{i=M-j+1}^{M-j+1} e_i \right) P_{M-j-1} P_{R(j-1)}
$$

$$
\cdots \left(\prod_{i=3}^{M-j+1} e_i \right) P_1 P_{R(j-1)} \quad \left. \left(\prod_{i=2}^{M-j+1} e_i \right) P_0 P_{R(j-1)} \right\}^{\mathrm{T}}
$$

$$
\tag{B6.5}
$$

References

Alam, S. and Baba, S. (1993) Robust active optimal control scheme including soil–structure interaction. *Journal of Structural Engineering*, **119**, 2533–2551.

Der Kiureghian, A. (1980) A response spectrum method for random vibrations. Report no. UCB/EERC 80-15, Earthquake Engineering Research Center, University of California, Berkeley, CA.

Lysmer, J. and Kuhlemeyer, R.L. (1969) Finite dynamic model for infinite media. *Journal of the Engineering Mechanics Division*, **94** (EM4), 859–877.

Schnabel, P.B., Lysmer, J., and Seed, H.B. (1972) SHAKE: a computer program for earthquake response analysis of horizontally layered sites. A computer program distributed by NISEE/Computer Applications, Berkeley.

Smith, H.A. and Wu, W.H. (1997) Effective optimal structural control of soil–structure interaction systems. *Earthquake Engineering and Structural Dynamics*, **26**, 549–570.

Smith, H.A., Wu, W.H., and Borja, R.I. (1994) Structural control considering soil–structure interaction effects. *Earthquake Engineering and Structural Dynamics*, **23**, 609–626.

Takewaki, I. (1998a) Optimal damper positioning in beams for minimum dynamic compliance. *Computer Methods in Applied Mechanics and Engineering*, **156** (1–4), 363–373.

Takewaki, I. (1998b) Remarkable response amplification of building frames due to resonance with the surface ground. *Soil Dynamics and Earthquake Engineering*, **17** (4), 211–218.

Takewaki, I. and Yoshitomi, S. (1998) Effects of support stiffnesses on optimal damper placement for a planar building frame. *The Structural Design of Tall Buildings*, **7** (4), 323–336.

Takewaki, I., Fujii, N., and Uetani, K. (2002a) Nonlinear surface ground analysis via statistical approach. *Soil Dynamics and Earthquake Engineering*, **22** (6), 499–509.

Takewaki, I., Fujii, N., and Uetani, K. (2002b) Simplified inverse stiffness design for nonlinear soil amplification. *Engineering Structures*, **24** (11), 1369–1381.

Wong, H.L. and Luco, J.E. (1991) Structural control including soil–structure interaction effects. *Journal of Engineering Mechanics*, ASCE, **117**, 2237–2250.

7

Optimal Sensitivity-based Design of Dampers in Bending-shear Buildings on Surface Ground under Earthquake Loading

7.1 Introduction

It is well recognized that passive control devices are effective tools for suppressing the structural responses with a small amount of cost (Kobori, 1996; Housner *et al.*, 1997). Seismic responses of building structures both with and without passive control devices are influenced greatly by soil–structure interaction. It is desired, therefore, to construct a theoretical basis for effective damper placement under the soil–structure interaction. The purpose of this chapter is to develop a systematic method for optimal viscous damper placement in shear-flexural building models taking into account response amplification due to the surface ground. It is well known that, in shear-flexural building models, story-installation-type dampers are effective for shear deformations only and design implications derived for shear-building models are not necessarily useful for the shear-flexural building models. The design implications for optimal damper placement in shear-flexural building models are discussed here. Nonlinear amplification of ground motion within the surface ground is described by a deterministic equivalent linear model, and local interaction with the surrounding soil is incorporated through horizontal and rotational springs and dashpots. Hysteretic damping of the surface ground and radiation damping into the semi-infinite ground are included in the model. While several results have been reported on active control of building structures including soil–structure interaction effects (Wong and Luco, 1991; Luco, 1998; Alam and Baba, 1993; Smith *et al.*, 1994; Smith and Wu, 1997), there are very few on optimal damper placement in a building structure, especially a shear-flexural building model, taking into account interaction with the surface ground.

The original steepest direction search algorithm (Takewaki, 1998a; Takewaki and Yoshitomi, 1998) explained in Chapters 4 and 5 for fixed-base structures is extended to an interaction model of a shear-flexural building with the surface ground as in Chapter 6. While resonant steady-state responses of structures with fixed bases were treated in Chapters 2–5 (Takewaki, 1998a; Takewaki and Yoshitomi, 1998), earthquake responses to random inputs are introduced here as controlled parameters as in Chapter 6. In comparison with the model in Chapter 6, a decomposed model is treated here where the amplification of ground motion within the surface ground is expressed first and that expression is introduced in the seismic response analysis of the super-building models. Optimal placement and the capacity of the viscous dampers are found automatically via the steepest direction search algorithm, which is based on successive satisfaction of the optimality conditions. Several examples for various surface ground properties are presented to demonstrate the effectiveness and validity of the present method and to examine the effects of surface ground characteristics on optimal damper placement.

There are many useful investigations on the optimal damper placement for fixed-base models (e.g., Zhang and Soong, 1992; Tsuji and Nakamura, 1996). Actually, the method due to Zhang and Soong (1992) is simple and may be applicable to a broad range of structural models. However, it does not appear that the previous approaches are applicable to soil–structure interaction models due to their limitations on modeling of structures and damping mechanisms or computational efficiency.

7.2 Building and Ground Model

7.2.1 Ground Model

The building structure is described by a shear-flexural building model introduced by Takewaki (2000) and the surface ground is represented by a shear beam model (see Figure 7.1). Local interaction of the shear-flexural building model with the surrounding soil is described by horizontal and rotational springs and dashpots between the two substructures. Let n and N denote the number of stories of the shear-flexural building model and the number of surface soil layers respectively. The random acceleration input is defined at the level just below the bedrock on which the layered surface ground lies.

Radiation damping into the semi-infinite visco-elastic ground is taken into account by a viscous boundary (Lysmer and Kuhlemeyer, 1969). Let $c_b = \rho_0 V_S A = \sqrt{\rho_0 G_0} A$ denote the total damping coefficient of the viscous boundary where ρ_0, V_S, and G_0 are respectively the mass density, the shear wave velocity, and the shear modulus of the semi-infinite visco-elastic ground and A is the governing area of the surface ground. Let l_i, ρ_i, G_i, and β_i respectively denote the layer thickness, the mass density,

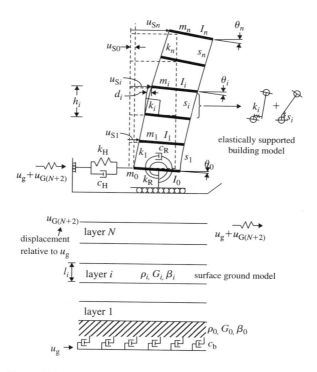

Figure 7.1 Bending-shear building model on surface ground.

the shear modulus, and the hysteretic damping ratio in layer i. A linear displacement function is used in evaluating the stiffness and consistent mass matrices in the ground FE model. Nonlinear amplification due to the surface ground is taken into account through an equivalent linear model (Schnabel *et al.*, 1972). The equivalent linear model used for the surface ground is almost equivalent to the SHAKE program. This model is a deterministic equivalent linear model and its accuracy has been demonstrated by many researchers. To evaluate the mean peak soil response of the equivalent linear model, the present method uses a peak factor (Der Kiureghian, 1980) multiplied by the standard deviation of the shear strain. The validity of this treatment in a linear model has been demonstrated by Der Kiureghian (1980). Once the surface ground is modeled by the equivalent linear model, the whole system consisting of a structure and the surface ground is a linear system. The random vibration theory for linear models is well established and its accuracy has been checked by many investigators.

Let k_H, k_R, c_H, and c_R denote the horizontal soil spring stiffness, the rotational soil spring stiffness, the horizontal soil dashpot damping coefficient, and the rotational soil dashpot damping coefficient respectively. These parameters k_H, k_R, c_H, and c_R are assumed to be prescribed for each surface ground model.

7.2.2 Building Model

A new shear-flexural building model is introduced. Let m_i, I_i, k_i, and s_i denote the floor mass, the floor mass moment of inertia, the story shear stiffness (stiffness with respect to shear deformation), and the story bending stiffness (stiffness with respect to interfloor rotation) in the ith story respectively. The height of the ith story is denoted by h_i. Let c_{Vi} denote the damping coefficient of the added damper in the ith story, which is effective for shear deformation only. The set $\{c_{Vi}\}$ is the design variable set. A member coordinate system and a system coordinate system are introduced to construct a system stiffness matrix from member stiffness equations. Let $\mathbf{F}_i^m = \{Q_i^m \quad M_i^m \quad Q_{i-1}^m \quad M_{i-1}^m\}^T$ and $\mathbf{u}_i^m = \{u_i^m \quad \theta_i^m \quad u_{i-1}^m \quad \theta_{i-1}^m\}^T$ denote the member-end forces and member-end displacements respectively in the member coordinate system in the ith story. On the other hand, let $\mathbf{F}_i^s = \{Q_i^s \quad M_i^s \quad Q_{i-1}^s \quad M_{i-1}^s\}^T$ and $\mathbf{u}_i^s = \{u_i^s \quad \theta_i^s \quad u_{i-1}^s \quad \theta_{i-1}^s\}^T$ denote the member-end forces and member-end displacements respectively in the system coordinate system in the ith story (see Figure 7.1).

Owing to the member equilibrium, $\mathbf{F}_i^s$ can be expressed in terms of $\mathbf{F}_i^m$ with the transformation matrix $\mathbf{T}_i$:

$$\begin{aligned}
\mathbf{F}_i^s &= \{Q_i^s \quad M_i^s \quad Q_{i-1}^s \quad M_{i-1}^s\}^T \\
&= \{Q_i^m \quad M_i^m \quad -Q_i^m \quad -(Q_i^m h_i + M_i^m)\}^T \\
&= \mathbf{T}_i^T \mathbf{F}_i^m
\end{aligned} \tag{7.1}$$

where

$$\mathbf{T}_i = \begin{bmatrix} 1 & 0 & -1 & -h_i \\ 0 & 1 & 0 & -1 \\ 0 & 0 & 0 & 0 \\ 0 & 0 & 0 & 0 \end{bmatrix} \tag{7.2}$$

Owing to the coordinate transformation, $\mathbf{u}_i^m$ can be expressed in terms of $\mathbf{u}_i^s$:

$$\begin{aligned}
\mathbf{u}_i^m &= \{u_i^m \quad \theta_i^m \quad u_{i-1}^m \quad \theta_{i-1}^m\}^T \\
&= \{u_i^s - u_{i-1}^s - h_i \theta_{i-1}^s \quad \theta_i^s - \theta_{i-1}^s \quad 0 \quad 0\}^T \\
&= \mathbf{T}_i \mathbf{u}_i^s
\end{aligned} \tag{7.3}$$

The element stiffness matrices associated with shear and bending deformations in the member coordinate system in the ith story may be described by

$$\boldsymbol{\kappa}_i^S = k_i \begin{bmatrix} 1 & 0 & -1 & 0 \\ 0 & 0 & 0 & 0 \\ -1 & 0 & 1 & 0 \\ 0 & 0 & 0 & 0 \end{bmatrix} \tag{7.4a}$$

$$\boldsymbol{\kappa}_i^B = s_i \begin{bmatrix} 0 & 0 & 0 & 0 \\ 0 & 1 & 0 & -1 \\ 0 & 0 & 0 & 0 \\ 0 & -1 & 0 & 1 \end{bmatrix} \tag{7.4b}$$

The element stiffness matrices in the system coordinate system in the ith story can then be derived as

$$\mathbf{T}_i^T \boldsymbol{\kappa}_i^S \mathbf{T}_i = k_i \begin{bmatrix} 1 & 0 & -1 & -h_i \\ 0 & 0 & 0 & 0 \\ -1 & 0 & 1 & h_i \\ -h_i & 0 & h_i & h_i^2 \end{bmatrix} \tag{7.5a}$$

$$\mathbf{T}_i^T \boldsymbol{\kappa}_i^B \mathbf{T}_i = s_i \begin{bmatrix} 0 & 0 & 0 & 0 \\ 0 & 1 & 0 & -1 \\ 0 & 0 & 0 & 0 \\ 0 & -1 & 0 & 1 \end{bmatrix} \tag{7.5b}$$

Let $\mathbf{u}_S = \{u_{Sn} \quad \theta_n \quad \cdots \quad u_{S1} \quad \theta_1 \quad u_{S0} \quad \theta_0\}^T$ denote a set of displacements in the system coordinate system. From Equations 7.5a and 7.5b, the system stiffness matrix of the elastically supported shear-flexural building model with respect to $\mathbf{u}_S$ can be expressed as

$$\mathbf{K}_S = \mathbf{K}_{SS} + \mathbf{K}_{SG} \tag{7.6}$$

where

$$\mathbf{K}_{SS} = \sum_{j=1}^{f} k_{f-j+1} \begin{bmatrix} & \vdots & & & \vdots & \\ \cdots & 1 & 0 & -1 & -h_{f-j+1} & \cdots & 2(j-1)+1 \\ & 0 & 0 & 0 & 0 & \\ & -1 & 0 & 1 & h_{f-j+1} & \\ \cdots & -h_{f-j+1} & 0 & h_{f-j+1} & h_{f-j+1}^2 & \cdots \\ & \vdots & & & \vdots & \end{bmatrix}$$

$$2(j-1)+1$$

$$+ \sum_{j=1}^{f} s_{f-j+1} \begin{bmatrix} & \vdots & & \vdots & \\ \cdots & 0 & 0 & 0 & 0 & \cdots & 2(j-1)+1 \\ & 0 & 1 & 0 & -1 & \\ & 0 & 0 & 0 & 0 & \\ \cdots & 0 & -1 & 0 & 1 & \cdots \\ & \vdots & & \vdots & \end{bmatrix}$$

$$2(j-1)+1 \tag{7.7a}$$

$$\mathbf{K}_{SG} = \text{diag}(0 \quad \cdots \quad 0 \quad k_H \quad k_R) \tag{7.7b}$$

The system damping matrix of the model with respect to $\dot{\mathbf{u}}_S$ can be described as

$$\mathbf{C}_S = \mathbf{C}_{SS} + \mathbf{C}_{SV} + \mathbf{C}_{SG} \tag{7.8}$$

where

$$\mathbf{C}_{SS} = (2h^s/\omega^s)\mathbf{K}_{SS} \tag{7.9a}$$

$$\mathbf{C}_{SV} = \sum_{j=1}^{f} c_{V(f-j+1)} \begin{bmatrix} & \vdots & & & & \vdots & \\ \cdots & 1 & 0 & -1 & -h_{f-j+1} & \cdots & 2(j-1)+1 \\ & 0 & 0 & 0 & 0 & & \\ & -1 & 0 & 1 & h_{f-j+1} & & \\ \cdots & -h_{f-j+1} & 0 & h_{f-j+1} & h_{f-j+1}^2 & \cdots & \\ & \vdots & & & & \vdots & \end{bmatrix}$$

$$2(j-1)+1 \tag{7.9b}$$

$$\mathbf{C}_{SG} = \mathrm{diag}(0 \quad \cdots \quad 0 \quad c_H \quad c_R) \tag{7.9c}$$

In Equation 7.9a, ω^s is the fundamental natural circular frequency of the building model and h^s is the lowest mode damping ratio. It should be noted that, since each damper is effective with respect to shear deformation only, the components associated with shear deformation velocities alone are taken into account in the system damping matrix.

7.3 Equations of Motion in Ground

Let $\mathbf{M}_G$, $\mathbf{C}_G$, $\mathbf{D}$, $\mathbf{K}_G$, $\mathbf{r}_G = \{1 \quad \cdots \quad 1\}^T$ denote the system mass matrix (consistent mass matrix), the system viscous damping matrix of the ground (viscous boundary), the system hysteretic damping matrix of the soil layers, the system stiffness matrix, and the influence coefficient vector respectively. An example of $\mathbf{M}_G$, $\mathbf{C}_G$, $\mathbf{D}$, $\mathbf{K}_G$ is shown in Appendix 7. A for a surface ground model with two soil layers. Equations of motion of the surface ground model in the frequency domain may be written as

$$(-\omega^2\mathbf{M}_G + i\omega\mathbf{C}_G + i\mathbf{D} + \mathbf{K}_G)\mathbf{U}_G(\omega) = -\mathbf{M}_G\mathbf{r}_G\ddot{U}_g(\omega) \tag{7.10}$$

where i is the imaginary unit. $\mathbf{U}_G(\omega)$ and $\ddot{U}_g(\omega)$ are the Fourier transforms of the horizontal displacements of the nodes relative to the base input and the Fourier transform of the horizontal input acceleration $\ddot{u}_g(t)$ defined at the level of l_0 below the bedrock. The horizontal input acceleration is assumed to be a stationary Gaussian random process with zero mean. Equation 7.10 can be reduced to the following form:

$$\mathbf{A}_G\mathbf{U}_G(\omega) = \mathbf{B}_G\ddot{U}_g(\omega) \tag{7.11}$$

where

$$\mathbf{A}_G = (-\omega^2 \mathbf{M}_G + i\omega \mathbf{C}_G + i\mathbf{D} + \mathbf{K}_G) \tag{7.12a}$$

$$\mathbf{B}_G = -\mathbf{M}_G \mathbf{r}_G \tag{7.12b}$$

Then $\mathbf{U}_G(\omega)$ can be expressed as

$$\mathbf{U}_G(\omega) = \mathbf{A}_G^{-1} \mathbf{B}_G \ddot{U}_g(\omega) \equiv \mathbf{H}_G(\omega) \ddot{U}_g(\omega) \tag{7.13}$$

The Fourier transform $U_{G(N+2)}(\omega)$ of the ground surface motion (horizontal displacement) may be obtained from

$$U_{G(N+2)}(\omega) = H_{G(N+2)}(\omega) \ddot{U}_g(\omega) \tag{7.14}$$

where $H_{G(N+2)}(\omega)$ is the $(N+2)$-th component of $\mathbf{H}_G(\omega)$.

7.4 Equations of Motion in Building and Seismic Response

As stated before, different from the model in Chapter 6, a decomposed model is treated here where the amplification of ground motion within the surface ground is expressed first (i.e., Equation 7.14) and that expression is introduced in the seismic response analysis of the super-building models (i.e., Equation 7.15).

Let $\mathbf{M}_S$, $\mathbf{C}_S$, $\mathbf{K}_S$, $\mathbf{r}_S = \{1 \quad 0 \quad \cdots \quad 1 \quad 0\}^T$ denote the system mass matrix (lumped mass matrix), the system viscous damping matrix, the system stiffness matrix of the elastically supported building model, and the building influence coefficient vector respectively. An example of $\mathbf{M}_S$, $\mathbf{K}_S$, and $\mathbf{C}_S$ is shown in Appendix 7.B for a two-story model. Equations of motion of the elastically supported building model in the frequency domain may be written as

$$(-\omega^2 \mathbf{M}_S + i\omega \mathbf{C}_S + \mathbf{K}_S) \mathbf{U}_S(\omega) = -\mathbf{M}_S \mathbf{r}_S \{\ddot{U}_g(\omega) + \ddot{U}_{G(N+2)}(\omega)\} \tag{7.15}$$

where $\ddot{U}_{G(N+2)}(\omega) = -\omega^2 U_{G(N+2)}(\omega)$ and $\mathbf{U}_S(\omega)$ are the Fourier transforms of the displacements $\mathbf{u}_S(t)$ consisting of the floor horizontal displacements relative to the free-field ground surface motion and the angles of floor rotation.

Substitution of Equation 7.14 into Equation 7.15 leads to the following expression of $\mathbf{U}_S(\omega)$:

$$\begin{aligned} \mathbf{U}_S(\omega) &= \mathbf{A}_S^{-1} \mathbf{B}_S \{1 - \omega^2 H_{G(N+2)}(\omega)\} \ddot{U}_g(\omega) \\ &= \mathbf{H}_S(\omega) \{1 - \omega^2 H_{G(N+2)}(\omega)\} \ddot{U}_g(\omega) \end{aligned} \tag{7.16}$$

where

$$\mathbf{A}_S = (-\omega^2 \mathbf{M}_S + i\omega \mathbf{C}_S + \mathbf{K}_S) \tag{7.17a}$$

$$\mathbf{B}_S = -\mathbf{M}_S \mathbf{r}_S \tag{7.17b}$$

$$\mathbf{H}_S(\omega) = \mathbf{A}_S^{-1} \mathbf{B}_S \tag{7.17c}$$

Let us define interstory drifts $\mathbf{d}(t) = \{d_i(t)\}$ of the shear-flexural building model and their Fourier transforms $\mathbf{\Delta}(\omega) = \{\Delta_i(\omega)\}$. The Fourier transforms $\mathbf{\Delta}(\omega)$ of interstory drifts are related to the Fourier transforms $\mathbf{U}_S(\omega)$ of nodal horizontal displacements by

$$\mathbf{\Delta}(\omega) = \mathbf{P}\mathbf{U}_S(\omega) \tag{7.18}$$

$\mathbf{P}$ is a constant rectangular matrix consisting of $1, -1, 0$ and story heights. Substitution of Equation 7.16 into Equation 7.18 leads to

$$\mathbf{\Delta}(\omega) = \mathbf{P}\mathbf{H}_S(\omega)\{1 - \omega^2 H_{G(N+2)}(\omega)\}\ddot{U}_g(\omega) \tag{7.19}$$

Equation 7.19 is simply expressed as

$$\mathbf{\Delta}(\omega) = \mathbf{H}_{\Delta}(\omega)\ddot{U}_g(\omega) \tag{7.20}$$

where $\mathbf{H}_{\Delta}(\omega) = \{H_{\Delta_i}(\omega)\}$ are the transfer functions of interstory drifts and described as

$$\mathbf{H}_{\Delta}(\omega) = \mathbf{P}\mathbf{H}_S(\omega)\{1 - \omega^2 H_{G(N+2)}(\omega)\} \tag{7.21}$$

Let $U_{Sn}(\omega)$ denote the Fourier transform of the top-floor horizontal displacement relative to the free-field horizontal ground surface motion. From Equation 7.16, the Fourier transform $\ddot{U}_g(\omega) + \ddot{U}_{G(N+2)}(\omega) + \ddot{U}_{Sn}(\omega)$ of the top-floor absolute acceleration may be expressed as

$$
\begin{aligned}
\ddot{U}_g(\omega) &+ \ddot{U}_{G(N+2)}(\omega) + \ddot{U}_{Sn}(\omega) \\
&= [1 - \omega^2 H_{G(N+2)}(\omega) - \omega^2 H_{S(top)}(\omega)\{1 - \omega^2 H_{G(N+2)}(\omega)\}]\ddot{U}_g(\omega) \\
&= \{1 - \omega^2 H_{S(top)}(\omega)\}\{1 - \omega^2 H_{G(N+2)}(\omega)\}\ddot{U}_g(\omega) \\
&\equiv H_{SA}(\omega)\ddot{U}_g(\omega)
\end{aligned} \tag{7.22}
$$

where $H_{S(top)}(\omega)$ is the component in $\mathbf{H}_S(\omega)$ corresponding to the top-floor horizontal displacement and

$$H_{SA}(\omega) = \{1 - \omega^2 H_{S(top)}(\omega)\}\{1 - \omega^2 H_{G(N+2)}(\omega)\} \tag{7.23}$$

Let $S_g(\omega)$ denote the PSD function of the input $\ddot{u}_g(t)$. Using the random vibration theory, the mean-square response of the ith interstory drift can be computed from

$$\sigma_{\Delta_i}^2 = \int_{-\infty}^{\infty} |H_{\Delta_i}(\omega)|^2 S_g(\omega)\,d\omega = \int_{-\infty}^{\infty} H_{\Delta_i}(\omega)H_{\Delta_i}^*(\omega)S_g(\omega)\,d\omega \tag{7.24}$$

where $(\ )^*$ denotes the complex conjugate. Similarly, the mean-square response of the ith-floor absolute acceleration may be evaluated by

$$\sigma_A^2 = \int_{-\infty}^{\infty} |H_{SA}(\omega)|^2 S_g(\omega)\,d\omega = \int_{-\infty}^{\infty} H_{SA}(\omega)H_{SA}^*(\omega)S_g(\omega)\,d\omega \tag{7.25}$$

where $H_{SA}(\omega)$ is defined in Equation 7.23.

7.5 Problem of Optimal Damper Placement and Optimality Criteria

The problem of optimal damper placement for soil–structure interaction (shear-flexural model) (PODP-SSI-SF) may be described as follows.

Problem 7.1 PODP-SSI-SF Given the ground properties, the building properties (story stiffnesses, masses, and structural damping), find the damping coefficients $c_V = \{c_{Vi}\}$ of added viscous dampers to minimize the sum of the mean squares of the interstory drifts

$$f = \sum_{i=1}^{n} \sigma_{\Delta_i}^2 \tag{7.26}$$

subject to the constraint on total damper capacity

$$\sum_{i=1}^{n} c_{Vi} = \overline{W} \tag{7.27}$$

and to the constraints on each damper capacity

$$0 \le c_{Vi} \le \overline{c}_{Vi} \quad (i = 1, \cdots, n) \tag{7.28}$$

where $\overline{W}$ is the specified total damper capacity and $\overline{c}_{Vi}$ is the upper bound of the damping coefficient of damper i.

7.5.1 Optimality Conditions

The generalized Lagrangian for Problem 7.1 may be defined as:

$$L(c_V, \lambda, \mu, \nu) = \sum_{i=1}^{n} \sigma_{\Delta_i}^2 + \lambda \left(\sum_{i=1}^{n} c_{Vi} - \overline{W} \right) + \sum_{i=1}^{n} \mu_i(0 - c_{Vi}) + \sum_{i=1}^{n} \nu_i(c_{Vi} - \overline{c}_{Vi}) \tag{7.29}$$

where $\mu = \{\mu_i\}$ and $\nu = \{\nu_i\}$. The principal optimality criteria for Problem 7.1 without active upper and lower bound constraints on damping coefficients may be derived from stationarity conditions of the generalized Lagrangian $L(\mu = 0, \nu = 0)$ with respect to c_V and λ:

$$f_{,j} + \lambda = 0 \quad \text{or} \quad 0 < c_{Vj} < \overline{c}_{Vj} \quad (j = 1, \cdots, n) \tag{7.30}$$

$$\sum_{i=1}^{n} c_{Vi} - \overline{W} = 0 \tag{7.31}$$

Here, and in the following, $(\cdot)_{,j}$ denotes the partial differentiation with respect to c_{Vj}. When the lower or upper bound of the constraint on damping coefficients is active, the optimality condition should be modified to the following forms:

$$f_{,j} + \lambda \ge 0 \quad \text{for } c_{Vj} = 0 \tag{7.32}$$

$$f_{,j} + \lambda \le 0 \quad \text{for } c_{Vj} = \overline{c}_{Vj} \tag{7.33}$$

7.6 Solution Algorithm

In the proposed procedure the model without added dampers, namely $c_{Vj} = 0$ $(j = 1, \cdots, n)$, is adopted as the initial model. The damping coefficients are added via an original steepest direction search algorithm. Let Δc_{Vi} and ΔW denote the increment of the damping coefficient of the ith added damper (damper in the ith story) and the increment of the sum of the damper damping coefficients respectively. Given ΔW, the problem is to determine the effective position and the amount of the increments of the damper damping coefficients. The first- and second-order sensitivities of the objective function with respect to a design variable are needed in this algorithm. Those quantities are derived by differentiating Equation 7.24 by the design variables.

The first derivatives are computed by

$$(\sigma_{\Delta_i}^2)^{,j} = \int_{-\infty}^{\infty} \{H_{\Delta_i}(\omega)\}_{,j} H_{\Delta_i}^*(\omega) S_g(\omega)\, d\omega + \int_{-\infty}^{\infty} H_{\Delta_i}(\omega) \{H_{\Delta_i}^*(\omega)\}_{,j} S_g(\omega)\, d\omega$$

(7.34)

The second derivatives are calculated by

$$(\sigma_{\Delta_i}^2)_{,jl} = \int_{-\infty}^{\infty} \{H_{\Delta_i}(\omega)\}_{,j} \{H_{\Delta_i}^*(\omega)\}_{,l} S_g(\omega)\, d\omega + \int_{-\infty}^{\infty} \{H_{\Delta_i}(\omega)\}_{,l} \{H_{\Delta_i}^*(\omega)\}_{,j} S_g(\omega)\, d\omega$$
$$+ \int_{-\infty}^{\infty} \{H_{\Delta_i}(\omega)\}_{,jl} H_{\Delta_i}^*(\omega) S_g(\omega)\, d\omega + \int_{-\infty}^{\infty} H_{\Delta_i}(\omega) \{H_{\Delta_i}^*(\omega)\}_{,jl} S_g(\omega)\, d\omega$$

(7.35)

where

$$\{H_{\Delta_i}(\omega)\}_{,j} = P_i (A_S^{-1})_{,j} B_S \{1 - \omega^2 H_{G(N+2)}(\omega)\}$$

(7.36)

$$\{H_{\Delta_i}(\omega)\}_{,jl} = P_i (A_S^{-1})_{,jl} B_S \{1 - \omega^2 H_{G(N+2)}(\omega)\}$$

(7.37)

$$\{H_{\Delta_i}^*(\omega)\}_{,j} = P_i (A_S^{-1})_{,j}^* B_S \{1 - \omega^2 H_{G(N+2)}^*(\omega)\}$$

(7.38)

$$\{H_{\Delta_i}^*(\omega)\}_{,jl} = P_i (A_S^{-1})_{,jl}^* B_S \{1 - \omega^2 H_{G(N+2)}^*(\omega)\}$$

(7.39)

In Equations 7.36–7.39, P_i is the ith row vector in the matrix P.

The first derivative of A_S^{-1} is computed by $(A_S^{-1})_{,j} = -A_S^{-1} A_{S,j} A_S^{-1}$. Because $A_{S,jl} = 0$, the second derivative of A_S^{-1} is obtained from

$$(A_S^{-1})_{,jl} = A_S^{-1} (A_{S,l} A_S^{-1} A_{S,j} + A_{S,j} A_S^{-1} A_{S,l}) A_S^{-1}$$

(7.40)

The first-order derivative of $(A^{-1})^*$ is computed as $\{(A_S^{-1})_{,j}\}^*$ and the second-order derivative of $(A^{-1})^*$ is found as $(A^{-1})_{,jl}^* = \{(A^{-1})_{,jl}\}^*$.

The solution algorithm in the case of $c_{Vj} < \bar{c}_{Vj}$ for all j may be described as follows:

Step 0 Initialize all the supplemental dampers as $c_{Vj} = 0$ $(j = 1, \cdots, n)$. Assume ΔW.
Step 1 Compute the first-order sensitivity $f_{,i}$ of the objective function by Equation 7.34.
Step 2 Find the index p such that

$$-f_{,p} = \max_i(-f_{,i}) \tag{7.41}$$

Step 3 Update the objective function f by the first-order approximation $f + f_{,p}\Delta c_{Vp}$, where $\Delta c_{Vp} = \Delta W$.
Step 4 Update the first-order sensitivity $f_{,i}$ of the objective function by the first-order approximation $f_{,i} + f_{,ip}\Delta c_{Vp}$ using Equation 7.35.
Step 5 If, in Step 4, there exists a damper of an index j such that the condition

$$-f_{,p} = \max_{j, j \neq p}(-f_{,j}) \tag{7.42}$$

is satisfied, then compute the corresponding increment $\Delta \tilde{c}_{Vp}$ of the damping coefficient and update the first-order sensitivity $f_{,i}$ of the objective function by the first-order approximation $f_{,i} + f_{,ip}\Delta \tilde{c}_{Vp}$ using Equation 7.35.
Step 6 Repeat from Step 2 to Step 5 until the constraint in Equation 7.27 (i.e., $\sum_{i=1}^{n} c_{Vi} = \overline{W}$) is satisfied.

In Step 2 and Step 3, the direction which reduces the objective function most effectively under the condition $\sum_{i=1}^{n} \Delta c_{Vi} = \Delta W$ is found and the design is updated in that direction. It is appropriate, therefore, to call the present algorithm "the steepest direction search algorithm," as in Chapters 4–6 (Takewaki, 1998a). A simple numerical example of damping sensitivity of the performance (sum of transfer function amplitudes of interstory drifts introduced in Chapter 2) for a two-story shear building model is presented in Appendix 2.A in Chapter 2. This sensitivity example just corresponds to the gradient direction of the performance function at the origin in the schematic diagram shown in Figure 7.2. This algorithm is similar to the well-known steepest descent method in mathematical programming (see Figure 7.2). However, while the conventional steepest descent method uses the gradient vector itself of the objective function as the direction and does not utilize optimality criteria, the present algorithm takes full advantage of the newly derived optimality criteria in Equations 7.30, 7.32, and 7.33 and does not adopt the gradient vector as the direction. In other words, the steepest direction search guarantees the successive satisfaction of the optimality criteria. For example, if Δc_{Vp} is added to the pth added damper in which Equation 7.41 is satisfied, then its damper $(c_p > 0)$ satisfies the optimality condition in Equation 7.30 and the other dampers $(c_j = 0, \; j \neq p)$ satisfy the optimality condition in

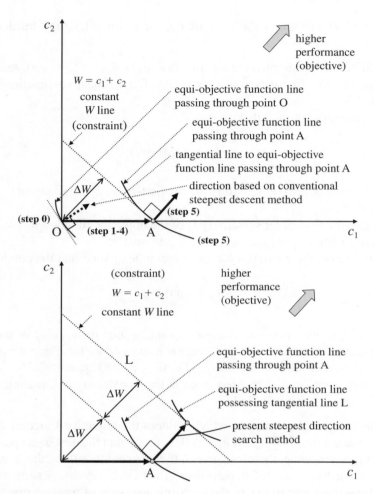

Figure 7.2 Comparison of the proposed steepest direction search method with the conventional steepest descent method ($\mathbf{c}_V$ is denoted here by $\mathbf{c}$ for simplicity of expression).

Equation 7.32. It should be noted that a series of subproblems is introduced here tentatively in which the damper level $\overline{W}$ is increased gradually by ΔW from zero through the specified value.

If multiple indices $p_1, \ldots, p_m$ exist in Step 2, then f and $f_{,j}$ have to be updated by

$$f \to f + \sum_{i=p_1}^{p_m} f_{,i} \Delta c_{Vi} \tag{7.43a}$$

$$f_{,j} \to f_{,j} + \sum_{i=p_1}^{p_m} f_{,ji} \Delta c_{Vi} \tag{7.43b}$$

Furthermore, the index p in Step 5 has to be replaced by the indices $p_1, \ldots, p_m$. The ratios among the magnitudes Δc_{Vi} must be determined so that the following relations are satisfied:

$$f_{,p_1} + \sum_{i=p_1}^{p_m} f_{,p_1 i} \Delta c_{Vi} = \cdots = f_{,p_m} + \sum_{i=p_1}^{p_m} f_{,p_m i} \Delta c_{Vi} \qquad (7.44)$$

Equation 7.44 requires that the optimality condition in Equation 7.30 continues to be satisfied in the dampers with the indices $p_1, \cdots, p_m$.

In the case where the damping coefficients of some added dampers attain their upper bounds, such constraints must be incorporated in the aforementioned algorithm. In that case, the increment Δc_{Vp} is added subsequently to the damper in which $-f_{,p}$ attains the maximum among all the dampers, except for those attaining the upper bound.

7.7 Numerical Examples

Consider a band-limited white noise as the base input motion. The PSD function of the input motion $\ddot{u}_g(t)$ is given by

$$S_g(\omega) = 0.01 \text{ m}^2/\text{s}^3 \quad (-2\pi \times 20 \le \omega \le -2\pi \times 0.2, \ 2\pi \times 0.2 \le \omega \le 2\pi \times 20)$$

$$S_g(\omega) = 0 \quad \text{otherwise}$$

Three different soil conditions are considered: very stiff ground, stiff ground, and soft ground. Every ground consists of three soil layers of identical thickness $l_i = 6$ m and these layers lie on a semi-infinite visco-elastic ground. The shear modulus, damping ratio, and mass density of the semi-infinite visco-elastic ground are $8.0 \times 10^8 \text{ N/m}^2$, 0.01, and $2.0 \times 10^3 \text{ kg/m}^3$ respectively. The input motion is defined as $l_0 = 6$ m below the boundary (bedrock) between the three soil layers and the semi-infinite visco-elastic ground. The mass densities of the three soil layers are identical over the very stiff, stiff, and soft grounds and have a value of $1.8 \times 10^3 \text{ kg/m}^3$. The shear moduli of the soil layers are identical within the three layers and have values of $1.6 \times 10^8 \text{ N/m}^2$ for the very stiff ground, $0.8 \times 10^8 \text{ N/m}^2$ for the stiff ground, and $0.2 \times 10^8 \text{ N/m}^2$ for the soft ground. The dependence of shear moduli and damping ratios on the strain level is taken into account via an equivalent linear model. Their relations are shown in Figure 7.3 for clay, sand, and gravel. It is assumed that layer 1 consists of gravel, layer 2 of clay, and layer 3 of sand. The effect of mean effective pressure is taken into account. Linear interpolation has been used in evaluating the intermediate values. The effective strain level (Schnabel *et al.*, 1972) has been evaluated from $2.5\sigma_{\Delta_i} \times 0.65$, where the coefficient 2.5 indicates the peak factor (Der Kiureghian, 1980) and 0.65 was introduced by Schnabel *et al.* (1972). The convergence of the stiffness reduction ratios of the shear moduli and damping ratios has been confirmed in a few cycles.

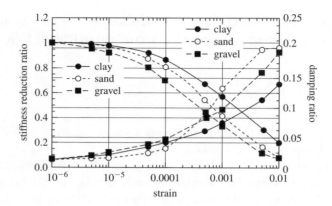

Figure 7.3 Strain-dependent stiffness reduction ratio and damping ratio of clay, sand, and gravel.

The interaction spring stiffnesses and the damping coefficients have been evaluated by the formula in Veletsos and Verbic (1974):

$$k_H + i\omega c_H = \frac{8Gr}{2 - \nu}[1 + i(b_1 a_0)]$$

$$k_R(\omega) + i\omega c_R(\omega) = \frac{8Gr^3}{3(1 - \nu)}\left[1 - b_1^* \frac{(b_2 a_0)^2}{1 + i b_2 a_0} - b_3 a_0^2\right]$$

where r is the equivalent radius of the base mat, G is the shear modulus, $a_0 = \omega r/V_S$, and $b_1 = 0.65$, $b_1^* = 0.5$, $b_2 = 0.8$, and $b_3 = 0$ for Poisson's ratio $\nu = 1/3$. For simplicity, $k_R = k_R(\omega_0)$, $c_R = c_R(\omega_0)$, and $\omega_0 = 2\pi/0.4$ have been adopted. The equivalent shear modulus of the top soil layer has been adopted as the shear modulus in Veletsos and Verbic (1974) and Poisson's ratio $\nu = 1/3$; the equivalent radius of the base mat is 3.57 m. The fundamental natural periods of the surface grounds at the small strain level are 0.239 s, 0.338 s, and 0.675 s for the very stiff ground, stiff ground, and soft ground respectively, and those computed from the equivalent stiffnesses are 0.322 s, 0.617 s, and 1.48 s for the very stiff ground, stiff ground, and soft grounds respectively.

Consider a three-story shear-building model. The floor masses are assumed to be $m_0 = 96 \times 10^3$ kg and $m_i = 32 \times 10^3$ kg $(i = 1, \ldots, 3)$ and the moments of inertia of the floor masses are $I_0 = 392 \times 10^3$ kg m^2 and $I_i = 131 \times 10^3$ kg m^2. The story shear stiffnesses are $k_i = 3.76 \times 10^7$ N/m$(i = 1, \ldots, 3)$ and the story bending stiffnesses are $s_i = 8.50 \times 10^9$ N m/rad$(i = 1, \ldots, 3)$. The fundamental natural periods of the elastically supported shear-flexural building model are 0.439 s, 0.467 s, and 0.639 s for the very stiff ground, stiff ground, and soft ground respectively. The structural viscous damping matrix of the shear-flexural building model has been given so that it is proportional to the stiffness matrix, and the lowest mode damping ratio is equal to 0.02. The increment ΔW of the sum of the added viscous dampers is 9.375 kN s/m.

Figure 7.4(a) shows the variation of damping coefficients in optimal damper placement for the very stiff ground with respect to the step number (total damper capacity level). It is observed that the damping coefficient is first added in the first story and then in the second story. Figures 7.4(b) and (c) illustrate the variations of damping coefficients in optimal damper placement for the stiff and soft grounds respectively. It can be understood from Figures 7.4(a)–(c) that optimal damper placement is strongly dependent on the surface ground properties. Resonance of the fundamental natural frequency of the building structure with the predominant frequency of the surface ground could sometimes cause amplified responses (Takewaki, 1998b). It seems that the relation of the fundamental natural frequency of the building structure with the predominant frequency of the surface ground (equivalent stiffness) can be a key parameter for characterizing the optimal damper placement.

Figure 7.5 shows the variations of the objective function, Equation 7.26, with respect to the step number (total damper capacity level) for the model on the very stiff, stiff, and soft grounds. It is observed that added viscous dampers are effective for stiffer grounds. Figure 7.6 illustrates the variations of the mean square, defined by Equation 7.25, of the top-floor absolute acceleration and the first-story interstory drift with respect to the step number for the model on the very stiff, stiff, and soft grounds. It is understood that added viscous dampers are also effective in the reduction of acceleration. Figure 7.7 shows the multicriteria plot obtained from Figures 7.5 and 7.6 with respect to the sum of mean-square interstory drifts and the mean-square top-floor acceleration. It is observed that the effectiveness of supplemental dampers is effective in a stiffer ground. Furthermore, the effectiveness of the optimal damper placement in the reduction of interstory drift (first story) is shown in Figure 7.8.

Figure 7.9 shows the lowest mode damping ratio with respect to total damper quantity for (a) very stiff ground, (b) stiff ground, and (c) soft ground. It is observed that a larger lowest mode damping ratio can be attained in the very stiff ground and supplemental viscous dampers are effective for stiffer ground.

Figure 7.10 illustrates the lowest mode shear deformation of the model on very stiff, stiff, and soft grounds. As the ground becomes stiffer, the ratio of shear deformation gets larger. Since the shear deformation is effective for supplemental dampers, this also supports the effectiveness of supplemental dampers in stiffer ground.

Figure 7.11 shows the optimal damper distribution of the three-story model in the different combination of the structure's fundamental natural period T_{S1} and the predominant period T_{G1} of ground for (a) $T_{S1} = T_{G1}$ and (b) $T_{S1} > T_{G1}$. It is observed that the optimal damper distribution depends largely on the relationship of the structure's fundamental natural period T_{S1} and the predominant period T_{G1} of ground.

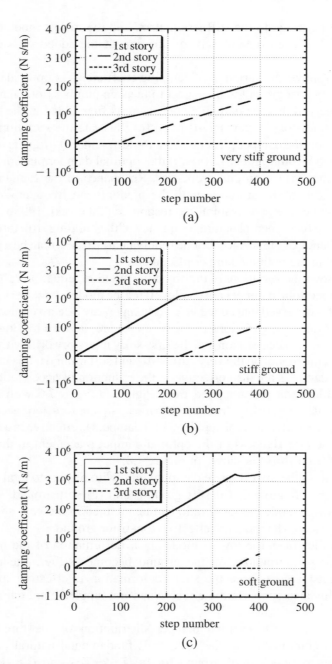

Figure 7.4 Optimal damper distribution of three-story model with respect to total damper quantity: (a) very stiff ground; (b) stiff ground; (c) soft ground.

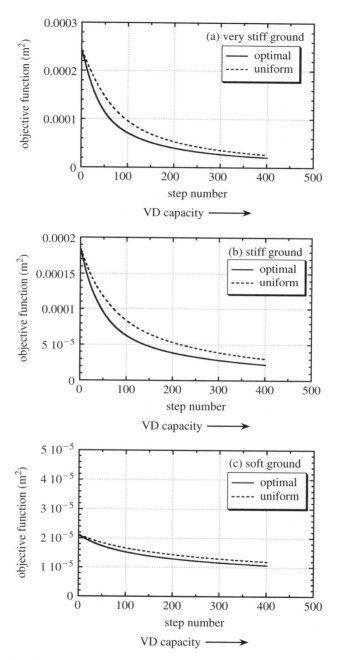

Figure 7.5 Objective function of three-story model with respect to total damper quantity: (a) very stiff ground; (b) stiff ground; (c) soft ground.

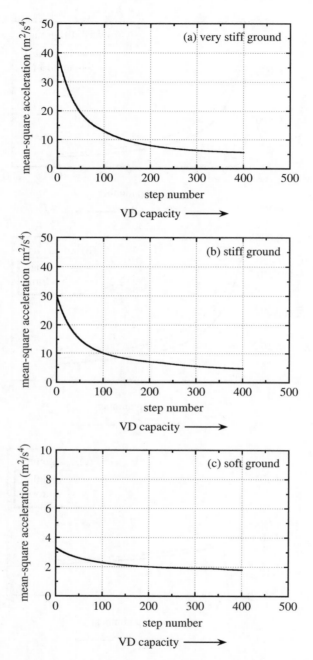

Figure 7.6 Mean-square acceleration at top floor of three-story model with respect to total damper quantity: (a) very stiff ground; (b) stiff ground; (c) soft ground.

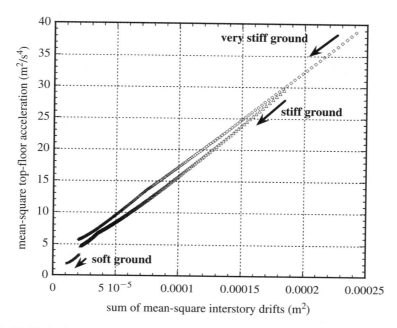

Figure 7.7 Multicriteria plot with respect to sum of mean-square interstory drifts and mean-square top-floor acceleration.

7.8 Summary

The results are summarized as follows.

1. Viscous damping of supplemental dampers in a shear-flexural building model, hysteretic damping in a surface ground, and a viscous damping at the ground viscous boundary can be taken into account exactly in the present formulation based on a steepest direction search algorithm. The shear-flexural building model can express the role of story-installation-type dampers in a deformed frame more exactly than the shear building model.

2. The effects of the surface ground properties on optimal damper placement in shear-flexural building models have been examined. It has been clarified that the optimal damper placement depends strongly on the surface ground characteristics, and the relation of the fundamental natural frequency of the elastically supported shear-flexural building model with the predominant frequency of the surface ground can be a key parameter for characterizing the optimal damper placement.

3. Supplemental viscous dampers are effective for stiffer ground. This may be because the effectiveness of dampers is related to the existence of stiff supports. Furthermore, supplemental viscous dampers are also effective in the reduction of acceleration.

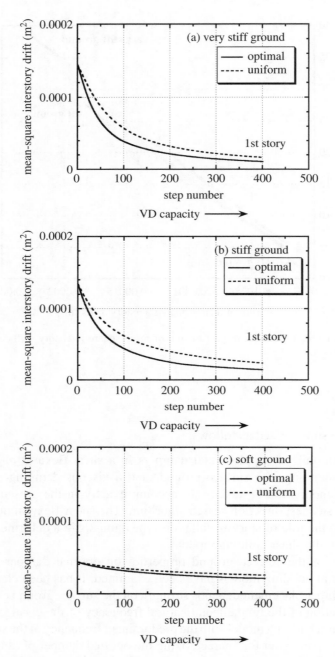

Figure 7.8 Mean-square interstory drift at first story of three-story model with respect to total damper quantity: (a) very stiff ground; (b) stiff ground; (c) soft ground.

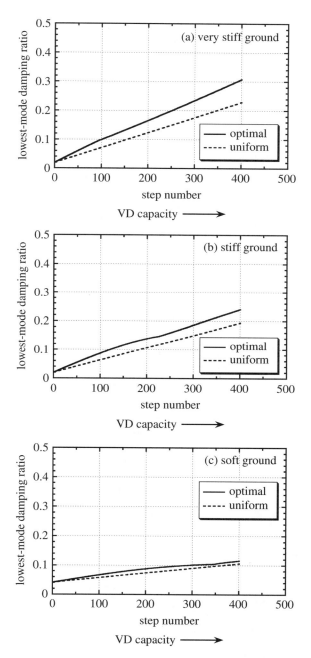

Figure 7.9 Lowest mode damping ratio with respect to total damper quantity: (a) very stiff ground; (b) stiff ground; (c) soft ground.

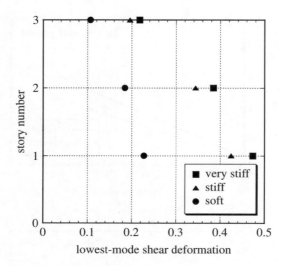

Figure 7.10 Lowest mode shear deformation of the model on very stiff ground, stiff ground, and soft ground.

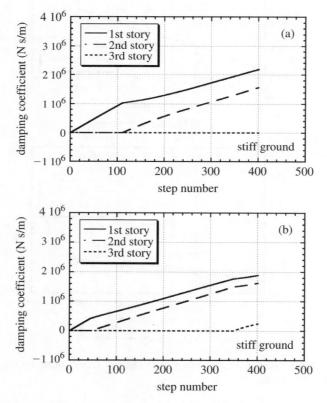

Figure 7.11 Optimal damper distribution of the three-story model with different combinations of the structure's fundamental natural period T_{S1} and predominant period of ground T_{G1}: (a) $T_{S1} = T_{G1}$; (b) $T_{S1} > T_{G1}$.

Appendix 7.A: System Mass, Damping, and Stiffness Matrices for a Surface Ground Model with Two Soil Layers

The system mass, damping, and stiffness matrices for a surface ground model with two soil layers may be defined by

$$\mathbf{M}_G = \begin{bmatrix} \mu_0/3 & \mu_0/6 & & \mathbf{0} \\ \mu_0/6 & (\mu_0 + \mu_1)/3 & \mu_1/6 & \\ & \mu_1/6 & (\mu_1 + \mu_2)/3 & \mu_2/6 \\ \mathbf{0} & & \mu_2/6 & \mu_2/3 \end{bmatrix} \tag{A7.1}$$

$$\mathbf{C}_G = \text{diag}(c_b \quad 0 \quad 0 \quad 0) \tag{A7.2}$$

$$\mathrm{i}\mathbf{D} + \mathbf{K}_G = \begin{bmatrix} K_0 & -K_0 & & \mathbf{0} \\ -K_0 & K_0 + K_1 & -K_1 & \\ & -K_1 & K_1 + K_2 & -K_2 \\ \mathbf{0} & & -K_2 & K_2 \end{bmatrix} \tag{A7.3}$$

where $\mu_i = \rho_i A l_i$ and $K_i = (G_i A / l_i)(1 + \mathrm{i}2\beta_i)$.

Appendix 7.B: System Mass, Damping, and Stiffness Matrices for a Two-Story Shear-Flexural Building Model

The system mass, damping, and stiffness matrices for a two-story shear-flexural building model may be defined by

$$\mathbf{M}_S = \text{diag}(m_2 \quad I_2 \quad m_1 \quad I_1 \quad m_0 \quad I_0) \tag{B7.1}$$

$$\mathbf{K}_S = \mathbf{K}_{SS} + \mathbf{K}_{SG} \tag{B7.2}$$

$$\mathbf{K}_{SS} = k_2 \begin{bmatrix} 1 & 0 & -1 & -h_2 & & \\ 0 & 0 & 0 & 0 & \mathbf{0} \\ -1 & 0 & 1 & h_2 & & \\ -h_2 & 0 & h_2 & h_2^2 & & \\ & \mathbf{0} & & & & \mathbf{0} \end{bmatrix} + k_1 \begin{bmatrix} \mathbf{0} & & & & \mathbf{0} & \\ & 1 & 0 & -1 & -h_1 \\ & 0 & 0 & 0 & 0 \\ \mathbf{0} & -1 & 0 & 1 & h_1 \\ & -h_1 & 0 & h_1 & h_1^2 \end{bmatrix}$$

$$+ s_2 \begin{bmatrix} 0 & 0 & 0 & 0 & & \\ 0 & 1 & 0 & -1 & \mathbf{0} \\ 0 & 0 & 0 & 0 & & \\ 0 & -1 & 0 & 1 & & \\ & \mathbf{0} & & & & \mathbf{0} \end{bmatrix} + s_1 \begin{bmatrix} \mathbf{0} & & & & \mathbf{0} & \\ & 0 & 0 & 0 & 0 \\ & 0 & 1 & 0 & -1 \\ \mathbf{0} & 0 & 0 & 0 & 0 \\ & 0 & -1 & 0 & 1 \end{bmatrix} \tag{B7.3}$$

$$\mathbf{K}_{SG} = \text{diag}(0 \quad \cdots \quad \cdots \quad 0 \quad k_H \quad k_R) \tag{B7.4}$$

$$\mathbf{C}_S = \mathbf{C}_{SS} + \mathbf{C}_{SV} + \mathbf{C}_{SG} \tag{B7.5}$$

$$\mathbf{C}_{SS} = (2h^s/\omega^s)\mathbf{K}_{SS} \tag{B7.6}$$

$$\mathbf{C}_{SV} = c_{V2}\begin{bmatrix} 1 & 0 & -1 & -h_2 & \\ 0 & 0 & 0 & 0 & \mathbf{0} \\ -1 & 0 & 1 & h_2 & \\ -h_2 & 0 & h_2 & h_2^2 & \\ & \mathbf{0} & & & \mathbf{0} \end{bmatrix} + c_{V1}\begin{bmatrix} \mathbf{0} & & & \mathbf{0} & \\ & 1 & 0 & -1 & -h_1 \\ & 0 & 0 & 0 & 0 \\ \mathbf{0} & -1 & 0 & 1 & h_1 \\ & -h_1 & 0 & h_1 & h_1^2 \end{bmatrix} \tag{B7.7}$$

$$\mathbf{C}_{SG} = \text{diag}(0 \quad \cdots \quad \cdots \quad 0 \quad c_H \quad c_R) \tag{B7.8}$$

References

Alam, S. and Baba, S. (1993) Robust active optimal control scheme including soil–structure interaction. *Journal of Structural Engineering*, **119**, 2533–2551.

Der Kiureghian, A. (1980) A response spectrum method for random vibrations, Report no. UCB/EERC 80-15, Earthquake Engineering Research Center, University of California, Berkeley, CA.

Housner, G., Bergmann, L.A., and Caughey, T.A. (1997) Structural control: past, present, and future (special issue). *Journal of Engineering Mechanics*, **123** (9), 897–971.

Kobori, T. (1996) Structural control for large earthquakes. Proceedings of the IUTAM Symposium, Kyoto, Japan, pp. 3–28.

Luco, J.E. (1998) A simple model for structural control including soil–structure interaction effects. *Earthquake Engineering and Structural Dynamics*, **27**, 225–242.

Lysmer, J. and Kuhlemeyer, R.L. (1969) Finite dynamic model for infinite media. *Journal of Engineering Mechanics*, **94** (EM4), 859–877.

Schnabel, P.B., Lysmer, J., and Seed, H.B. (1972) SHAKE: a computer program for earthquake response analysis of horizontally layered sites. A computer program distributed by NISEE/Computer Applications, Berkeley.

Smith, H.A., Wu, W.H., and Borja, R.I. (1994) Structural control considering soil-structure interaction effects. *Earthquake Engineering and Structural Dynamics*, **23**, 609–626.

Smith, H.A. and Wu, W.H. (1997) Effective optimal structural control of soil-structure interaction systems. *Earthquake Engineering and Structural Dynamics*, **26**, 549–570.

Takewaki, I. (1998a) Optimal damper positioning in beams for minimum dynamic compliance. *Computer Methods in Applied Mechanics and Engineering*, **156** (1–4), 363–373.

Takewaki, I. (1998b) Remarkable response amplification of building frames due to resonance with the surface ground. *Soil Dynamics and Earthquake Engineering*, **17** (4), 211–218.

Takewaki, I. (2000) *Dynamic Structural Design: Inverse Problem Approach*, WIT Press.

Takewaki, I. and Yoshitomi, S. (1998) Effects of support stiffnesses on optimal damper placement for a planar building frame. *The Structural Design of Tall Buildings*, **7** (4), 323–336.

Tsuji, M. and Nakamura, T. (1996) Optimum viscous dampers for stiffness design of shear buildings. *The Structural Design of Tall Buildings*, **5**, 217–234.

Veletsos, A.S. and Verbic, B. (1974) Basic response functions for elastic foundations. *Journal of Engineering Mechanics*, **100** (EM2), 189–202.

Wong, H.L. and Luco, J.E. (1991) Structural control including soil-structure interaction effects. *Journal of Engineering Mechanics*, **117**, 2237–2250.

Zhang, R.H. and Soong, T.T. (1992). Seismic design of viscoelastic dampers for structural applications. *Journal of Structural Engineering*, **118** (5), 1375–1392.

8

Optimal Sensitivity-based Design of Dampers in Shear Buildings with TMDs on Surface Ground under Earthquake Loading

8.1 Introduction

As explained in Chapter 6, the soil or ground under a structure greatly influences the structural vibration properties. It is important, therefore, to develop a seismic-resistant design method for such an interaction model. Passive control devices have recently been recognized as effective tools for suppressing the structural responses and upgrading the structural performances. In building structures with passive energy dissipation systems, the property mentioned above is also true and the theory for passive dampers should be developed for those interaction models.

The purpose of this chapter is to explain a systematic method for optimal viscous damper placement in building structures with a TMD. Refer to Appendix 8.A for the fundamentals of TMD systems. Amplification of ground motion within the surface ground is also taken into account. It is frequently pointed out that, while a TMD system is effective in the response reduction of the tuned mode (usually a fundamental mode), supplemental viscous dampers installed between consecutive floors are effective for all the natural modes (Housner et al., 1997). For this reason, the combined passive damper system of the TMD and the viscous dampers effective for story deformation is discussed in this chapter.

Nonlinear amplification of ground motion within the surface ground will be described by an equivalent linear model, and local interaction with the surrounding soil will be incorporated through a set of horizontal spring and dashpot. Both hysteretic damping of the surface ground and radiation damping into the semi-infinite ground are

Building Control with Passive Dampers   Izuru Takewaki
© 2009 John Wiley & Sons (Asia) Pte Ltd

included in the model. While several useful investigations have been reported on active control of building structures including soil–structure interaction effects (Wong and Luco, 1991; Luco, 1998; Alam and Baba, 1993; Smith *et al.*, 1994; Smith and Wu, 1997), it should be remarked that there are a very few on optimal damper placement in a building structure taking into account interaction with the surface ground. In particular, simultaneous use of a TMD at the roof and supplemental viscous dampers between consecutive floors is of great interest from the viewpoints of both dampers' effective collaboration in the high-performance response reduction.

The unique steepest direction search algorithm (Takewaki, 1998a; Takewaki and Yoshitomi, 1998) explained in Chapters 4 and 5 for fixed-base structures has been extended to a building–ground system in Chapters 6 and 7. This algorithm is utilized and extended to a TMD structure–surface ground interaction model in this chapter. While resonant steady-state responses of structures with fixed bases were treated in Chapters 4 and 5 (Takewaki, 1998a; Takewaki and Yoshitomi, 1998), earthquake responses to random inputs are introduced here as controlled parameters, as in Chapters 6 and 7. It is shown that closed-form expressions of the inverse of the tri-diagonal coefficient matrix in the governing equations (equations of motion) lead to drastic reduction of computational time of mean-square responses of the TMD building structure to the random earthquake input and their derivatives with respect to the design variables (damping coefficients of supplemental viscous dampers). The optimal placement and the quantity of supplemental viscous dampers are found simultaneously and automatically via the explained steepest direction search algorithm, which requires successive approximate satisfaction of the optimality conditions. Several examples with and without a TMD for various surface ground properties are presented to demonstrate the effectiveness and validity of the present method and to examine the effects of surface ground characteristics on optimal damper placement.

As for optimal damper placement for fixed-base models, several useful algorithms have been developed (e.g., Zhang and Soong, 1992; Tsuji and Nakamura, 1996). Actually, the method due to Zhang and Soong (1992) is simple and may be applicable to a broad range of structural models. However, it does not appear that the previous approaches are applicable to the soil–structure interaction models with a TMD, due to their limitations on modeling of structures and damping mechanisms or computational efficiency.

8.2 Building with a TMD and Ground Model

In this chapter, a building structure with a TMD system at the roof is assumed to rest on a surface ground. For simple and clear presentation, the building structure is described by a shear-building model and the surface ground is represented by a shear-beam model (see Figure 8.1). Local interaction of the building structure with the surrounding soil is represented by a horizontal spring and a dashpot between these two substructures (building and surface ground), as in Chapter 7. Let n and N denote

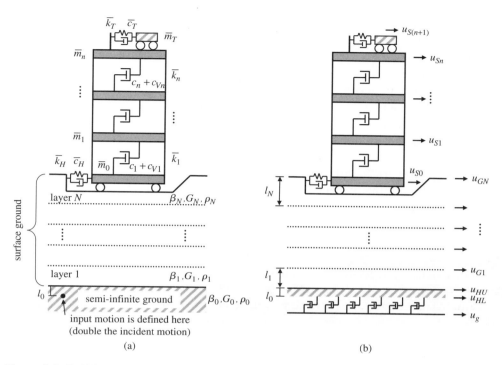

Figure 8.1 Building structure with a TMD at the roof supported by a surface ground. (Originally published in I. Takewaki, "Soil–structure random response reduction via TMD-MD simultaneous use," *Computer Methods in Applied Mechanics and Engineering*, **190**, no. 5–7, 677–690, 2000, Elsevier B.V.).

the number of stories of the shear building model and the number of surface soil layers respectively. The random horizontal acceleration input is defined at the level just below the bedrock on which the layered surface ground lies.

Radiation damping from the surface ground into the semi-infinite visco-elastic ground is taken into account by using a viscous boundary (Lysmer and Kuhlemeyer, 1969) at the engineering bedrock surface. This treatment is exact in a one-dimensional model because the incident angle of the wave to the surface is $\pi/2$. Let $c_b = \rho_0 V_S A = \sqrt{\rho_0 G_0} A$ denote the total damping coefficient of the viscous boundary, where ρ_0, V_S, and G_0 respectively denote the mass density, the shear wave velocity, and the shear modulus of the semi-infinite visco-elastic ground, and A is the horizontal governing area of the surface ground. It has been confirmed that a single-input model at the engineering bedrock surface can simulate the seismic response of a multi-input model within a reasonable accuracy by adopting a fairly large area A (Takewaki and Nakamura, 1995). In this multi-input model, the input is from the engineering bedrock surface and various points in the surface ground. This fact of simulation accuracy will also be demonstrated later in Section 8.7.

Let l_i, ρ_i, G_i, and β_i respectively denote the thickness, the mass density, the shear modulus, and the hysteretic damping ratio in layer i. As in Chapters 6 and 7, a linear displacement function is used in evaluating the stiffness and consistent mass matrices in the ground FE model. Nonlinear amplification of ground motion in the surface ground is taken into account by using an equivalent linear model (Schnabel *et al.*, 1972). The equivalent linear model introduced for the surface ground is almost equivalent to the well-known SHAKE program (one-dimensional wave propagation theory including an equivalent linear model). This model is a deterministic equivalent linear model and its accuracy has been demonstrated by many researchers (see Takewaki *et al.*, 2002a, 2002b). In order to evaluate the mean peak soil response of the equivalent linear model, the present method uses a peak factor (Der Kiureghian, 1980) multiplied by the standard deviation of the shear strain, as in Chapters 6 and 7. The validity of this treatment in a linear model has been demonstrated by Der Kiureghian (1980). Once the surface ground is represented by the equivalent linear model, the whole system consisting of a structure and the surface ground is a linear system. The random vibration theory for such linear models is well established and its accuracy has been checked and investigated by many investigators.

As for the building structure, various parameters are prescribed. Let $\overline{m}_i$ and $\overline{k}_i$ denote the floor mass and the story stiffness in the ith story respectively. Furthermore, let c_i and c_{Vi} denote the structural damping coefficient in the ith story and the damping coefficient of the supplemental damper in the ith story respectively. The set $\{c_{Vi}\}$ of damping coefficients of supplemental dampers is the design variable set in this chapter. As for the surrounding soil model and the TMD system, let $\overline{k}_H$ and $\overline{c}_H$ denote the soil interaction spring stiffness and the damping coefficient respectively, and let $\overline{k}_T$, $\overline{c}_T$, and $\overline{m}_T$ denote the TMD spring stiffness, damping coefficient, and mass respectively. These parameters $\overline{k}_H$, $\overline{c}_H$, $\overline{k}_T$, $\overline{c}_T$, and $\overline{m}_T$ are assumed to be prescribed in each surface ground model.

8.3 Equations of Motion and Seismic Response

Different from the model (decomposed model) in Chapter 7, a whole model is treated here where the super-building model and the surface ground model are dealt with simultaneously as one model. The comparison of the whole model with the decomposed model will be shown in Section 8.7.

Assume that this shear-building–surface-ground model with a TMD system at the roof is subjected to a random horizontal acceleration $\ddot{u}_g(t)$ at a level l_0 below the bedrock. Let $\mathbf{M}$, $\mathbf{C}$, $\mathbf{D}$, $\mathbf{K}$, and $\mathbf{r}$ denote the system mass matrix (combination of a lumped mass matrix and a consistent mass matrix), the system viscous damping matrix of the TMD building on the surrounding soil and the viscous boundary, the system hysteretic damping matrix of the soil layers, the system stiffness matrix (building, TMD system, surrounding soil, soil layer), and the influence coefficient vector respectively. An example of $\mathbf{M}$, $\mathbf{C}$, $\mathbf{D}$, and $\mathbf{K}$ is shown in Appendix 8.B for a two-story TMD shear-building model supported by two soil layers.

The equations of motion of the TMD building–ground interaction system in the frequency domain may be expressed by

$$(-\omega^2 \mathbf{M} + i\omega \mathbf{C} + i\mathbf{D} + \mathbf{K})\mathbf{U}(\omega) = -\mathbf{Mr}\ddot{U}_g(\omega) \tag{8.1}$$

where i is the imaginary unit. As stated in Chapters 6 and 7, it should be noted that, because the present model includes a hysteretic damping, the equations of motion in the time domain for the random horizontal acceleration $\ddot{u}_g(t)$ cannot be expressed and only the frequency-domain ones can be derived. In Equation 8.1, $\mathbf{U}(\omega)$ and $\ddot{U}_g(\omega)$ are the Fourier transforms of the horizontal displacements of the nodes relative to the base input $\ddot{u}_g(t)$ and the Fourier transform of the horizontal input acceleration $\ddot{u}_g(t)$ defined at the level l_0 below the bedrock. The horizontal input acceleration $\ddot{u}_g(t)$ is assumed to be a stationary Gaussian random process with zero mean. Equation 8.1 can then be reduced to the following compact form:

$$\mathbf{A}\mathbf{U}(\omega) = \mathbf{B}\ddot{U}_g(\omega) \tag{8.2}$$

where $\mathbf{A}$ and $\mathbf{B}$ are the coefficient matrix and the vector expressed by

$$\mathbf{A} = (-\omega^2 \mathbf{M} + i\omega \mathbf{C} + i\mathbf{D} + \mathbf{K}) \tag{8.3a}$$

$$\mathbf{B} = -\mathbf{Mr} \tag{8.3a}$$

$\mathbf{A}$ is a tri-diagonal matrix, as seen in Appendix 8.B for a two-story shear-building model supported by two soil layers. This property will be fully utilized in the efficient computation of its inverse.

Because the interstory drift can be a good indicator of the overall stiffness of the building, its control plays an important role in the stiffness design of the building. For this reason the interstory drift is treated here as the controlled parameter.

Let us define the time-domain interstory drifts $\mathbf{d}(t) = \{d_i(t)\}$ of the shear-building model and consider their Fourier transforms $\boldsymbol{\Delta}(\omega) = \{\Delta_i(\omega)\}$. $\boldsymbol{\Delta}(\omega)$ are related to $\mathbf{U}(\omega)$ by the use of the aforementioned transformation matrix $\mathbf{T}$:

$$\boldsymbol{\Delta}(\omega) = \mathbf{T}\mathbf{U}(\omega) \tag{8.4}$$

The transformation matrix $\mathbf{T}$ is a rectangular matrix and consists of 1, -1, and 0 because the total number of degrees of freedom of the model and the number of stories in the building are different. Substitution of Equation 8.2 into Equation 8.4 provides

$$\boldsymbol{\Delta}(\omega) = \mathbf{T}\mathbf{A}^{-1}\mathbf{B}\ddot{U}_g(\omega) \tag{8.5}$$

Equation 8.5 is expressed compactly as

$$\boldsymbol{\Delta}(\omega) = \mathbf{H}_\Delta(\omega)\ddot{U}_g(\omega) \tag{8.6}$$

where $\mathbf{H}_\Delta(\omega) = \{H_{\Delta_i}(\omega)\}$ are the transfer functions of the interstory drifts with respect to the input acceleration $\ddot{U}_g(\omega)$ and are described as

$$\mathbf{H}_\Delta(\omega) = \mathbf{TA}^{-1}\mathbf{B} \tag{8.7}$$

From Equation 8.2, the Fourier transforms $\ddot{U}(\omega)$ of the floor accelerations $\ddot{u}(t)$ relative to the input acceleration $\ddot{u}_g(t)$ are expressed in terms of $\ddot{U}_g(\omega)$ by

$$\ddot{U}(\omega) = -\omega^2 U(\omega) = -\omega^2 \mathbf{A}^{-1}\mathbf{B}\ddot{U}_g(\omega) \tag{8.8}$$

The Fourier transforms $\ddot{U}_A(\omega)$ of the absolute floor accelerations are then expressed by

$$\ddot{U}_A(\omega) = \ddot{U}(\omega) + \mathbf{1}\ddot{U}_g(\omega) = (\mathbf{1} - \omega^2 \mathbf{A}^{-1}\mathbf{B})\ddot{U}_g(\omega) \equiv \mathbf{H}_A(\omega)\ddot{U}_g(\omega) \tag{8.9}$$

where $\mathbf{1} = \{1 \cdots 1\}^T$ and $\mathbf{H}_A(\omega)$ is the acceleration transfer function with respect to the input acceleration $\ddot{U}_g(\omega)$.

Note again that the coefficient matrix $\mathbf{A}$ is a tri-diagonal matrix. Therefore, its inverse can be obtained in closed form (see Appendix 8.C). This property enables one to compute the deformation and acceleration transfer functions $\mathbf{H}_\Delta(\omega)$ and $\mathbf{H}_A(\omega)$ very efficiently. In particular, as the number of soil layers and building stories increases, this advantage may be remarkable.

The statistical characteristic of stationary random signals can be described by the PSD function. Let $S_g(\omega)$ denote the PSD function of the horizontal base input $\ddot{u}_g(t)$. Using the random vibration theory, the mean-square response of the ith interstory drift can be computed by

$$\sigma_{\Delta_i}^2 = \int_{-\infty}^{\infty} |H_{\Delta_i}(\omega)|^2 S_g(\omega)\mathrm{d}\omega = \int_{-\infty}^{\infty} H_{\Delta_i}(\omega)H_{\Delta_i}^*(\omega)S_g(\omega)\mathrm{d}\omega \tag{8.10}$$

where $(\ )^*$ denotes the complex conjugate. Similarly, the mean-square response of the ith-floor absolute acceleration may be evaluated by

$$\sigma_{A_i}^2 = \int_{-\infty}^{\infty} |H_{A_i}(\omega)|^2 S_g(\omega)\mathrm{d}\omega = \int_{-\infty}^{\infty} H_{A_i}(\omega)H_{A_i}^*(\omega)S_g(\omega)\mathrm{d}\omega \tag{8.11}$$

where $H_{A_i}(\omega)$ is an ith component of the transfer function vector $\mathbf{H}_A(\omega)$ defined in Equation 8.9.

8.4 Problem of Optimal Damper Placement and Optimality Criteria

In this section, the problem of optimal damper placement in the super-building is formulated. The problem of optimal damper placement for soil–structure interaction models with a TMD (PODP-SSI-TMD) system may be described as:

Problem 8.1 PODP-SSI-TMD   Given the properties of the surface ground, the building (story stiffnesses, masses, and structural damping) and the TMD system, find the damping coefficients $c_V = \{c_{Vi}\}$ of supplemental viscous dampers in the building to minimize the sum of the mean squares of the interstory drifts

$$f = \sum_{i=1}^{n} \sigma_{\Delta_i}^2 \qquad (8.12)$$

subject to the constraint on total capacity of supplemental dampers

$$\sum_{i=1}^{n} c_{Vi} = \overline{W} \qquad (8.13)$$

and to the constraints on each supplemental damper capacity

$$0 \leq c_{Vi} \leq \overline{c}_{Vi} \quad (i = 1, \cdots, n) \qquad (8.14)$$

where $\overline{W}$ is the specified total supplemental damper capacity and $\overline{c}_{Vi}$ is the upper bound of the damping coefficient of the supplemental damper i.

8.4.1 Optimality Conditions

It is straightforward to use the Lagrange multiplier method for solving constrained optimization problems. The constrained optimization problem stated above can be formulated appropriately by the generalized Lagrangian formulation. The generalized Lagrangian for Problem 8.1 may be defined as

$$L(c_V, \lambda, \mu, \nu) = \sum_{i=1}^{n} \sigma_{\Delta_i}^2 + \lambda \left(\sum_{i=1}^{n} c_{Vi} - \overline{W} \right) + \sum_{i=1}^{n} \mu_i (0 - c_{Vi}) + \sum_{i=1}^{n} \nu_i (c_{Vi} - \overline{c}_{Vi}) \qquad (8.15)$$

In Equation 8.15, $\mu = \{\mu_i\}$, $\nu = \{\nu_i\}$, and λ are the Lagrange multipliers. The principal (or major) optimality criteria for Problem 8.1 without active upper and lower bound constraints on damping coefficients of supplemental dampers may be derived from the stationarity conditions of the generalized Lagrangian L ($\mu = 0$, $\nu = 0$) with respect to each components of c_V and λ:

$$f_{,j} + \lambda = 0 \quad \text{for } 0 < c_{Vj} < \overline{c}_{Vj} \quad (j = 1, \cdots, n) \qquad (8.16)$$

$$\sum_{i=1}^{n} c_{Vi} - \overline{W} = 0 \tag{8.17}$$

Equation 8.16 has been derived by the differentiation of Equation 8.15 with respect to c_{Vj} and Equation 8.17 has been obtained by the differentiation of Equation 8.15 with respect to λ.

Here, and in the following, the mathematical symbol $(\cdot)_{,j}$ indicates the partial differentiation with respect to the damping coefficient c_{Vj} of the supplemental damper in the jth story. When the lower or upper bound of the constraint on damping coefficients of supplemental dampers is active, the optimality condition should be modified to the following two forms:

$$f_{,j} + \lambda \geq 0 \quad \text{for} \quad c_{Vj} = 0 \tag{8.18}$$

$$f_{,j} + \lambda \leq 0 \quad \text{for} \quad c_{Vj} = \bar{c}_{Vj} \tag{8.19}$$

8.5 Solution Algorithm

A solution algorithm for the problem stated above is explained in this section. In the solution algorithm, the model without supplemental viscous dampers, namely $c_{Vj} = 0$ ($j = 1, \cdots, n$), is used as the initial model. This treatment is well suited to the situation where a structural designer is starting the allocation and placement of added supplemental viscous dampers at appropriate positions. The damping coefficients of supplemental dampers are increased gradually based on the optimality criteria stated above. This algorithm is called a steepest direction search algorithm, as in Chapters 4–7.

Let Δc_{Vi} and ΔW denote the increment of the damping coefficient of the ith added supplemental damper (damper in the ith story) and the increment of the sum of the damping coefficients of added supplemental dampers respectively. Once ΔW is given, the problem is reduced to determining simultaneously the effective position and amount of the increments of the damping coefficients of added supplemental dampers. In order to develop this algorithm, the first- and second-order sensitivities of the objective function with respect to the design variable c_{Vj} are needed. Those quantities are derived by differentiating successively Equation 8.10 by the design variables.

First-order derivative of mean-square interstory drift:

$$(\sigma_{\Delta_i}^2)_{,j} = \int_{-\infty}^{\infty} \{H_{\Delta_i}(\omega)\}_{,j} H_{\Delta_i}^*(\omega) S_g(\omega) d\omega + \int_{-\infty}^{\infty} H_{\Delta_i}(\omega) \{H_{\Delta_i}^*(\omega)\}_{,j} S_g(\omega) d\omega \tag{8.20}$$

Second-order derivative of mean-square interstory drift:

$$(\sigma_{\Delta_i}^2)_{,jl} = \int_{-\infty}^{\infty} \{H_{\Delta_i}(\omega)\}_{,j} \{H_{\Delta_i}^*(\omega)\}_{,l} S_g(\omega) d\omega + \int_{-\infty}^{\infty} \{H_{\Delta_i}(\omega)\}_{,l} \{H_{\Delta_i}^*(\omega)\}_{,j} S_g(\omega) d\omega$$
$$+ \int_{-\infty}^{\infty} \{H_{\Delta_i}(\omega)\}_{,jl} H_{\Delta_i}^*(\omega) S_g(\omega) d\omega + \int_{-\infty}^{\infty} H_{\Delta_i}(\omega) \{H_{\Delta_i}^*(\omega)\}_{,jl} S_g(\omega) d\omega. \tag{8.21}$$

In Equations 8.20 and 8.21, the first- and second-order derivatives of the deformation transfer functions may be expressed as

$$\{H_{\Delta_i}(\omega)\}_{,j} = \mathbf{T}_i \mathbf{A}_{,j}^{-1} \mathbf{B} \tag{8.22}$$

$$\{H_{\Delta_i}(\omega)\}_{,jl} = \mathbf{T}_i \mathbf{A}_{,jl}^{-1} \mathbf{B} \tag{8.23}$$

$$\{H_{\Delta_i}^*(\omega)\}_{,j} = \mathbf{T}_i \mathbf{A}_{,j}^{-1*} \mathbf{B} \tag{8.24}$$

$$\{H_{\Delta_i}^*(\omega)\}_{,jl} = \mathbf{T}_i \mathbf{A}_{,jl}^{-1*} \mathbf{B} \tag{8.25}$$

In Equations 8.22–8.25, $\mathbf{T}_i$ is the ith row vector in the transformation matrix $\mathbf{T}$ introduced above.

The first-order derivative of the inverse $\mathbf{A}^{-1}$ of the coefficient matrix is computed by $(\mathbf{A}^{-1})_{,j} = -\mathbf{A}^{-1}\mathbf{A}_{,j}\mathbf{A}^{-1}$, derived by differentiating the identity $\mathbf{A}\mathbf{A}^{-1} = \mathbf{I}$. Because the components in the coefficient matrix $\mathbf{A}$ are linear functions of design variables, the relation $\mathbf{A}_{,jl} = \mathbf{0}$ holds. Then, the second-order derivative of the inverse $\mathbf{A}^{-1}$ is obtained from

$$(\mathbf{A}^{-1})_{,jl} = \mathbf{A}^{-1}(\mathbf{A}_{,l}\mathbf{A}^{-1}\mathbf{A}_{,j} + \mathbf{A}_{,j}\mathbf{A}^{-1}\mathbf{A}_{,l})\mathbf{A}^{-1} \tag{8.26}$$

The first-order derivative of the complex conjugate $\mathbf{A}^{-1*}$ of the inverse $\mathbf{A}^{-1}$ can be computed as $\{\mathbf{A}_{,j}^{-1}\}^*$ and the second-order derivative of $\mathbf{A}^{-1*}$ can be found as $\mathbf{A}_{,jl}^{-1*} = \{\mathbf{A}_{,jl}^{-1}\}^*$.

The solution algorithm in the case satisfying the conditions $c_{Vj} < \bar{c}_{Vj}$ for all j may be described as follows:

Step 0   Initialize all the added supplemental viscous dampers as $c_{Vj} = 0$ ($j = 1, \cdots, n$). In the initial design stage, the structural damping alone exists in the shear-building model. Assume the quantity ΔW.

Step 1   Compute the first-order derivative $f_{,i}$ of the objective function by Equation 8.20.

Step 2   Find the index p satisfying the condition

$$-f_{,p} = \max_i(-f_{,i}) \tag{8.27}$$

Step 3   Update the objective function f by the linear approximation $f + f_{,p}\Delta c_{Vp}$, where $\Delta c_{Vp} = \Delta W$. This is because the supplemental damper is added only in the pth story in the initial design stage.

Step 4   Update the first-order sensitivity $f_{,i}$ of the objective function by the linear approximation $f_{,i} + f_{,ip}\Delta c_{Vp}$ using Equation 8.21.

Step 5   If, in Step 4, there exists a supplemental damper of an index j such that the
condition

$$-f_{,p} = \max_{j, j \neq p}(-f_{,j})$$

is satisfied, then stop and compute the increment $\Delta \tilde{c}_{\mathrm{V}p}$ of the damping
coefficient of the corresponding damper. At this stage, update the first-
order sensitivity $f_{,i}$ of the objective function by the linear approximation
$f_{,i} + f_{,ip}\Delta \tilde{c}_{\mathrm{V}p}$ using Equation 8.21.

Step 6   Repeat the procedure from Step 2 to Step 5 until the constraint in Equation
8.13 (i.e., $\sum_{i=1}^{n} c_{\mathrm{V}i} = \overline{W}$) is satisfied.

In Step 2 and Step 3, the direction which decreases the objective function most
effectively under the condition $\sum_{i=1}^{n} \Delta c_{\mathrm{V}i} = \Delta W$ is found and the design (the quantity of supplemental dampers) is updated in that direction. It is suitable, therefore,
to call the present algorithm "the steepest direction search algorithm," as in Chapters 4–7 (Takewaki, 1998a). As explained before, this algorithm is similar to the
conventional steepest descent method in the mathematical programming (see Figure 7.2 in Chapter 7 to understand the concept). However, while the conventional
steepest descent method uses the gradient vector itself of the objective function as the
direction and does not utilize optimality criteria, the present algorithm takes advantage of the newly derived optimality criteria expressed by Equations 8.16, 8.18, and
8.19 and does not adopt the gradient vector as its direction. More specifically, the
explained steepest direction search guarantees the successive and approximate satisfaction of the optimality criteria. For example, if the increment $\Delta c_{\mathrm{V}p}$ is added to
the pth supplemental damper in which Equation 8.27 is satisfied, then its damper
($c_p > 0$) satisfies the optimality condition in Equation 8.16 and the other dampers
($c_j = 0$, $j \neq p$) alternatively satisfy the optimality condition in Equation 8.18. It is
important to note that a series of subproblems is introduced here tentatively in which
the total damper capacity level $\overline{W}$ is increased gradually by ΔW from zero through the
specified value.

It is necessary to investigate other possibilities. If multiple indices $p_1, \cdots, p_m$ exist
in Step 2, then the objective function f and its derivative $f_{,j}$ have to be updated by the
following rules:

$$f \rightarrow f + \sum_{i=p_1}^{p_m} f_{,i}\Delta c_{\mathrm{V}i} \tag{8.29a}$$

$$f_{,j} \rightarrow f_{,j} + \sum_{i=p_1}^{p_m} f_{,ji}\Delta c_{\mathrm{V}i} \tag{8.29b}$$

Furthermore, the index p defined in Step 5 has to be replaced by the multiple indices $p_1, \cdots, p_m$. The ratios among the magnitudes Δc_{Vi} have to be determined so that the following relations are satisfied:

$$f_{,p_1} + \sum_{i=p_1}^{p_m} f_{,p_1 i} \Delta c_{Vi} = \cdots = f_{,p_m} + \sum_{i=p_1}^{p_m} f_{,p_m i} \Delta c_{Vi} \tag{8.30}$$

Equation 8.30 requires that the optimality condition in Equation 8.16 continues to be satisfied in the supplemental dampers with the indices $p_1, \cdots, p_m$.

It may be the case in realistic situations that the maximum quantity of supplemental dampers is limited by the requirements of building design and planning. In the case where the damping coefficients of some supplemental dampers attain their upper bounds, such constraints must be incorporated in the aforementioned algorithm. In that case, the increment Δc_{Vp} of the supplemental dampers is added subsequently to the supplemental damper in which $-f_{,p}$ attains the maximum among all the supplemental dampers, except for those attaining the upper bound.

8.6 Numerical Examples

A band-limited white noise is often used as a test input of random disturbance. In this section, a band-limited white noise with the following PSD function is input to the building–ground system as the input acceleration $\ddot{u}_g(t)$:

$$S_g(\omega) = 0.01 \text{ m}^2/\text{s}^3 \quad (-2\pi \times 16 \le \omega \le -2\pi \times 0.2, \; 2\pi \times 0.2 \le \omega \le 2\pi \times 16)$$

$$S_g(\omega) = 0 \quad \text{otherwise}$$

As stated before, the soil under building structures influences greatly the seismic response of the building structures. In order to investigate in detail the influence of soil conditions on the seismic response and optimal distributions of supplemental dampers, three different soil conditions are considered. These three soil types are specified as very stiff ground, stiff ground, and soft ground. Every ground model consists of three soil layers of identical thickness $l_i = 6$ m and these layers rest on a semi-infinite visco-elastic ground as the engineering bedrock. The shear modulus, damping ratio, and mass density of the semi-infinite visco-elastic ground as the engineering bedrock are given by 8.0×10^8 N/m^2, 0.01, and 2.0×10^3 kg/m^3 respectively.

The governing area of the surface ground is $2500 (m^2)$. The input acceleration is defined at $l_0 = 6$ m below the boundary (bedrock) between the three soil layers and the semi-infinite visco-elastic ground. The mass densities of the three soil layers are identical over the very stiff ground, stiff ground, and soft ground and take the value 1.8×10^3 kg/m^3.

The shear moduli of the soil layers are also identical within each of the three layers and take the values of 1.6×10^8 N/m^2 for the very stiff ground, 0.8×10^8 N/m^2 for the stiff ground, and 0.2×10^8 N/m^2 for the soft ground. The shear moduli and damping

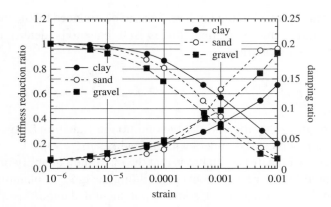

Figure 8.2 Dependence of shear moduli and damping ratios on strain level. (Originally published in I. Takewaki, "Soil–structure random response reduction via TMD-MD simultaneous use," *Computer Methods in Applied Mechanics and Engineering*, **190**, no. 5–7, 677–690, 2000, Elsevier B.V.).

ratios of the soil layers are dependent on the strain level, and its property is taken into account via the equivalent linear model explained above. Their relations are shown in Figure 8.2 for clay, sand, and gravel. It is assumed that layer 1 consists of gravel, layer 2 consists of clay, and layer 3 consists of sand. The effect of mean effective pressure on these relations is sometimes important and is taken into account here. Linear interpolation has been used in evaluating the intermediate values between the specified points. The effective strain level (Schnabel *et al.*, 1972) has been evaluated by $2.5\sigma_{\Delta_i} \times 0.65$, where the coefficient 2.5 indicates the peak factor (Der Kiureghian, 1980) and 0.65 as the effective strain ratio was introduced by Schnabel *et al.*, (1972) and Takewaki *et al.*, (2002a, 2002b). The convergence of stiffness reduction ratios of the shear moduli and damping ratios has been confirmed in a few cycles. The stiffness reduction ratios and damping ratios are shown in Tables 8.1 and 8.2. This evaluation of equivalent values has been performed for the soil–structure interaction system without supplemental dampers and a TMD system.

The interaction spring stiffness and the damping coefficient between the structure and the surface ground have been evaluated by the conventional formula by Veletsos and Verbic (1974). The equivalent shear modulus of the top soil layer was adopted as the shear modulus by Veletsos and Verbic (1974). In addition, Poisson's ratio was 1/3 and the equivalent radius of the base mat was 3.57 m. The fundamental natural periods of the surface grounds at the small strain level are 0.239 s, 0.338 s, and 0.675 s for the very stiff, stiff, and soft grounds, respectively, and those computed from the equivalent stiffnesses are 0.322 s, 0.617 s, and 1.48 s for the very stiff ground, stiff ground, and soft ground respectively.

As for super-buildings, consider a three-story shear-building model. The floor masses are assumed to be $\overline{m}_0 = 96 \times 10^3$ kg and $\overline{m}_i = 32 \times 10^3$ kg $(i = 1, 2, 3)$ and the

Table 8.1 Stiffness reduction ratios via an equivalent linear model. (Originally published in I. Takewaki, "Soil–structure random response reduction via TMD-MD simultaneous use," *Computer Methods in Applied Mechanics and Engineering*, **190**, no. 5–7, 677–690, 2000, Elsevier B.V.).

Soil layer no.	Very stiff ground	Stiff ground	Soft ground
3	0.783	0.688	0.615
2	0.730	0.678	0.400
1	0.480	0.228	0.162

Table 8.2 Damping ratios via an equivalent linear model. (Originally published in I. Takewaki, "Soil–structure random response reduction via TMD-MD simultaneous use," *Computer Methods in Applied Mechanics and Engineering*, **190**, no. 5–7, 677–690, 2000, Elsevier B.V.).

Soil layer no.	Very stiff ground	Stiff ground	Soft ground
3	0.0378	0.0562	0.0606
2	0.0552	0.0600	0.0977
1	0.0761	0.126	0.124

story stiffnesses are $\bar{k}_i = 3.76 \times 10^7$ N/m ($i = 1, 2, 3$). Then, the fundamental natural period of the structure with a fixed base is 0.412 s. The fundamental natural periods of the elastically supported shear-building model are 0.414 s, 0.418 s, and 0.449 s for the very stiff ground, stiff ground, and soft ground respectively. In the TMD system, $\bar{m}_T$ is given as 1.0×10^3 kg and $\bar{k}_T$ and $\bar{c}_T$ are determined for each surface ground model so that the TMD system is tuned with the fundamental natural frequency of the elastically supported shear-building model and the TMD system (as an SDOF system) has a damping ratio of 0.1. In this case, the fundamental natural periods of the elastically supported shear-building models are located between that of the very stiff ground and that of the stiff ground. The structural viscous damping matrix of the shear-building model has been given so that it is proportional to the stiffness matrix, and the lowest mode damping ratio is equal to 0.02. The increment ΔW of the sum of the supplemental viscous dampers is 9.375 kN s/m.

Figure 8.3(a) shows the variation of damping coefficients of supplemental viscous dampers in optimal damper placement for the very stiff ground with respect to the step number (total damper capacity level). It is observed that the damping coefficient of the supplemental dampers is first added in the first story and then in the second story. Figures 8.3(b) and (c) illustrate the variations of damping coefficients of supplemental viscous dampers in optimal damper placement for the stiff and soft grounds. It can be understood from Figures 8.3(a)–(c) that optimal damper placement is strongly

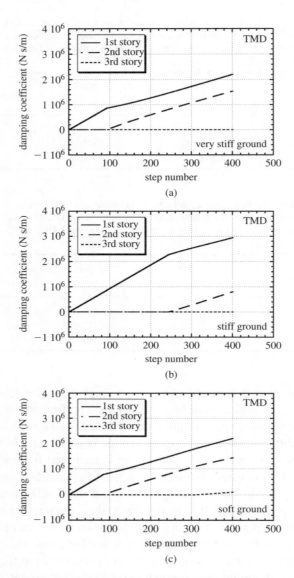

Figure 8.3 Variation of damping coefficients in optimal damper placement with respect to the step number (total damper capacity level). (Originally published in I. Takewaki, "Soil–structure random response reduction via TMD-MD simultaneous use," *Computer Methods in Applied Mechanics and Engineering*, **190**, no. 5–7, 677–690, 2000, Elsevier B.V.).

dependent on the surface ground properties. Resonance of the fundamental natural frequency of the building structure with the predominant frequency of the surface ground could sometimes cause amplified responses (Takewaki, 1998b). It seems that the relation of the fundamental natural frequency of the building structure with the

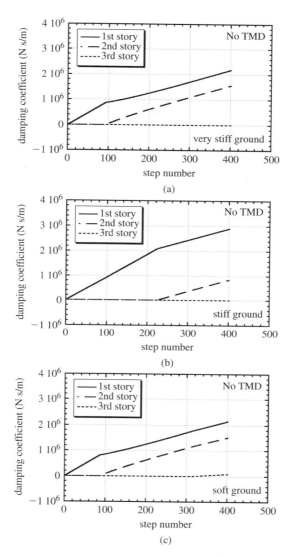

Figure 8.4 Variation of damping coefficients in optimal damper placement with respect to the step number without TMD. (Originally published in I. Takewaki, "Soil–structure random response reduction via TMD-MD simultaneous use," *Computer Methods in Applied Mechanics and Engineering*, **190**, no. 5–7, 677–690, 2000, Elsevier B.V.).

predominant frequency of the surface ground (equivalent stiffness) is a key parameter for characterizing the optimal damper placement.

Figures 8.4(a)–(c) illustrate the variations of damping coefficients of supplemental viscous dampers in optimal damper placement for the very stiff ground, stiff ground,

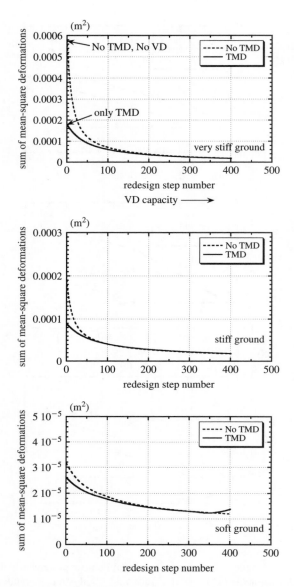

Figure 8.5 Variation of sum of mean-square deformations with respect to step number (total damper capacity level) for models with and without TMD on very stiff, stiff, and soft grounds. (Originally published in I. Takewaki, "Soil–structure random response reduction via TMD-MD simultaneous use," *Computer Methods in Applied Mechanics and Engineering*, **190**, no. 5–7, 677–690, 2000, Elsevier B.V.).

and soft ground respectively without a TMD system. This investigation was conducted to check the effect of the TMD system on the optimal damper placement. It is observed that the TMD system does not affect the optimal distribution of the supplemental viscous dampers much.

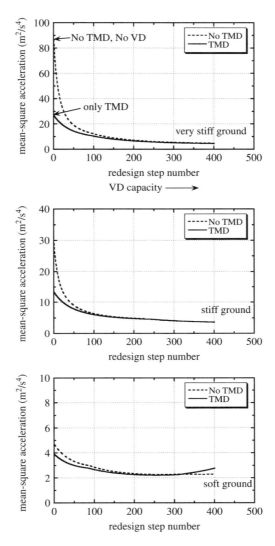

Figure 8.6 Variations of mean-square acceleration with respect to step number for models with and without a TMD on very stiff, stiff, and soft grounds. (Originally published in I. Takewaki, "Soil–structure random response reduction via TMD-MD simultaneous use," *Computer Methods in Applied Mechanics and Engineering*, **190**, no. 5–7, 677–690, 2000, Elsevier B.V.).

Figure 8.5 shows the variations of the objective function, Equation 8.12, with respect to the step number (total damper capacity level) for the model with and without a TMD system on the very stiff, stiff, and soft grounds. It is observed that the TMD system is effective in response reduction when the lowest mode of the structure is amplified; that is, the very stiff and stiff grounds in this case. This implies that a careful examination of the characteristics of surface grounds is absolutely necessary in using a TMD system.

Figure 8.6 indicates the variations of the mean square, defined by Equation 8.11, of the top-floor absolute acceleration with respect to the step number for the model with and without a TMD system on the very stiff, stiff, and soft grounds. It is observed that there exists a case where excessive introduction of supplemental viscous dampers is not necessarily effective in acceleration response reduction. It is often and repeatedly reported that this phenomenon can also be observed in base-isolated buildings. It can further be seen that appropriate introduction of supplemental viscous dampers is effective both in deformation and acceleration reductions even in the case where a TMD system is not effective, namely in the soft ground in this case. This implies that appropriate selection of a TMD system and supplemental viscous dampers is very important and essential in response reduction depending on the surface ground characteristics.

Figure 8.7(a) shows the multicriteria plot obtained from Figures 8.5 and 8.6 with respect to the sum of mean-square interstory drifts and the mean-square top-floor acceleration for models without a TMD; Figure 8.7(b) illustrates this for models with a TMD. It can be seen that the multicriteria plot is very useful in understanding the usefulness of supplemental viscous dampers and a TMD system in the reduction of deformation and acceleration under various ground conditions.

8.7 Whole Model and Decomposed Model

Although the whole model as shown in Figure 8.1 was adopted in this chapter, it may be meaningful to discuss the comparison between the whole model and the decomposed model. The decomposed model is a model where the super-building and the surface ground are treated independently, as shown in Chapter 7. It is well known that the structural responses evaluated by the whole model depend on the mass ratio between the structure and the ground (Nakamura *et al.*, 1996; Takewaki, 1998b). In the case where multiple buildings exist in urban areas, an appropriate mass ratio has to be chosen (see Chapter 6). On the other hand, in the case where only a building exists on a field, a fairly large mass ratio (ground mass to building mass) must be chosen.

As stated above, there is another model, as shown in Figure 8.8, referred to as "a decomposed model," which can represent the coupling of a structure and the ground. In this model, the surface ground motion in the free-field ground is computed first through wave propagation theory or vibration theory and the computed motions at several selected points are input simultaneously into the elastically supported building. It is noted that the surface ground motion in the free-field ground is not influenced by the response of the building.

Figure 8.9 shows the comparison of the mean squares of the interstory drifts for the whole models used in the foregoing sections and the decomposed models on the very stiff, stiff, and soft grounds. From this figure, small differences can be observed for the very stiff ground. However, these differences are negligible, because the mean peak

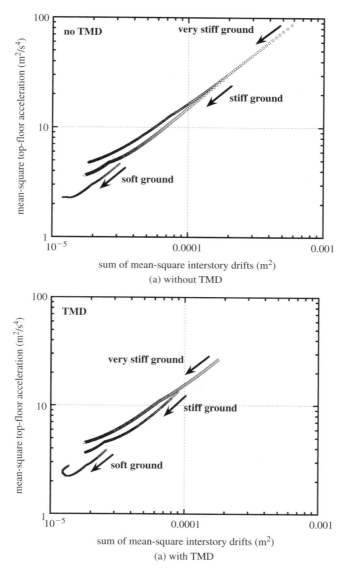

Figure 8.7 Multicriteria plot with respect to sum of mean-square interstory drifts and mean-square top-floor acceleration. (Originally published in I. Takewaki, "Soil–structure random response reduction via TMD-MD simultaneous use," *Computer Methods in Applied Mechanics and Engineering*, **190**, no. 5–7, 677–690, 2000, Elsevier B.V.).

responses are related to the square roots of the mean-square responses. Furthermore, Figure 8.10 illustrates the comparison of the amplitude of the transfer function of third-story drift between the whole model and the decomposed model. It is confirmed numerically that, as the mass ratio (ground mass to building mass) becomes larger,

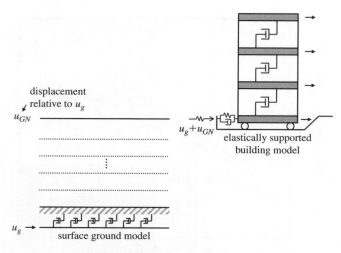

Figure 8.8   Decomposed model.

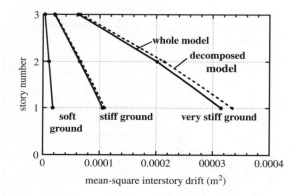

Figure 8.9   Mean squares of interstory drifts for the whole models.

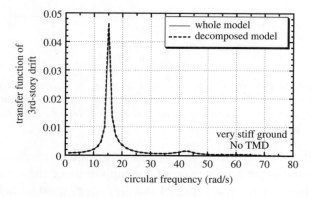

Figure 8.10   Amplitude of transfer function of third-story drift.

the responses of the whole model converge to those of the corresponding decomposed model. However, the numerical instability caused by the large difference of masses and stiffnesses between the structure and the ground has to be kept in mind carefully in using the whole model.

8.8 Summary

The results in this chapter may be summarized as follows.

1. All the structural and supplemental viscous damping in a TMD building structure, hysteretic damping in a surface ground, and viscous damping at the ground viscous boundary can be taken into account exactly in the present unified formulation based on a frequency-domain approach. A so-called steepest direction search algorithm has been developed and explained in detail.
2. Because deformation and acceleration transfer functions can be obtained in closed form even for MDOF systems with nonproportional damping owing to the tridiagonal property of the coefficient matrix, the mean-square deformations of the TMD building structure to the random earthquake input at the engineering bedrock surface and their derivatives with respect to the design variables (damper damping coefficients) can be computed very efficiently.
3. The effects of surface ground properties on optimal supplemental damper placement in TMD building structures have been examined. It has been clarified numerically that the relation of the fundamental natural frequency of the elastically supported TMD building structure with the predominant frequency of the surface ground (equivalent stiffness) is a key parameter for characterizing the optimal supplemental damper placement. Appropriate selection of a TMD system and supplemental viscous dampers is very important in the response reduction, depending on the surface ground characteristics.
4. The TMD system is effective in the response reduction when the lowest mode of the structure is amplified; that is, the very stiff and stiff grounds in this case. A case exists where excessive introduction of supplemental viscous dampers is not effective in acceleration response reduction. It is often reported that this phenomenon can also be observed in base-isolated buildings. Appropriate introduction of supplemental viscous dampers is effective both in deformation and acceleration reductions even in the case where a TMD system is not effective; that is, the soft ground in this case. This means that the TMD system and the supplemental viscous damper system can help each other, and there exists the possibility to take full advantage of these two systems.

Appendix 8.A: Fundamentals of TMD Systems

Consider the SDOF structure with a TMD system shown in Figure 8.11. The masses of the structure and the additional system are denoted by M and m and the stiffnesses of the

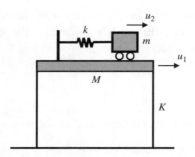

Figure 8.11 SDOF structure with TMD system without viscous damper.

structure and the additional system by K and k. Let u_1 and u_2 denote the displacement of the structural mass relative to the ground and that of the additional mass relative to the ground respectively. The objective of this appendix is to investigate whether it is possible for the structure to remain undeformed; that is, the displacement u_1 of the structural mass relative to the ground is zero when subjected to a sinusoidal ground motion $\ddot{u}_g = A \cos pt$.

The equations of motion for this system may be expressed by

$$\begin{bmatrix} M & 0 \\ 0 & m \end{bmatrix} \begin{Bmatrix} \ddot{u}_1 \\ \ddot{u}_2 \end{Bmatrix} + \begin{bmatrix} K+k & -k \\ -k & k \end{bmatrix} \begin{Bmatrix} u_1 \\ u_2 \end{Bmatrix} = - \begin{Bmatrix} M \\ m \end{Bmatrix} A \cos pt \qquad (A8.1)$$

Consider only the steady-state response and let us assume

$$u_1 = U_1 \cos pt \qquad (A8.2a)$$

$$u_2 = U_2 \cos pt \qquad (A8.2b)$$

Substitution of Equations A8.2a and A8.2b into Equation 8.1 and comparison of the coefficients on the terms including $\cos pt$ provide

$$-p^2 \begin{bmatrix} M & 0 \\ 0 & m \end{bmatrix} \begin{Bmatrix} U_1 \\ U_2 \end{Bmatrix} + \begin{bmatrix} K+k & -k \\ -k & k \end{bmatrix} \begin{Bmatrix} U_1 \\ U_2 \end{Bmatrix} = - \begin{Bmatrix} M \\ m \end{Bmatrix} A \qquad (A8.3)$$

Using the notation $m/M = \mu$, Equation A8.3 can be solved as follows:

$$\begin{aligned}
\begin{Bmatrix} U_1 \\ U_2 \end{Bmatrix} &= \left(p^2 \begin{bmatrix} M & 0 \\ 0 & m \end{bmatrix} - \begin{bmatrix} K+k & -k \\ -k & k \end{bmatrix} \right)^{-1} \begin{Bmatrix} M \\ m \end{Bmatrix} A \\
&= \left(p^2 M \begin{bmatrix} 1 & 0 \\ 0 & \mu \end{bmatrix} - \begin{bmatrix} K+k & -k \\ -k & k \end{bmatrix} \right)^{-1} \begin{Bmatrix} M \\ \mu M \end{Bmatrix} A \qquad (A8.4) \\
&= \frac{1}{\det \Delta} \begin{bmatrix} p^2 M \mu - k & -k \\ -k & p^2 M - (K+k) \end{bmatrix} \begin{Bmatrix} M \\ \mu M \end{Bmatrix} A
\end{aligned}$$

where Δ is the coefficient matrix of Equation (A8.3) with minus sign. The condition that the displacement u_1 of the structural mass relative to the ground is zero can be expressed as

$$U_1 = \frac{A}{\det \Delta}\{(p^2 M\mu - k)M - k\mu M\} = 0 \tag{A8.5}$$

Equation A8.5 leads to

$$p^2 M\mu - k - k\mu = 0 \tag{A8.6}$$

Assume here that the frequency of the input excitation coincides with the fundamental natural circular frequency of the structure, which can be expressed by

$$p = \sqrt{K/M} \tag{A8.7}$$

Substitution of Equation A8.7 into Equation A8.6 provides

$$\frac{K}{M}m - k(1 + \mu) = 0 \tag{A8.8}$$

Equation A8.8 can be rewritten as

$$\frac{K}{M}\frac{m}{k} = 1 + \mu \tag{A8.9}$$

Denoting the fundamental natural period of the structure by $T_1 = 2\pi\sqrt{M/K}$ and that of the additional mass-spring system by $T_2 = 2\pi\sqrt{m/k}$, Equation A8.9 may be re-expressed as

$$\frac{T_2}{T_1} = \frac{\sqrt{m/k}}{\sqrt{M/K}} = \sqrt{1 + \mu} \tag{A8.10}$$

In the case of $\mu = 0.01$, $T_2/T_1 \cong 1.005$.

For further study, refer to appropriate textbooks; for example, Connor and Klink (1996).

Appendix 8.B: System Mass, Damping, and Stiffness Matrices for a Two-story Shear-building Model Supported by Two Soil Layers

The system mass, damping, and stiffness matrices for a two-story shear building model supported by two soil layers may be expressed by

$$\mathbf{M} = \begin{bmatrix} \overline{m}_T & & & & & & \mathbf{0} \\ & \overline{m}_2 & & & & & \\ & & \overline{m}_1 & & & & \\ & & & \overline{m}_0 & & & \\ & & & & \mu_2/3 & \mu_2/6 & \\ & & & & \mu_2/6 & (\mu_2 + \mu_1)/3 & \mu_1/6 \\ & & & & & \mu_1/6 & (\mu_1 + \mu_0)/3 & \mu_0/6 \\ \mathbf{0} & & & & & & \mu_0/6 & \mu_0/3 \end{bmatrix} \tag{B8.1}$$

$$\mathbf{C} = \begin{bmatrix} \bar{c}_T & -\bar{c}_T & & & & & & & \mathbf{0} \\ -\bar{c}_T & \bar{c}_T + C_2 & -C_2 & & & & & \\ & -C_2 & C_2 + C_1 & -C_1 & & & & \\ & & -C_1 & C_1 + \bar{c}_H & -\bar{c}_H & & & \\ & & & -\bar{c}_H & \bar{c}_H & & & \\ & & & & & 0 & & \\ & & & & & & 0 & \\ \mathbf{0} & & & & & & & c_b \end{bmatrix} \tag{B8.2}$$

$$\mathbf{iD} + \mathbf{K} = \begin{bmatrix} \bar{k}_T & -\bar{k}_T & & & & & & & \mathbf{0} \\ -\bar{k}_T & \bar{k}_T + \bar{k}_2 & -\bar{k}_2 & & & & & \\ & -\bar{k}_2 & \bar{k}_2 + \bar{k}_1 & -\bar{k}_1 & & & & \\ & & -\bar{k}_1 & \bar{k}_1 + \bar{k}_H & -\bar{k}_H & & & \\ & & & -\bar{k}_H & \bar{k}_H + K_2 & -K_2 & & \\ & & & & -K_2 & K_2 + K_1 & -K_1 & \\ & & & & & -K_1 & K_1 + K_0 & -K_0 \\ \mathbf{0} & & & & & & -K_0 & K_0 \end{bmatrix}$$
$$\tag{B8.3}$$

where $\mu_i = \rho_i A l_i$, $C_i = c_i + c_{Vi}$, and $K_i = (G_i A / l_i)(1 + i2\beta_i)$.

Appendix 8.C: Closed-form Expression of the Inverse of a Tri-diagonal Matrix

Consider the following symmetric tri-diagonal matrix of $M \times M$:

$$\mathbf{A} = \begin{bmatrix} d_M & -e_M & & & & 0 \\ -e_M & \ddots & & \ddots & & \\ & \ddots & \ddots & \ddots & & \\ & & \ddots & \ddots & -e_3 & \\ & & & -e_3 & d_2 & -e_2 \\ 0 & & & & -e_2 & d_1 \end{bmatrix} \tag{C8.1}$$

Let us define the following principal minors:

$$P_0 = 1, \ P_1 = d_1, \ P_2 = \begin{vmatrix} d_2 & -e_2 \\ -e_2 & d_1 \end{vmatrix}, \cdots, P_M = \det \mathbf{A} \tag{C8.2}$$

$$P_{R0} = 1, \ P_{R1} = d_M, \ P_{R2} = \begin{vmatrix} d_M & -e_M \\ -e_M & d_{M-1} \end{vmatrix}, \cdots, P_{RM} = \det \mathbf{A} \tag{C8.3}$$

The principal minors satisfy the following recurrence formula:

$$P_{j-1} = d_{j-1} P_{j-2} - e_{j-1}^2 P_{j-3} \quad (j = 3, \cdots, M) \tag{C8.4}$$

The jth column of $\mathbf{A}^{-1}$ may be expressed as

$$
\frac{1}{\det \mathbf{A}} \left\{ \left(\prod_{i=M-j+2}^{M} e_i \right) P_{M-j} P_{R0} \quad \left(\prod_{i=M-j+2}^{M-1} e_i \right) P_{M-j} P_{R1} \quad \cdots \right.
$$
$$
\left(\prod_{i=M-j+2}^{M-j+2} e_i \right) P_{M-j} P_{R(j-2)} \quad P_{M-j} P_{R(j-1)} \quad \left(\prod_{i=M-j+1}^{M-j+1} e_i \right) P_{M-j-1} P_{R(j-1)}
$$
$$
\left. \cdots \quad \left(\prod_{i=3}^{M-j+1} e_i \right) P_1 P_{R(j-1)} \quad \left(\prod_{i=2}^{M-j+1} e_i \right) P_0 P_{R(j-1)} \right\}^{\mathrm{T}} \tag{C8.5}
$$

References

Alam, S. and Baba, S. (1993) Robust active optimal control scheme including soil–structure interaction. *Journal of Structural Engineering*, **119**, 2533–2551.

Connor, J.J. and Klink, B.S.A. (1996) *Introduction to Motion-Based Design*, WIT Press.

Der Kiureghian, A. (1980) A response spectrum method for random vibrations. Report No. UCB/EERC 80-15. Earthquake Engineering Research Center, University of California, Berkeley, CA.

Housner, G., Bergmann, L.A., and Caughey, T.A. (1997) Structural control: past, present, and future (special issue). *Journal of Engineering Mechanics*, **123** (9), 897–971.

Luco, J.E. (1998) A simple model for structural control including soil–structure interaction effects. *Earthquake Engineering and Structural Dynamics*, **27**, 225–242.

Lysmer, J. and Kuhlemeyer, R.L. (1969) Finite dynamic model for infinite media. *Journal of Engineering Mechanics*, **94** (EM4), 859–877.

Nakamura, T., Takewaki, I., and Asaoka, Y. (1996) Sequential stiffness design for seismic drift ranges of a shear building–pile–soil system. *Earthquake Engineering and Structural Dynamics*, **25** (12), 1405–1420.

Schnabel, P.B., Lysmer, J., and Seed, H.B. (1972) SHAKE: a computer program for earthquake response analysis of horizontally layered sites. A computer program distributed by NISEE/Computer Applications, Berkeley, CA.

Smith, H.A. and Wu, W.H. (1997) Effective optimal structural control of soil–structure interaction systems. *Earthquake Engineering and Structural Dynamics*, **26**, 549–570.

Smith, H.A., Wu, W.H., and Borja, R.I. (1994) Structural control considering soil–structure interaction effects. *Earthquake Engineering and Structural Dynamics*, **23**, 609–626.

Takewaki, I. (1998a) Optimal damper positioning in beams for minimum dynamic compliance. *Computer Methods in Applied Mechanics and Engineering*, **156** (1–4), 363–373.

Takewaki, I. (1998b) Remarkable response amplification of building frames due to resonance with the surface ground. *Soil Dynamics and Earthquake Engineering*, **17** (4), 211–218.

Takewaki, I. (2000) *Dynamic Structural Design: Inverse Problem Approach*, WIT Press.

Takewaki, I. and Nakamura, T. (1995) Stiffness design of a shear building–pile system supported by a ground with a specified coefficient of variation of stiffness for specified seismic deformation. *Journal of Structural Engineering B*, **41**, 129–138 (in Japanese).

Takewaki, I. and Yoshitomi, S. (1998) Effects of support stiffnesses on optimal damper placement for a planar building frame. *The Structural Design of Tall Buildings*, **7** (4), 323–336.

Takewaki, I., Fujii, N., and Uetani, K. (2002a) Nonlinear surface ground analysis via statistical approach. *Soil Dynamics and Earthquake Engineering*, **22** (6), 499–509.

Takewaki, I., Fujii, N., and Uetani, K. (2002b) Simplified inverse stiffness design for nonlinear soil amplification. *Engineering Structures*, **24** (11), 1369–1381.

Tsuji, M. and Nakamura, T. (1996) Optimum viscous dampers for stiffness design of shear buildings. *The Structural Design of Tall Buildings*, **5**, 217–234.

Veletsos, A.S. and Verbic, B. (1974) Basic response functions for elastic foundations. *Journal of Engineering Mechanics*, **100** (EM2), 189–202.

Wong, H.L. and Luco, J.E. (1991) Structural control including soil–structure interaction effects. *Journal of Engineering Mechanics*, **117**, 2237–2250.

Zhang, R.H. and Soong, T.T. (1992) Seismic design of viscoelastic dampers for structural applications. *Journal of Structural Engineering*, **118** (5), 1375–1392.

9

Design of Dampers in Shear Buildings with Uncertainties

9.1 Introduction

Sources of uncertainties in structural engineering usually come from input earthquake ground motions and parameter variability in structures. It is well accepted in earthquake-prone countries that the former uncertainties govern the principal design stage. However, the latter uncertainties are also important in the design decision stage. In this chapter, both uncertainties are treated and a few approaches to tackling these uncertainties simultaneously are explained.

As for the former uncertainties, the method of critical excitation was proposed by Drenick (1970). This method is aimed at finding the excitation producing the maximum response from a class of allowable inputs. By using the variational formulation, Drenick (1970) showed that the critical excitation for a given structural system is its impulse response function reversed in time. Just after the work by Drenick (1970), Shinozuka (1970) discussed the same problem in the frequency domain and presented a narrower upper bound of the maximum response. Fruitful practical application and extension of critical excitation methods have been made (Yang and Heer, 1971; Iyengar, 1972; Drenick, 1973, 1977; Wang *et al.*, 1978; Ahmadi, 1979; Wang and Yun, 1979; Pirasteh *et al.*, 1988; Srinivasan *et al.*, 1991; Baratta *et al.*, 1998). Iyengar and Manohar (1987), Srinivasan *et al.* (1992), Manohar and Sarkar (1995), and Sarkar and Manohar (1998) extended the concept of critical excitation to stochastic problems. Ben-Haim and Elishakoff (1990), Ben-Haim *et al.* (1996), and Pantelides and Tzan (1996) presented several interesting convex models.

One of the purposes of this chapter is to explain a probabilistic critical excitation method and a systematic method for optimal supplemental damper placement (Zhang and Soong, 1992; Tsuji and Nakamura, 1996) in building structures subjected to the critical excitation. In contrast to most of the conventional critical excitation methods,

Building Control with Passive Dampers   Izuru Takewaki
© 2009 John Wiley & Sons (Asia) Pte Ltd

a stochastic response index is treated as the objective function to be maximized. The energy and the intensity of the excitations are fixed and the critical excitation is found under these restrictions. The critical excitation method explained in this chapter is applicable to MDOF structural systems with nonproportional damping. The steepest direction search algorithm (Takewaki, 1998; Takewaki and Yoshitomi, 1998) explained in Chapters 4–8 is extended to a structure subjected to the critical excitation. While resonant steady-state responses of structures were treated in Chapters 4 and 5 (Takewaki, 1998; Takewaki and Yoshitomi, 1998), earthquake responses (mean-square responses) to random inputs are introduced here as controlled parameters, as in Chapters 6–8. It is shown that closed-form expressions of the inverse of the tri-diagonal coefficient matrix in the governing equations lead to drastic reduction of computational time of mean-square responses of the building structure to the random earthquake input and their derivatives with respect to the design variables (damping coefficients of supplemental dampers). Optimal placement and the capacity of the supplemental dampers are found automatically via the steepest direction search algorithm, which is based on successive satisfaction of the optimality conditions. A numerical example of a 6-DOF shear-building model is presented to demonstrate the effectiveness and validity of the method explained here.

The second purpose of this chapter is to explain a new evaluation method of robustness of passive control systems under both structural model and load uncertainties. Info-gap theory due to Ben-Haim (2001, 2006) is introduced and it is shown how the maximum robust placement of supplemental dampers can be attained under uncertain environments of damper and load properties.

9.2 Equations of Motion and Mean-square Response

For a simple and clear presentation of the theory, consider an n-story shear building model, as shown in Figure 9.1, subjected to a horizontal base acceleration input $\ddot{u}_g(t)$. Let $\mathbf{M}$, $\mathbf{C}$, $\mathbf{K}$, and $\mathbf{r} = \{1 \cdots 1\}^T$ denote the system mass, the viscous damping (original frame damping plus supplemental viscous damping) and stiffness matrices, and the influence coefficient vector respectively. Simple examples of $\mathbf{M}$, $\mathbf{C}$, and $\mathbf{K}$ are shown in Appendix 9.A for a two-story shear building model.

Because the formulation in the frequency domain is suitable for the response analysis under random vibration, it is treated in the first part of this chapter. Equations of motion of the building in the frequency domain may be expressed by

$$(-\omega^2 \mathbf{M} + i\omega \mathbf{C} + \mathbf{K})\mathbf{U}(\omega) = -\mathbf{M}\mathbf{r}\ddot{U}_g(\omega) \tag{9.1}$$

$\mathbf{U}(\omega)$ and $\ddot{U}_g(\omega)$ are the Fourier transforms of the horizontal displacements of the floors and the Fourier transform of the horizontal input acceleration $\ddot{u}_g(t)$. The horizontal input acceleration is assumed to be a stationary Gaussian random process with zero mean. Equation 9.1 can be reduced to the following compact form:

$$\mathbf{A}\mathbf{U}(\omega) = \mathbf{B}\ddot{U}_g(\omega) \tag{9.2}$$

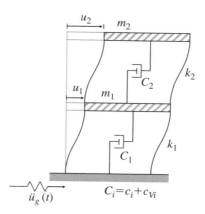

Figure 9.1 Shear building model. (Originally published in I. Takewaki, "Optimal damper placement for critical excitation," *Probabilistic Engineering Mechanics*, **15**, no. 4, 317–325, 2000, Elsevier B.V.).

where the coefficient matrix $\mathbf{A}$ and the vector $\mathbf{B}$ in the right-hand side are defined by

$$\mathbf{A} = (-\omega^2\mathbf{M} + i\omega\mathbf{C} + \mathbf{K}) \tag{9.3a}$$

$$\mathbf{B} = -\mathbf{Mr} \tag{9.3b}$$

Because the interstory drift can be a good indicator of the overall stiffness of the building, its control plays an important role in the stiffness design of the building. For this reason, the interstory drift is treated here as the controlled parameter. Let us define time-domain interstory drifts $\mathbf{d}(t) = \{d_i(t)\}$ of the shear building model and their Fourier transforms $\mathbf{D}(\omega) = \{D_i(\omega)\}$. $\mathbf{D}(\omega)$ can be related to $\mathbf{U}(\omega)$ by

$$\mathbf{D}(\omega) = \mathbf{T}\mathbf{U}(\omega) \tag{9.4}$$

The transformation matrix $\mathbf{T}$ is a constant matrix consisting of 1, -1, and 0. Substitution of a modified expression of Equation 9.2 into Equation 9.4 provides

$$\mathbf{D}(\omega) = \mathbf{T}\mathbf{A}^{-1}\mathbf{B}\ddot{U}_g(\omega) \tag{9.5}$$

Equation 9.5 is simply expressed as

$$\mathbf{D}(\omega) = \mathbf{H}_D(\omega)\ddot{U}_g(\omega) \tag{9.6}$$

In Equation 9.6, $\mathbf{H}_D(\omega) = \{H_{D_i}(\omega)\}$ are the transfer functions of interstory drifts and are described as

$$\mathbf{H}_D(\omega) = \mathbf{T}\mathbf{A}^{-1}\mathbf{B} \tag{9.7}$$

Note that $\mathbf{A}$ is a tri-diagonal matrix and its inverse can be obtained in closed form (see Appendix 9.B). This property enables one to compute the transfer functions of interstory drifts very efficiently.

The statistical characteristic of stationary random signals can be described by the PSD function. Let $S_g(\omega)$ denote the PSD function of the input acceleration $\ddot{u}_g(t)$. Using the random vibration theory, the mean-square response of the ith interstory drift can then be computed from

$$\sigma_{D_i}^2 = \int_{-\infty}^{\infty} |H_{D_i}(\omega)|^2 S_g(\omega) d\omega = \int_{-\infty}^{\infty} H_{D_i}(\omega) H_{D_i}^*(\omega) S_g(\omega) d\omega \qquad (9.8)$$

where ( )* denotes the complex conjugate.

9.3 Critical Excitation

The problem of critical excitation is explained in this section. As a system flexibility measure, the sum of the mean squares of the interstory drifts is introduced here. By using Equation 9.8, the sum of the mean squares of the interstory drifts can be expressed as

$$f = \sum_{i=1}^{n} \sigma_{D_i}^2 = \int_{-\infty}^{\infty} F(\omega) S_g(\omega) d\omega \qquad (9.9)$$

where n denotes the number of stories of the shear building model as stated above and the function $F(\omega)$ can be described by

$$F(\omega) = \sum_{i=1}^{n} |H_{D_i}(\omega)|^2 \qquad (9.10)$$

The problem of defining a critical excitation for the model without added dampers may be stated as follows.

Problem 9.1 Critical Excitation   Given the floor masses, story stiffnesses, and original structural damping of the n-story shear building model, find the critical PSD function $\tilde{S}_g(\omega)$ to maximize the sum f of the mean squares of the interstory drifts defined by Equation 9.9 subject to

$$\int_{-\infty}^{\infty} S_g(\omega) d\omega \leq \overline{S} \quad (\overline{S}: \text{given energy limit}) \qquad (9.11)$$

$$\sup S_g(\omega) \leq \overline{s} \quad (\overline{s}: \text{given PSD amplitude limit}) \qquad (9.12)$$

Equation 9.11 constrains the power of the excitation (Shinozuka, 1970; Iyengar and Manohar, 1987) and Equation 9.12 is introduced to keep the present excitation model physically realistic by avoiding the PDF function from attaining an extremely large value.

It is well known that a PSD function, a Fourier amplitude spectrum, and an undamped velocity response spectrum of an earthquake have a certain relationship. If the time

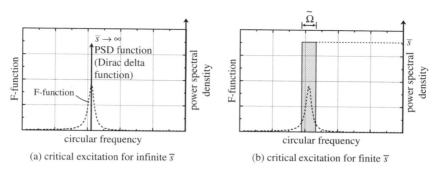

(a) critical excitation for infinite $\bar{s}$ (b) critical excitation for finite $\bar{s}$

Figure 9.2 Solution procedure: (a) without PSD bound; (b) with PSD bound. (Originally published in I. Takewaki, "Optimal damper placement for critical excitation," *Probabilistic Engineering Mechanics*, 15, no. 4, 317–325, 2000, Elsevier B.V.).

duration of the earthquake is fixed, then the PSD function corresponds to the Fourier amplitude spectrum and almost corresponds to the undamped velocity response spectrum (Hudson, 1962). Therefore, the present limitation on the peak of the PSD function indicates approximately the introduction of a bound on the undamped velocity response spectrum.

It is meaningful to discuss the problem structure of Problem 9.1 by focusing on the attainability of the maximum value of the input-motion PSD function. In the case where $\bar{s}$ is infinite, $\tilde{S}_g(\omega)$ becomes the Dirac delta function (see Figure 9.2(a)) and the value f takes the following quantity:

$$f = \bar{S}F(\omega_M) \tag{9.13}$$

where

$$F(\omega_M) = \max_{\omega} F(\omega) \tag{9.14}$$

This implies that the critical excitation is almost resonant to the fundamental natural frequency of the structural model.

When $\bar{s}$ is finite, the critical PSD function $\tilde{S}_g(\omega)$ becomes a constant value of $\bar{s}$ in a finite interval $\tilde{\Omega} = \bar{S}/\bar{s}$ (see Figure 9.2(b)). The optimization procedure to be implemented in Problem 9.1 is very simple because of the positive definiteness of the functions $F(\omega)$ and $S_g(\omega)$ in Equation 9.9; it is sufficient to find the finite interval $\tilde{\Omega}$, which can be searched by decreasing a horizontal line in the figure of the function $F(\omega)$ until the interval length becomes $\bar{S}/\bar{s}$ and finding the intersection. This finite interval $\tilde{\Omega}$ may not necessarily be a continuous interval (see Figure 9.3) (i.e., decomposed ones are possible) and can be defined as the interval with a fixed interval length $\bar{S}/\bar{s}$ to maximize

$$G(\Omega) = \int_{\omega \in \Omega} F(\omega) d\omega \tag{9.15}$$

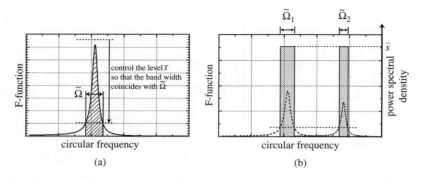

Figure 9.3 Continuous and uncontinuous intervals as solutions depending on PSD bound: (a) continuous; (b) uncontinuous. (Originally published in I. Takewaki, "Optimal damper placement for critical excitation," *Probabilistic Engineering Mechanics*, **15**, no. 4, 317–325, 2000, Elsevier B.V.).

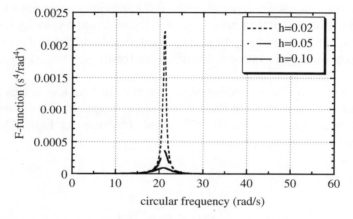

Figure 9.4 Example of the function $F(\omega)$ for two-DOF shear building models with three different damping ratios. (Originally published in I. Takewaki, "Optimal damper placement for critical excitation," *Probabilistic Engineering Mechanics*, **15**, no. 4, 317–325, 2000, Elsevier B.V.).

The function $G(\Omega)$ defined by Equation 9.15 indicates the area of the function $F(\omega)$ in the interval Ω, and maximizing the function $G(\Omega)$ means maximizing the area of the function $F(\omega)$ in the interval Ω with a constraint on fixed interval length $\overline{S}/\overline{s}$. Note that Ω is a finite interval (not necessarily a continuous interval as stated above) and $G(\Omega)$ is not an ordinary function; that is, not a function of a variable. In this case, the value f takes the following form:

$$f = \overline{s} \int_{\omega \in \tilde{\Omega}} F(\omega) d\omega \qquad (9.16)$$

Figure 9.4 shows an example of the function $F(\omega)$ for two-DOF shear building models with three different lowest mode damping ratios. Figure 9.5 illustrates the

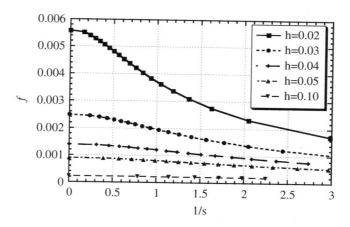

Figure 9.5 Variation of the function f with respect to $1/\bar{s}$ under a constant energy limit $\bar{S}$. (Originally published in I. Takewaki, "Optimal damper placement for critical excitation," *Probabilistic Engineering Mechanics*, **15**, no. 4, 317–325, 2000, Elsevier B.V.).

variation of the function f defined by Equation 9.16 with respect to $1/\bar{s}$ under a constant energy limit $\bar{S}$. Five lowest mode damping ratios are utilized. The case of $1/\bar{s}=0$ corresponds to $\bar{s}$ attaining infinity, and its value indicates the value given by Equation 9.13.

It is remarkable to note that the probabilistic critical excitation method explained here is applicable to structural systems with nonproportional damping and the objective function Equation 9.9 can easily be replaced by other response indices; for example, the top-floor absolute acceleration. For a more detailed explanation of critical excitation methods in earthquake engineering, Takewaki (2006) may be appropriate.

9.4 Conservativeness of Bounds (Recorded Ground Motions)

In this section, the level of conservativeness of the critical excitation explained in Section 9.3 is investigated through comparison with the results for recorded earthquakes. Three representative recorded earthquake ground motions are considered: El Centro NS 1940 (Imperial Valley), Taft EW 1952 (Kern County), and Kobe University NS 1995 (Hyogoken-Nanbu). For a simple and essential investigation and comparison, an SDOF elastic model is taken as the structural model. The interstory drift of the model is chosen as the response parameter to be compared. Two SDOF models with different damping ratios of 0.02 and 0.10 are investigated. The model with a rather high damping ratio represents passive control systems or base-isolated structures.

Figure 9.6 shows the PSD functions for these three ground motions. Although earthquake ground motions are well expressed by a nonstationary random process and possess a nonstationary nature, the PSD function in the relaxed sense is employed

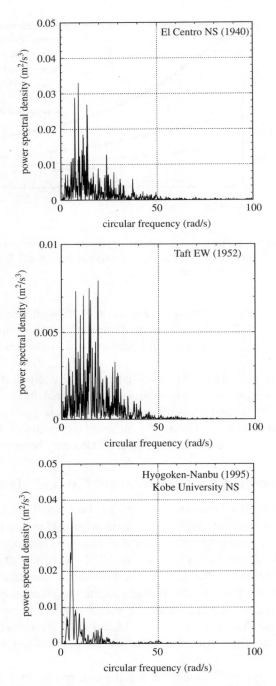

Figure 9.6 PSD functions for three recorded ground motions. (Originally published in I. Takewaki, "Optimal damper placement for critical excitation," *Probabilistic Engineering Mechanics*, **15**, no. 4, 317–325, 2000, Elsevier B.V.).

here. It can be observed that a rather sharp peak exists around the frequency of 5 rad/s (period $\cong 1.2$ s) in Kobe University NS 1995.

Figure 9.7 illustrates the standard deviation of the interstory drift of the SDOF model with the damping ratio of 0.02, plotted with respect to the undamped natural frequency of the SDOF model, to each recorded ground motion (the PSD of input motion is substituted in Equation 9.8) and that to the corresponding critical excitation. In evaluating the critical PSD function in Figure 9.7, the area of the PSD function and the peak value of the PSD function have been computed for each recorded ground motion. These values are: $\overline{S}=0.278$ m^2/s^4 and $\overline{s}=0.0330$ m^2/s^3 for El Centro NS 1940; $\overline{S}=0.0901$ m^2/s^4 and $\overline{s}=0.007\,92$ m^2/s^3 for Taft EW 1952; and $\overline{S}=0.185$ m^2/s^4 and $\overline{s}=0.0364$ m^2/s^3 for Kobe University NS 1995. It can be understood from Figure 9.7 that, while the level of conservativeness is about 2 or 3 in the natural frequency range of interest in El Centro NS 1940 and Taft EW 1952, a closer coincidence can be found around the frequency of 5 rad/s in Kobe University NS 1995. This means that Kobe University NS 1995 has a predominant frequency around 5 rad/s and the resonant characteristic of this ground motion can be well represented by the critical excitation explained in Section 9.3.

Figure 9.8 shows the standard deviation of the interstory drift of the SDOF model with a damping ratio of 0.10. A similar tendency can also be observed from Figure 9.8.

9.5 Design of Dampers in Shear Buildings under Uncertain Ground Motions

Consider a problem to incorporate story-installation-type viscous dampers into the shear building model to reduce the seismic response. This problem is the same as in Chapters 2–8. However, the properties of the input ground motion are not specified completely in the present problem. Therefore, the methods explained in Chapters 2–8 cannot be used in the present problem. On the other hand, if the properties of the building structure are given, then the method of finding the critical input explained in the previous section can be utilized here. However, the structural properties are not specified completely in the present problem because supplemental viscous dampers can be incorporated in any story of the buildings.

In order to overcome this difficulty, the problem of optimal damper placement is discussed here such that an optimal damper placement is found for the corresponding critical input. This problem is a highly nonlinear problem.

Let c_{Vi} denote the damping coefficient of the supplemental viscous damper in the ith story and let $\mathbf{c}_V = \{c_{Vi}\}$ denote the set of these damping coefficients. A general formulation is to regard the function f in Equation 9.9 as a functional of the PSD function $S_g(\omega)$ of the input and the damping coefficient $\mathbf{c}_V$ of the supplemental viscous dampers; that is, the expression $f(S_g(\omega):\mathbf{c}_V)$ may be appropriate in this problem (see Figure 9.9). The design objective is to find the most effective distribution $\mathbf{c}_V$ for the worst input $S_g(\omega)$ which maximizes the function f under the given conditions of

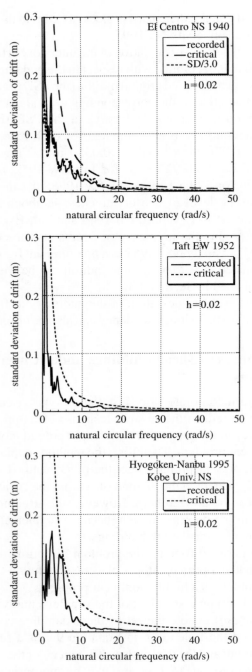

Figure 9.7 Standard deviation of the interstory drift of the SDOF model with a damping ratio of 0.02 for each recorded ground motion and the corresponding critical excitation plotted with respect to the undamped natural frequency of the SDOF model. (Originally published in I. Takewaki, "Optimal damper placement for critical excitation," *Probabilistic Engineering Mechanics*, **15**, no. 4, 317–325, 2000, Elsevier B.V.).

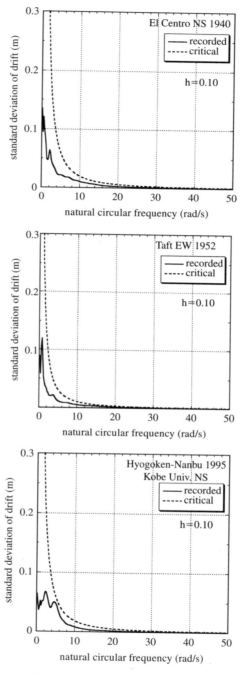

Figure 9.8 Standard deviation of the interstory drift of the SDOF model with a damping ratio of 0.10 for each recorded ground motion and the corresponding critical excitation plotted with respect to the undamped natural frequency of the SDOF model. (Originally published in I. Takewaki, "Optimal damper placement for critical excitation," *Probabilistic Engineering Mechanics*, **15**, no. 4, 317–325, 2000, Elsevier B.V.).

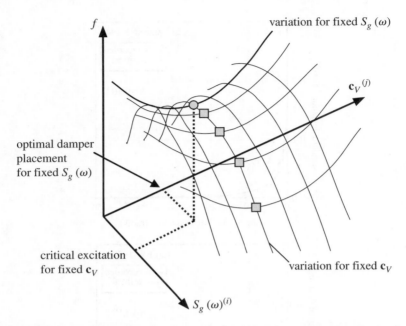

Figure 9.9 Design-dependent critical excitation and effective damper placement for critical excitation. (Originally published in I. Takewaki, "Optimal damper placement for critical excitation," *Probabilistic Engineering Mechanics*, **15**, no. 4, 317–325, 2000, Elsevier B.V.).

Equations 9.11 and 9.12. Since the worst input $S_g(\omega)$ is dependent on $\mathbf{c}_V$, the exact treatment of this problem is complex and can be expressed by

$$\min_{\mathbf{c}_V} \; \max_{S_g(\omega)} \; f(S_g(\omega) : \mathbf{c}_V)$$

Furthermore, since story stiffnesses are fixed in the present problem, it is expected that the PSD function $S_g(\omega)$ of the worst input is insensitive to the change of $\mathbf{c}_V$. For example, Figure 9.10 shows the distributions of the function $F(\omega)$ defined by Equation 9.10 for two-DOF models with three different damper distributions (installed in both stories, installed only in the first story and installed only in the second story). It can be observed that the frequencies attaining the peak value of the function $F(\omega)$ are almost the same irrespective of the damper distribution, and the worst input $S_g(\omega)$ is insensitive to the change of $\mathbf{c}_V$. For this reason, the worst input is to be found for the model without supplemental viscous dampers here. Then, the optimal damper placement is found for the fixed worst input. In the case of viscoelastic dampers, this problem becomes serious, and the dependency of the worst input on the damper's properties will have to be considered appropriately. This is a challenging problem in the future.

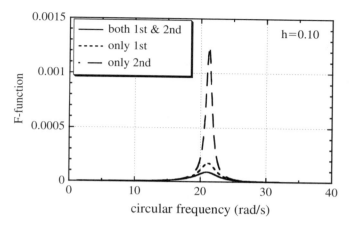

Figure 9.10 Distributions of the function $F(\omega)$ for two-DOF models with three different damper distributions. (Originally published in I. Takewaki, "Optimal damper placement for critical excitation," *Probabilistic Engineering Mechanics*, **15**, no. 4, 317–325, 2000, Elsevier B.V.).

In the following formulation, the PSD function of the input motion is fixed as that of the critical excitation obtained in the previous section for the case of a *finite* PSD amplitude $\bar{s}$. The mean squares of the interstory drifts are evaluated by Equation 9.8 for that critical excitation. The problem of optimal damper placement for that critical excitation (PODP-CE) may be described as follows.

Problem 9.2 PODP-CE Given the floor masses, story stiffnesses, and structural damping of the shear building model and the PSD function of the input motion, find the optimal distribution $\mathbf{c}_V$ of supplemental viscous dampers to minimize the sum of the mean squares of the interstory drifts

$$f = \sum_{i=1}^{n} \sigma_{D_i}^2 \tag{9.17}$$

subject to the constraint on total damper capacity

$$\sum_{i=1}^{n} c_{Vi} = \overline{W} \quad (\overline{W}: \text{specified total damper capacity}) \tag{9.18}$$

and to the constraints on each damper capacity

$$0 \le c_{Vi} \le \bar{c}_{Vi} \quad (i = 1, \cdots, n) \quad (\bar{c}_{Vi} : \text{ upper bound of damping coefficient}) \tag{9.19}$$

9.5.1 Optimality Conditions

It is straightforward to use the Lagrange multiplier method for solving constrained optimization problems. The constrained optimization problem stated above can be formulated mathematically by the generalized Lagrangian formulation. The generalized Lagrangian L for Problem 9.2 may be expressed by

$$L(\mathbf{c_V}, \lambda, \boldsymbol{\mu}, \boldsymbol{v}) = \sum_{i=1}^{n} \sigma_{D_i}^2 + \lambda \left(\sum_{i=1}^{n} c_{Vi} - \overline{W} \right) + \sum_{i=1}^{n} \mu_i (0 - c_{Vi}) + \sum_{i=1}^{n} v_i (c_{Vi} - \bar{c}_{Vi})$$

(9.20)

In Equation 9.20, $\boldsymbol{\mu} = \{\mu_i\}$ and $\boldsymbol{v} = \{v_i\}$ are the Lagrange multipliers together with λ. The principal (or major) optimality conditions for Problem 9.2 without active upper and lower bound constraints on damping coefficients of the supplemental viscous dampers may be derived from the stationarity conditions of $L(\boldsymbol{\mu} = \mathbf{0}, \ \boldsymbol{v} = \mathbf{0})$ with respect to $\mathbf{c_V}$ and λ:

$$f_{,j} + \lambda = 0 \quad \text{for } 0 < c_{Vj} < \bar{c}_{Vj} \quad (j = 1, \cdots, n)$$

(9.21)

$$\sum_{i=1}^{n} c_{Vi} - \overline{W} = 0$$

(9.22)

Equation 9.13 has been derived by the differentiation with respect to c_{Vj} and Equation 9.14 has been obtained by the differentiation with respect to λ.

Here, and in the following, the mathematical symbol $(\cdot)_{,j}$ indicates the partial differentiation with respect to the damping coefficient c_{Vj} of the supplemental damper in the jth story. When the lower or upper bound of the constraint on damping coefficients of supplemental dampers is active, the optimality condition should be modified to the following forms:

$$f_{,j} + \lambda \geq 0 \quad \text{for } c_{Vj} = 0$$

(9.23)

$$f_{,j} + \lambda \leq 0 \quad \text{for } c_{Vj} = \bar{c}_{Vj}$$

(9.24)

9.5.2 Solution Algorithm

A solution algorithm for the problem stated above is explained here. In the aforementioned problem, the total damper capacity $\overline{W}$ is a fixed given value. From the practical structural design's point of view, however, it may be more useful if the optimal damper placement is available for multiple total damper capacity levels. In this section, a procedure is presented for finding such a set of optimal damper placements.

The model without supplemental viscous dampers, namely $c_{Vj} = 0 \ (j = 1, \cdots, n)$, is employed as the initial model. This treatment is well suited to the situation where a structural designer is starting the allocation and placement of added supplemental viscous dampers at appropriate positions. The damping coefficients of added supplemental dampers are *increased gradually* based on the optimality criteria stated above. This algorithm is called the steepest direction search algorithm, as in Chapters 4–8.

Let Δc_{Vi} and ΔW denote the increment of the damping coefficient of the ith added damper (damper in the ith story) and the increment of the sum of the damping coefficients of added supplemental dampers respectively. Once ΔW is given, the problem is to determine simultaneously the effective position and amount of the increments of the damping coefficients of added supplemental dampers. In order to develop this algorithm, the first- and second-order sensitivities of the objective function with respect to a design variable are needed. Those quantities are derived by differentiating successively Equation 9.8 by the design variables.

First-order derivative of mean-square interstory drift:

$$(\sigma_{D_i}^2)_{,j} = \int_{-\infty}^{\infty} \{H_{D_i}(\omega)\}_{,j} H_{D_i}^*(\omega) S_g(\omega)\, d\omega + \int_{-\infty}^{\infty} H_{D_i}(\omega)\{H_{D_i}^*(\omega)\}_{,j} S_g(\omega)\, d\omega \quad (9.25)$$

Second-order derivative of mean-square interstory drift:

$$(\sigma_{D_i}^2)_{,jl} = \int_{-\infty}^{\infty} \{H_{D_i}(\omega)\}_{,j}\{H_{D_i}^*(\omega)\}_{,l} S_g(\omega) d\omega + \int_{-\infty}^{\infty} \{H_{D_i}(\omega)\}_{,l}\{H_{D_i}^*(\omega)\}_{,j} S_g(\omega) d\omega$$

$$+ \int_{-\infty}^{\infty} \{H_{D_i}(\omega)\}_{,jl} H_{D_i}^*(\omega) S_g(\omega) d\omega + \int_{-\infty}^{\infty} H_{D_i}(\omega)\{H_{D_i}^*(\omega)\}_{,jl} S_g(\omega) d\omega$$

$$(9.26)$$

In Equations 9.25 and 9.26, the first and second-order derivatives of transfer functions may be expressed as

$$\{H_{D_i}(\omega)\}_{,j} = \mathbf{T}_i \mathbf{A}_{,j}^{-1} \mathbf{B} \quad (9.27)$$

$$\{H_{D_i}(\omega)\}_{,jl} = \mathbf{T}_i \mathbf{A}_{,jl}^{-1} \mathbf{B} \quad (9.28)$$

$$\{H_{D_i}^*(\omega)\}_{,j} = \mathbf{T}_i \mathbf{A}_{,j}^{-1*} \mathbf{B} \quad (9.29)$$

$$\{H_{D_i}^*(\omega)\}_{,jl} = \mathbf{T}_i \mathbf{A}_{,jl}^{-1*} \mathbf{B} \quad (9.30)$$

In Equations 9.27–9.30, $\mathbf{T}_i$ is the ith row vector in the transformation matrix $\mathbf{T}$ defined in Equation 9.4.

The first derivative of the inverse $\mathbf{A}^{-1}$ of the coefficient matrix can be computed by $(\mathbf{A}^{-1})_{,j} = -\mathbf{A}^{-1}\mathbf{A}_{,j}\mathbf{A}^{-1}$ obtained by differentiating $\mathbf{A}\mathbf{A}^{-1} = \mathbf{I}$. Because the components in the coefficient matrix $\mathbf{A}$ are linear functions of design variables, $\mathbf{A}_{,jl} = \mathbf{0}$. Then the second-order derivative of the inverse $\mathbf{A}^{-1}$ can be obtained from

$$\mathbf{A}_{,jl}^{-1} = \mathbf{A}^{-1}(\mathbf{A}_{,l}\mathbf{A}^{-1}\mathbf{A}_{,j} + \mathbf{A}_{,j}\mathbf{A}^{-1}\mathbf{A}_{,l})\mathbf{A}^{-1} \quad (9.31)$$

The first-order derivative of the complex conjugate A^{-1*} of the inverse can be computed as $\{A^{-1}_{,j}\}^*$ and the second-order derivative of A^{-1*} can be found as $A^{-1*}_{,jl} = \{A^{-1}_{,jl}\}^*$.

The solution algorithm in the case satisfying the conditions $c_{Vj} < \bar{c}_{Vj}$ for all j may be summarized as follows:

Step 0 Initialize all the damping coefficients of supplemental viscous dampers as $c_{Vj} = 0$ $(j = 1, \cdots, n)$. In the initial design stage, the structural damping alone exists in the shear building model.

Step 1 Find the critical excitation $S_g(\omega)$ for the model without supplemental dampers.

Step 2 Assume ΔW.

Step 3 Compute the first-order derivative $f_{,i}$ of the objective function by Equation 9.25.

Step 4 Find the index p satisfying the condition

$$-f_{,p} = \max_i (-f_{,i}) \tag{9.32}$$

Step 5 Update the objective function f by the linear approximation $f + f_{,p}\Delta c_{Vp}$, where $\Delta c_{Vp} = \Delta W$. This is because the supplemental damper is added only in the pth story in the initial design stage.

Step 6 Update the first-order sensitivity $f_{,i}$ of the objective function by the linear approximation $f_{,i} + f_{,ip}\Delta c_{Vp}$ using Equation 9.26.

Step 7 If, in Step 6, there exists a supplemental damper of an index j such that the condition

$$-f_{,p} = \max_{j, j \neq p}(-f_{,j}) \tag{9.33}$$

is satisfied, then stop and compute the increment $\Delta \tilde{c}_{Vp}$ of the damping coefficient of the corresponding damper. At this stage, update $f_{,i}$ by $f_{,i} + f_{,ip}\Delta \tilde{c}_{Vp}$ using Equation 9.26.

Step 8 Repeat the procedure from Step 4 to Step 7 until the constraint in Equation 9.18 (i.e., $\sum_{i=1}^{n} c_{Vi} = \overline{W}$) is satisfied.

In Step 4 and Step 5, the direction which decreases the objective function most effectively under the condition $\sum_{i=1}^{n} \Delta c_{Vi} = \Delta W$ is found and the design (the quantity of supplemental dampers) is updated in that direction. It is appropriate, therefore, to call the present algorithm "the steepest direction search algorithm," as in Chapters 4–8. As explained before, this algorithm is similar to the conventional steepest descent method in mathematical programming (see Figure 7.2 in Chapter 7 to understand the concept). However, while the conventional steepest descent method uses the gradient vector itself of the objective function as the direction and does not utilize optimality criteria, the present algorithm takes advantage of the newly derived optimality criteria expressed by Equations 9.21, 9.23, and 9.24 and does not adopt the gradient vector as the direction. More specifically, the explained steepest direction search guarantees the successive and approximate satisfaction of the optimality criteria. For example, if Δc_{Vp} is added to the pth added viscous damper in which Equation 9.32 is satisfied, then its damper ($c_{Vp} > 0$) satisfies the optimality condition in Equation 9.21 and the other dampers ($c_{Vj} = 0$, $j \neq p$) alternatively satisfy the optimality condition in Equation 9.23. It is important to note that a series of subproblems is introduced here tentatively in which the total damper level $\overline{W}$ is increased gradually by ΔW from zero through the specified value.

It is necessary to investigate other possibilities. If multiple indices $p_1, \cdots, p_m$ exist in Step 4, then the objective function f and its derivative f_j have to be updated by the following rules:

$$f \to f + \sum_{i=p_1}^{p_m} f_{,i} \Delta c_{Vi} \tag{9.34a}$$

$$f_j \to f_j + \sum_{i=p_1}^{p_m} f_{,ji} \Delta c_{Vi} \tag{9.34b}$$

Furthermore, the index p defined in Step 7 has to be replaced by the multiple indices $p_1, \cdots, p_m$. The ratios among the magnitudes Δc_{Vi} have to be determined so that the following relations are satisfied:

$$f_{,p_1} + \sum_{i=p_1}^{p_m} f_{,p_1 i} \Delta c_{Vi} = \cdots = f_{,p_m} + \sum_{i=p_1}^{p_m} f_{,p_m i} \Delta c_{Vi} \tag{9.35}$$

Equation 9.35 requires that the optimality condition (9.21) continues to be satisfied in the supplemental dampers with the indices $p_1, \cdots, p_m$.

It may be the case in realistic situations that the maximum quantity of supplemental dampers is limited by the requirements of building design and planning. In the case where the damping coefficients of some added supplemental dampers attain their upper bounds, such constraints must be incorporated in the aforementioned algorithm. In that case the increment Δc_{Vp} is added subsequently to the supplemental damper in which $-f_{,p}$ attains the maximum among all the supplemental dampers, except for those attaining the upper bound.

9.6 Numerical Examples I

For a simple and clear presentation of the method explained above, consider a six-story shear-building model. The energy limit of the excitation is given by $\overline{S} = 0.05 \, \mathrm{m^2/s^4}$ and the amplitude limit of the excitation PSD is set to $\overline{s} = 0.01 \, \mathrm{m^2/s^3}$. From the method in Section 9.3, the critical excitation turns out to be the band-limited white noise, and its PSD function is given by

$$S_g(\omega) = 0.01 \, \mathrm{m^2/s^3} \quad (-9.40 \, \mathrm{rad/s} \leq \omega \leq -6.90, \; 6.90 \leq \omega \leq 9.40)$$

$$S_g(\omega) = 0 \quad \text{otherwise}$$

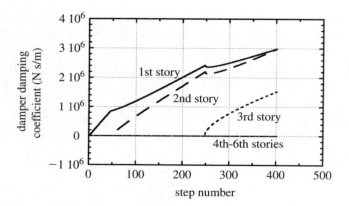

Figure 9.11 Variation of damping coefficients for optimal damper placement with respect to the varied total damper capacity. (Originally published in I. Takewaki, "Optimal damper placement for critical excitation," *Probabilistic Engineering Mechanics*, **15**, no. 4, 317–325, 2000, Elsevier B.V.).

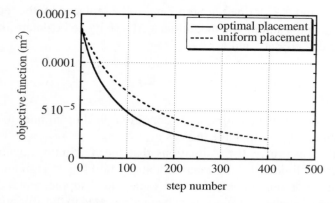

Figure 9.12 Variations of the objective function for optimal placement and for uniform placement with the same total damper capacity. (Originally published in I. Takewaki, "Optimal damper placement for critical excitation," *Probabilistic Engineering Mechanics*, **15**, no. 4, 317–325, 2000, Elsevier B.V.).

The floor masses of the shear building model are given by $m_i = 32 \times 10^3$ kg $(i = 1, \cdots, 6)$ and the story stiffnesses are specified as $k_i = 3.76 \times 10^7$ N/m $(i = 1, \cdots, 6)$. Then, the fundamental natural period of the structure attains 0.760 s. The structural viscous damping matrix of the shear building model has been given so that it is proportional to the stiffness matrix, and the lowest mode damping ratio is equal to 0.05. The increment ΔW of the total damper capacity for finding a series of optimal damper placements with respect to varied total damper capacity is set to $\Delta W = 1.875 \times 10^4$ N s/m.

Figure 9.11 shows the variation of damping coefficients of the supplemental viscous dampers for the optimal damper placement with respect to the varied total damper capacity. The damping coefficients of the supplemental viscous dampers for the optimal placement are added first in the first story and then in the second and third stories successively. It is useful to compare the performance of the shear building model with the optimal placement with that of the shear building model with the uniform distribution of dampers. Figure 9.12 illustrates the variation of the objective function for the optimal placement and that for the uniform placement with the same total damper capacity. It is observed from this figure that the optimal placement can reduce the objective function more effectively than the uniform placement. It is noted that a series of optimal designs is found in the method explained above with respect to total damper capacity, and the step number in Figures 9.11 and 9.12 does not mean the redesign step number in the conventional numerical optimization algorithms. It has been found that the closed-form expressions of the inverse of the coefficient matrix actually reduced the CPU time drastically from 130 seconds to 15 seconds.

9.7 Approach Based on Info-gap Uncertainty Analysis

As stated at the beginning of this chapter, load uncertainties and structural model uncertainties are two major sources of actual uncertainties encountered in the design of structures. Critical difficulties may be caused by the situation that load uncertainties and structural model uncertainties are independent in some cases and are dependent in a complicated manner in other cases. Analysis of such complicated dependency itself is an important research subject (e.g., Schueller, 2008; Tsompanakis *et al.*, 2008). While simultaneous consideration of both the load and structural model uncertainties is very important and challenging, as stated above, only a limited number of publications have accumulated (Igusa and Der Kiureghian, 1988; Ghanem and Spanos, 1991; Jensen and Iwan, 1992; Cherng and Wen, 1994a, 1994b; Koyluoglu *et al.*, 1995; Katafygiotis and Papadimitriou, 1996; Jensen, 2000; Qiu and Wang, 2003; Schueller, 2008; Tsompanakis *et al.*, 2008).

Because civil engineering structures, unlike mechanical products, are not mass produced, and because the occurrence rate of large earthquakes and other severe disturbances critical to their safety design is very low, the probabilistic representation of the effect of these disturbances on structural systems seems to be difficult in most cases. This implies the difficulty of applying structural reliability theory to such problems.

The method of critical excitation is one of the powerful strategies for overcoming difficulties arising in the modeling of the nonprobabilistic load uncertainty, and many useful investigations have been made (Drenick, 1970; Shinozuka, 1970; Westermo, 1985; Ben-Haim and Elishakoff, 1990; Takewaki, 2001a, 2001b, 2002a, 2002b; 2004, 2006). In most of these critical excitation methods, except Westermo (1985) and Takewaki (2004, 2006), deformation or displacement parameters were treated as response performance functions defining the criticality of the loads. On the contrary, the earthquake input energy to passively controlled structures is introduced in this chapter as a new measure of structural performance. This is because some control devices have a limitation on energy dissipation capacity and their modeling. It is acknowledged in general that, while the structural properties and mechanical performance of ordinary structural systems are well recognized through a lot of experiences and databases, those of control devices installed in those ordinary structural systems are not necessarily well defined and reliable. In this situation, it may be reasonable to take into account the uncertainties of damping coefficients of supplemental viscous dampers and to use the earthquake input energy to those passively controlled structures as a new measure of structural performance.

The purpose of this section is to introduce and explain a new structural design concept which combines uncertainties in both the load and the structural parameters. For this goal, it is absolutely necessary to identify the critical load (excitation) theoretically, if possible, and the corresponding critical set of structural model parameters. As mentioned above, it is well recognized that the critical load (excitation) depends on the structural model parameters, and it is extremely difficult to deal with load uncertainties and structural model parameter uncertainties simultaneously. In order to tackle these difficult problems, the info-gap models of uncertainty (nonprobabilistic uncertainty models) by Ben-Haim (1996, 2001, 2005, 2006) are used. This concept enables one to represent uncertainties which exist in the Fourier amplitude spectrum of the load (input ground acceleration) and in parameters of the vibration model of the structure.

As a simple example, let us consider a vibration model with viscous damping systems in addition to masses and springs. It is well understood in the field of structural control and health monitoring that viscous damping coefficients c_i in a vibration model are very uncertain in comparison with masses and stiffnesses. By using a method for describing such uncertainty, the uncertain viscous damping coefficient can be expressed in terms of the nominal value $\tilde{c}_i$ and the unknown uncertainty level α, as shown in Figure 9.13 (Takewaki and Ben-Haim, 2005).

$$C(\alpha, \tilde{\mathbf{c}}) = \left\{ \mathbf{c} : \left| \frac{c_i - \tilde{c}_i}{\tilde{c}_i} \right| \leq \alpha, \ i = 1, \cdots, N \right\} \quad \alpha \geq 0 \qquad (9.36a)$$

The inequality in Equation 9.36a can be rewritten as

$$(1 - \alpha)\tilde{c}_i \leq c_i \leq (1 + \alpha)\tilde{c}_i \qquad (9.36b)$$

There is another method for describing such uncertainty of viscous damping. The energy transfer function introduced in the following section or the deformation (or acceleration) transfer function to the input acceleration can be one of the dynamic structural performances which a vibration model or system possesses. Let $F(\omega, \mathbf{c}, \mathbf{k})$ denote the "energy transfer function" defined in the following section. The energy transfer function is a function such that, when it is multiplied by the squared Fourier amplitude spectrum of the input and integrated in the frequency range, it provides

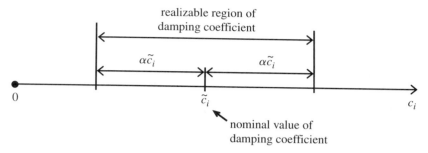

Figure 9.13 Uncertain damping coefficient with unknown horizon of uncertainty α. (© 2008 from *Structural Design Optimization Considering Uncertainties* by Y. Tsompanakis, N.D. Lagaros, and M. Papadrakakis (eds). Reproduced by permission of Taylor and Francis Group, LLC, a division of Informa plc.).

the earthquake input energy to the vibration model. The energy transfer function of a vibration model is a function of the viscous damping coefficients c_i, and the following info-gap model may be introduced in terms of the nominal function $\tilde{F}$ corresponding to the nominal viscous damping coefficients $\tilde{c}_i$:

$$F(\alpha, \tilde{F}) = \left\{ F(\omega, \mathbf{c}, \mathbf{k}) : \left| \frac{c_i - \tilde{c}_i}{\tilde{c}_i} \right| \leq \alpha, \ i = 1, \cdots, N \right\} \quad \alpha \geq 0 \qquad (9.37)$$

While the energy transfer functions are regarded as functions of the viscous damping coefficients c_i, as shown in Equation 9.37, the following family of sets of functions may also be considered for the definition of the info-gap uncertainty model:

$$F^*(\alpha, \tilde{F}) = \{ F(\omega, \mathbf{k}) : \ |F(\omega, \mathbf{k}) - \tilde{F}(\omega, \mathbf{k})| \ \leq \alpha \} \quad \alpha \geq 0 \qquad (9.38a)$$

$$F^{**}(\alpha, \tilde{F}) = \{ F(\omega, \mathbf{k}) : \ |F(\omega, \mathbf{k}) - \tilde{F}(\omega, \mathbf{k})| \ \leq \alpha \psi(\omega) \} \quad \alpha \geq 0 \qquad (9.38b)$$

Inequalities in Equations 9.38a and 9.38b can be expressed by

$$\tilde{F}(\omega, \mathbf{k}) - \alpha \leq F(\omega, \mathbf{k}) \leq \tilde{F}(\omega, \mathbf{k}) + \alpha \qquad (9.39a)$$

$$\tilde{F}(\omega, \mathbf{k}) - \alpha \psi(\omega) \leq F(\omega, \mathbf{k}) \leq \tilde{F}(\omega, \mathbf{k}) + \alpha \psi(\omega) \qquad (9.39b)$$

It is noted that, while the nominal function $\tilde{F}$ in Equation 9.37 has been introduced as the function corresponding to the nominal viscous damping coefficients $\tilde{c}_i$, the function $\tilde{F}(\omega, \mathbf{k})$ in Equations 9.38a and 9.38b may not necessarily be the function corresponding to the nominal viscous damping coefficients $\tilde{c}_i$. This treatment enables one to incorporate uncertainties other than the viscous damping coefficients.

9.7.1 Info-gap Robustness Function

The info-gap uncertainty analysis was introduced by Ben-Haim (2001) for measuring the robustness of a structure subjected to external loads. Simply speaking, the info-gap robustness is the greatest horizon of uncertainty α up to which the performance function $f(\mathbf{c}, \mathbf{k})$ does not exceed a critical value f_C. The performance function may be a peak displacement, peak stress, earthquake input energy, and so on.

Let us define the following info-gap robustness function corresponding to the info-gap uncertainty model represented by Equation 9.36a:

$$\hat{\alpha}(\mathbf{k}, f_C) = \max \left\{ \alpha : \{ \max_{\mathbf{c} \in C(\alpha, \tilde{\mathbf{c}})} f(\mathbf{c}, \mathbf{k}) \} \leq f_C \right\} \tag{9.40}$$

Another info-gap robustness function corresponding to the info-gap uncertainty model represented by Equation 9.37 may be introduced by

$$\hat{\alpha}(\mathbf{k}, f_C) = \max \left\{ \alpha : \{ \max_{F \in \mathcal{F}(\alpha, \tilde{F})} f(\mathbf{c}, \mathbf{k}) \} \leq f_C \right\} \tag{9.41}$$

Let us put $f_{C0} = f(\tilde{\mathbf{c}}, \mathbf{k})$ for the nominal damping coefficients. Then one can show that $\hat{\alpha}(\mathbf{k}, f_{C0}) = 0$ for the specific value f_{C0}, as shown in Figure 9.14. Furthermore, let us define $\hat{\alpha}(\mathbf{k}, f_C) = 0$ if $f_C \leq f_{C0}$ (see Figure 9.14). This means that, when the performance requirement is too small, we cannot satisfy the performance requirement for any admissible damping coefficients. The definitions in Equations 9.40 and 9.41 also imply that the robustness is the maximum level of the structural model parameter uncertainty α satisfying the performance requirement $f(\mathbf{c}, \mathbf{k}) \leq f_C$ for all admissible variation of the structural model parameter represented by Equation 9.36a or 9.37.

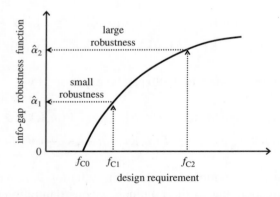

Figure 9.14 Info-gap robustness function $\hat{\alpha}$ with respect to design requirement f_C. (© 2008 from *Structural Design Optimization Considering Uncertainties* by Y. Tsompanakis, N.D. Lagaros, and M. Papadrakakis (eds). Reproduced by permission of Taylor and Francis Group, LLC, a division of Informa plc.).

9.7.2 Earthquake Input Energy to an SDOF System

The earthquake input energy is now introduced as the performance function of a vibration model. Many investigations have accumulated on the topic of the earthquake input energy since the pioneering work by Housner (1959). Housner (1959) pointed out that the input energy, although this definition is somewhat ambiguous, can be related to the velocity response spectrum of the input ground motion and the constant input energy criterion may hold approximately not only in elastic structures but also in elastic–plastic structures. Most research in this field has been conducted using the time-domain approach, which enables the treatment of even inelastic structures. In contrast to most of these previous studies (e.g., Housner, 1959; Akiyama, 1985; Uang and Bertero, 1990), the earthquake input energy is formulated here in the frequency domain (Page, 1952; Lyon, 1975; Ordaz et al., 2003; Takewaki, 2004, 2006) to facilitate the derivation of a bound of the earthquake input energy. Although the structures to be treated are restricted to elastic structures, this bound analysis may be useful and meaningful in uncertainty analysis of ground motions.

For a simple but essential presentation of the theory, consider a damped linear SDOF system of mass m, stiffness k, and damping coefficient c. Let $\Omega = \sqrt{k/m}$, $h = c/(2\Omega m)$ and x denote the undamped natural circular frequency, the damping ratio, and the displacement of the mass relative to the ground respectively. The time derivative is denoted by an over-dot. The earthquake input energy to this SDOF system by a unidirectional horizontal ground acceleration $\ddot{u}_g(t) = a(t)$ from $t = 0$ to $t = t_0$ (end of input) can be defined by the work of the ground on this SDOF structural system and is expressed in time domain by

$$E_I = \int_0^{t_0} m(\ddot{u}_g + \ddot{x})\dot{u}_g dt \qquad (9.42)$$

In the system shown in Figure 9.15, the term $-m(\ddot{u}_g + \ddot{x})$ is modified from the term in Equation 9.42 with a minus sign to indicate the inertial force on the mass at an

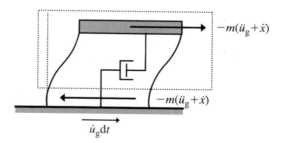

Figure 9.15 Free-body diagram for defining input energy. (© 2008 from *Structural Design Optimization Considering Uncertainties* by Y. Tsompanakis, N.D. Lagaros, and M. Papadrakakis (eds). Reproduced by permission of Taylor and Francis Group, LLC, a division of Informa plc.).

arbitrary time t, which is equal to the sum of the restoring force kx and the damping force $c\dot{x}$. Integration by parts of Equation 9.42 provides

$$E_I = \int_0^{t_0} m(\ddot{x} + \ddot{u}_g)\dot{u}_g dt = \int_0^{t_0} m\ddot{x}\dot{u}_g dt + \left[\frac{1}{2}m\dot{u}_g^2\right]_0^{t_0}$$

$$= [m\dot{x}\dot{u}_g]_0^{t_0} - \int_0^{t_0} m\dot{x}\ddot{u}_g dt + \left[\frac{1}{2}m\dot{u}_g^2\right]_0^{t_0} \tag{9.43}$$

If the relative velocity satisfies $\dot{x} = 0$ at $t = 0$ and the velocity of ground motion satisfies $\dot{u}_g = 0$ at $t = 0$ and $t = t_0$, then the earthquake input energy can be reduced simply to the following form:

$$E_I = -\int_0^{t_0} m\ddot{u}_g \dot{x} dt \tag{9.44a}$$

Equation 9.44a is alternatively defined by multiplying the relative velocity $\dot{x}$ on both sides of the equation of motion and integrating from time $t = 0$ to $t = t_0$ as follows:

$$\int_0^{t_0} m\ddot{x}\dot{x} dt + \int_0^{t_0} c\dot{x}\dot{x} dt + \int_0^{t_0} kx\dot{x} dt = -\int_0^{t_0} m\ddot{u}_g \dot{x} dt \tag{9.44b}$$

This also provides the following energy balance:

$$\left[\frac{1}{2}m\dot{x}^2\right]_0^{t_0} + \int_0^{t_0} c\dot{x}^2 dt + \left[\frac{1}{2}kx^2\right]_0^{t_0} = -\int_0^{t_0} m\ddot{u}_g \dot{x} dt \tag{9.44c}$$

As an actual recorded ground motion, consider the motion of El Centro NS 1940 (Imperial Valley) shown in Figure 9.16. The time history of the earthquake input energy per unit mass is shown in Figure 9.17. This can be computed using Equation 9.44a by regarding t_0 as t.

It is known (Page, 1952; Lyon, 1975; Ordaz et al., 2003; Takewaki, 2004, 2006) that, once the Fourier transformation is applied to $\ddot{u}_g$ and $\dot{x}$, the earthquake input energy expressed by Equation 9.44a can also be expressed in the frequency domain:

$$\frac{E_I}{m} = -\int_{-\infty}^{\infty} \dot{x}a dt = -\int_{-\infty}^{\infty} \left(\frac{1}{2\pi}\int_{-\infty}^{\infty} \dot{X}e^{i\omega t} d\omega\right) a dt$$

$$= -\frac{1}{2\pi}\int_{-\infty}^{\infty} \left(\int_{-\infty}^{\infty} ae^{i\omega t} dt\right) \{H_V(\omega; \Omega, h)A(\omega)\} d\omega$$

$$= -\frac{1}{2\pi}\int_{-\infty}^{\infty} A(-\omega)\{H_V(\omega; \Omega, h)A(\omega)\} d\omega \tag{9.45}$$

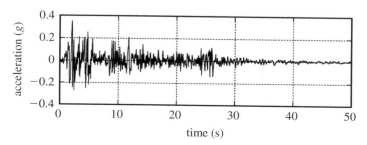

Figure 9.16 Ground motion of El Centro NS 1940 (Imperial Valley). (© 2008 from *Structural Design Optimization Considering Uncertainties* by Y. Tsompanakis, N.D. Lagaros, and M. Papadrakakis (eds). Reproduced by permission of Taylor and Francis Group, LLC, a division of Informa plc.).

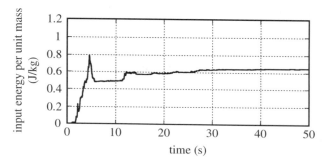

Figure 9.17 Time history of earthquake input energy under El Centro NS 1940 (Imperial Valley). (© 2008 from *Structural Design Optimization Considering Uncertainties* by Y. Tsompanakis, N.D. Lagaros, and M. Papadrakakis (eds). Reproduced by permission of Taylor and Francis Group, LLC, a division of Informa plc.).

where $H_V(\omega; \Omega, h)$ is the velocity transfer function defined by $\dot{X}(\omega) = H_V(\omega; \Omega, h)A(\omega)$ and is expressed by $H_V(\omega; \Omega, h) = -i\omega(\Omega^2 - \omega^2 + 2ih\Omega\omega)$. The functions $\dot{X}$ and $A(\omega)$ are the Fourier transforms of the relative velocity $\dot{x}$ and input acceleration $\ddot{u}_g(t) = a(t)$ respectively. The symbol i denotes the imaginary unit. Since the imaginary part of the velocity transfer function $H_V(\omega; \Omega, h)$ is an odd function of ω, Equation 9.45 can be simplified to

$$\frac{E_I}{m} = \int_0^\infty |A(\omega)|^2 \left\{ -\frac{1}{\pi} \mathrm{Re}[H_V(\omega; \Omega, h)] \right\} d\omega$$

$$\equiv \int_0^\infty |A(\omega)|^2 F(\omega) d\omega \tag{9.46}$$

where $F(\omega)$ is the energy transfer function expressed by $F(\omega) = -\mathrm{Re}[H_V(\omega; \Omega, h)]/\pi$. Equation 9.46 implies that the earthquake input energy to damped linear elastic SDOF

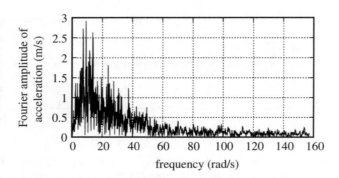

Figure 9.18 Fourier amplitude spectrum of ground acceleration of El Centro NS 1940 (Imperial Valley). (© 2008 from *Structural Design Optimization Considering Uncertainties* by Y. Tsompanakis, N.D. Lagaros, and M. Papadrakakis (eds). Reproduced by permission of Taylor and Francis Group, LLC, a division of Informa plc.).

systems does not depend on the phase of input motions, as pointed out by Page (1952), Lyon (1975), Ordaz *et al.* (2003), and Takewaki (2004).

Figure 9.18 shows the Fourier amplitude spectrum $|A(\omega)|$ of El Centro NS 1940.

9.7.3 Earthquake Input Energy to an MDOF System

Consider next a damped linear elastic MDOF shear building model of mass matrix $[M]$ subjected to a unidirectional horizontal ground acceleration $\ddot{u}_g(t) = a(t)$. The method explained here can be applied to both proportionally damped and nonproportionally damped structures.

Let $\{x\}$ denote a set of the horizontal floor displacements of the MDOF shear building model relative to the ground. The time derivative is denoted by an over-dot. As in SDOF models, the earthquake input energy to this MDOF system by the ground motion from $t = 0$ to $t = t_0$ (end of input) can be defined by the work of the ground on this MDOF system and is expressed by

$$E_I = \int_0^{t_0} \{1\}^T [M](\{1\}\ddot{u}_g + \{\ddot{x}\})\dot{u}_g dt \qquad (9.47)$$

In Equation 9.47, the vector $\{1\}$ is the influence coefficient vector denoted by $\{1\} = \{1 \quad \cdots \quad 1\}^T$. The term $\{1\}^T[M](\{1\}\ddot{u}_g + \{\ddot{x}\})$ in Equation 9.47 indicates the sum of the horizontal inertial forces with a minus sign acting on this system shown in Figure 9.19.

As in SDOF models, integration by parts of Equation 9.47 provides

$$E_I = \left[\frac{1}{2}\{1\}^T[M]\{1\}\dot{u}_g^2\right]_0^{t_0} + \left[\{\dot{x}\}^T[M]\{1\}\dot{u}_g\right]_0^{t_0} - \int_0^{t_0} \{\dot{x}\}^T[M]\{1\}\ddot{u}_g dt \qquad (9.48)$$

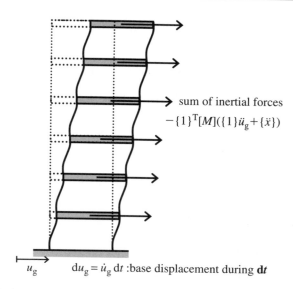

sum of inertial forces
$-\{1\}^T[M](\{1\}\ddot{u}_g+\{\ddot{x}\})$

u_g $du_g = \dot{u}_g \, dt$:base displacement during dt

Figure 9.19 Free-body diagram for defining earthquake input energy to MDOF model. (© 2008 from *Structural Design Optimization Considering Uncertainties* by Y. Tsompanakis, N.D. Lagaros, and M. Papadrakakis (eds). Reproduced by permission of Taylor and Francis Group, LLC, a division of Informa plc.).

If the relative velocity satisfies $\{\dot{x}\} = \{0\}$ at $t = 0$ and the velocity of ground motion satisfies $\dot{u}_g = 0$ at $t = 0$ and $t = t_0$, then the earthquake input energy can be reduced simply to the following form:

$$E_I = -\int_0^{t_0} \{\dot{x}\}^T[M]\{1\}\ddot{u}_g dt \qquad (9.49)$$

The earthquake input energy can also be expressed in the frequency domain as in the SDOF system. Let $\{\dot{X}\}$ denote the Fourier transform of the relative velocity $\{\dot{x}\}$. Application of the Fourier inverse transformation of the relative velocities $\{\dot{x}\}$ to Equation 9.49 provides

$$E_I = -\int_{-\infty}^{\infty} \left[\frac{1}{2\pi} \int_{-\infty}^{\infty} \{\dot{X}\}^T e^{i\omega t} d\omega \right] [M]\{1\}\ddot{u}_g dt$$

$$= -\frac{1}{2\pi} \int_{-\infty}^{\infty} \{\dot{X}\}^T[M]\{1\} \left[\int_{-\infty}^{\infty} \ddot{u}_g e^{i\omega t} \, dt \right] d\omega \qquad (9.50)$$

$$= -\frac{1}{2\pi} \int_{-\infty}^{\infty} \{\dot{X}\}^T[M]\{1\}A(-\omega)d\omega$$

In Equation 9.50, $A(\omega)$ is the Fourier transform of ground acceleration $\ddot{u}_g(t) = a(t)$ as defined above.

By using the Fourier transform of the equations of motion, the Fourier transform $\dot{X}(\omega)$ of the relative velocities can be expressed by

$$\{\dot{X}(\omega)\} = -i\omega(-\omega^2[M] + i\omega[C] + [K])^{-1}[M]\{1\}A(\omega) \qquad (9.51)$$

After substitution of Equation 9.51 into Equation 9.50, the earthquake input energy may be evaluated by

$$E_I = \int_0^\infty F_M(\omega)|A(\omega)|^2 d\omega \qquad (9.52)$$

where $F_M(\omega)$ denotes the following energy transfer function for MDOF models:

$$F_M(\omega) = \frac{1}{\pi} \text{Re}[i\omega\{1\}^T[M]^T[Y(\omega)][M]\{1\}] \qquad (9.53a)$$

In Equation 9.53a, the matrix $[Y(\omega)]$ is given by

$$[Y(\omega)] = (-\omega^2[M] + i\omega[C] + [K])^{-1} \qquad (9.53b)$$

9.7.4 Critical Excitation Problem for Acceleration Power

This section demonstrates that a critical excitation method for the earthquake input energy as a performance index can provide upper bounds on earthquake input energy. This is common in the field of critical excitation; that is, an upper bound of the objective function in the critical excitation problem can always be derived. Over two decades ago, Westermo (1985) discussed a similar problem for the maximum input energy to an SDOF system subjected to external forces. His solution of critical excitation is interesting but restrictive, because it is of the form including the velocity response quantity containing the solution itself implicitly. To remove this restriction, a more general solution procedure will be explained here.

In the field of earthquake engineering, the scaling of ground motions based on an appropriate measure is quite important from the viewpoint of risk-based design. The capacity or intensity of ground motions is often defined in terms of the time integral of squared ground acceleration $a(t)^2$ (Arias, 1970; Housner and Jennings, 1975; Takewaki, 2004, 2006). This quantity is well known as the Arias intensity measure, except for a difference in the coefficient. The constraint on this quantity can be expressed by

$$\int_{-\infty}^\infty a(t)^2 \, dt = \frac{1}{\pi} \int_0^\infty |A(\omega)|^2 \, d\omega = \overline{C}_A \qquad (9.54)$$

where $\overline{C}_A$ is the specified value of the time integral of squared ground acceleration and $A(\omega)$ is the Fourier transform of $a(t)$. It is also natural to assume that the maximum value of the Fourier amplitude spectrum of input ground acceleration is finite because the duration is finite. The infinite Fourier amplitude spectrum may correspond to a perfect harmonic function with infinite duration or that multiplied by an exponential function (Drenick, 1970), which is unrealistic as an actual ground motion. The constraint on this property may be described by

$$|A(\omega)| \leq \overline{A} \tag{9.55}$$

where $\overline{A}$ is a specified upper bound of the Fourier amplitude spectrum.

The critical excitation problem for the MDOF system may be stated as follows.

Problem 9.3   Find the Fourier amplitude spectrum $|A(\omega)|$ of ground acceleration that maximizes the earthquake input energy expressed by Equation 9.52 subject to the constraints in Equations 9.54 and 9.55 on ground acceleration.

It is known and clear from the work by Takewaki (2001a, 2001b, 2002b) or the explanation in Section 9.3 on PSD functions that, if the upper bound $\overline{A}$ is infinite, $|A(\omega)|^2$ turns out to be the Dirac delta function which has a nonzero value at the point maximizing $F(\omega)$. On the other hand, if the upper bound $\overline{A}$ is finite, then $|A(\omega)|^2$ yields a rectangular function attaining $\overline{A}^2$ in a certain finite frequency range. The band-width of the frequency can be obtained as $\Delta\omega = \pi \overline{C}_A / \overline{A}^2$. The position of the rectangular function (i.e., the lower and upper frequency limits) can be computed by maximizing the function $\overline{A}^2 \int_{\omega_L}^{\omega_U} F(\omega) \, d\omega$. It is noted that the lower and upper frequency limits satisfy the condition $\omega_U - \omega_L = \Delta\omega$. It can be shown that a good and simple approximation can be obtained from $(\omega_U + \omega_L)/2 = \Omega$, where Ω is the undamped fundamental natural circular frequency of the MDOF system. The essential feature of the solution procedure explained in this section is shown in the schematic diagram in Figure 9.20. It is interesting to note that the above-mentioned periodic solution of Westermo (1985) may correspond to the case of infinite $\overline{A}$.

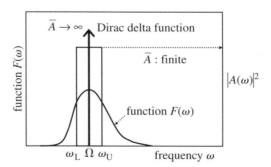

Figure 9.20   Schematic diagram of solution procedure for critical excitation problem. (© 2008 from *Structural Design Optimization Considering Uncertainties* by Y. Tsompanakis, N.D. Lagaros, and M. Papadrakakis (eds). Reproduced by permission of Taylor and Francis Group, LLC, a division of Informa plc.).

9.8 Evaluation of Robustness of Shear Buildings with Uncertain Damper Properties under Uncertain Ground Motions

9.8.1 Load Uncertainty Representation in Terms of Info-gap Models

As mentioned above, the simultaneous consideration of load and structural model uncertainties is not easy because both uncertainties have different dimensions and characters. To explain this simultaneous consideration, consider first an uncertainty model of load which is expressed in terms of a Fourier amplitude spectrum of the input acceleration.

Let $\tilde{A}$ and α_s denote the nominal Fourier amplitude spectrum of the uncertain input acceleration and its uncertainty level. The nominal Fourier amplitude spectrum may be interpreted as a design spectrum and the uncertainty level may be regarded as the degree of variation resulting from various factors. An info-gap model (Ben-Haim, 2001) of load $\mathcal{A}(\alpha_s, \tilde{A})$ for $\alpha_s \geq 0$ is introduced here to represent uncertainty in the Fourier amplitude spectrum of the input acceleration. This info-gap model of load may be defined by

$$\mathcal{A}[\alpha_s, \tilde{A}^2(\omega; \Delta\omega, C_A)] = \left\{ \left| A^2(\omega) \right| = s^* \tilde{A}^2 \left(\omega; \frac{\Delta\omega}{s^*}, C_A \right) : \right.$$

$$\left. s^* = \frac{s}{\tilde{s}}, \left| \frac{s - \tilde{s}}{\tilde{s}} \right| \leq \alpha_s \right\} \quad \alpha_s \geq 0 \tag{9.56a}$$

The inequality in Equation 9.56a can be rewritten as

$$(1 - \alpha_s)\tilde{A}^2 \leq |A^2(\omega)| \leq (1 + \alpha_s)\tilde{A}^2 \tag{9.56b}$$

In the past, Takewaki and Ben-Haim (2005) proposed a similar info-gap model for PSD functions.

The graphical and schematic expression of $\mathcal{A}(\alpha_s, \tilde{A}^2)$ can be found in Figure 9.21. Note that the quantity C_A is related to the power or intensity of the input acceleration and is assumed to be constant here. This leads to the constant area of the critical rectangular function of the squared Fourier amplitude spectrum. For this reason, as the amplitude changes uncertainly, the band-width varies correspondingly.

In order to explain in more detail the physical meaning of variation of the squared Fourier amplitude spectrum shown in Figure 9.21, consider the two finite-duration sinusoidal waves shown in Figures 9.22 and 9.23. These two finite-duration sinusoidal waves have the same acceleration power C_A. While Figure 9.22 represents a short-duration ground acceleration (representative of near-field ground motion) which has an intensive input during a short duration, Figure 9.23 presents a long-duration ground acceleration and simulates approximately a far-field ground motion which has a rather small input over a long duration. Figure 9.24 is the Fourier amplitude spectrum of the sinusoidal wave shown in Figure 9.22 (near-field ground motion) and Figure 9.25 is the Fourier amplitude spectrum of the sinusoidal wave shown in Figure 9.23 (far-field

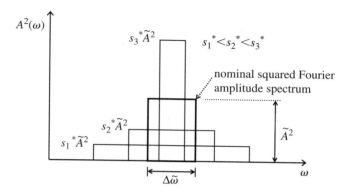

Figure 9.21 Variation of critical rectangular function of the squared Fourier amplitude spectrum of input acceleration. (© 2008 from *Structural Design Optimization Considering Uncertainties* by Y. Tsompanakis, N.D. Lagaros, and M. Papadrakakis (eds). Reproduced by permission of Taylor and Francis Group, LLC, a division of Informa plc.).

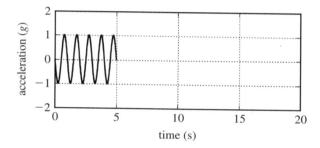

Figure 9.22 Short-duration sinusoidal motion. (© 2008 from *Structural Design Optimization Considering Uncertainties* by Y. Tsompanakis, N.D. Lagaros, and M. Papadrakakis (eds). Reproduced by permission of Taylor and Francis Group, LLC, a division of Informa plc.).

ground motion). From these figures, it may be said that a smaller level and wider range in squared Fourier amplitude spectrum represents a variation to a short-duration ground acceleration (representative of near-field ground motion) and a larger level and narrower range of squared Fourier amplitude spectrum assumes a variation to be a long-duration ground motion (representative of far-field ground motion).

9.8.2 Info-gap Robustness Function for Load and Structural Uncertainties

As explained before, the info-gap model for uncertainty in the dynamic model can be expressed by $\mathcal{F}(\alpha_m, \tilde{F})$ for $\alpha_m \geq 0$ in compliance with the definition (9.37). It should be noted that two different uncertainty parameters α_m and α_s are eventually used here simultaneously. The parameter α_s for load uncertainty has been introduced just above and the parameter α_m for model uncertainty is defined here by including the expression of α_s.

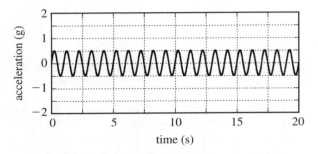

Figure 9.23 Long-duration sinusoidal motion. (© 2008 from *Structural Design Optimization Considering Uncertainties* by Y. Tsompanakis, N.D. Lagaros, and M. Papadrakakis (eds). Reproduced by permission of Taylor and Francis Group, LLC, a division of Informa plc.).

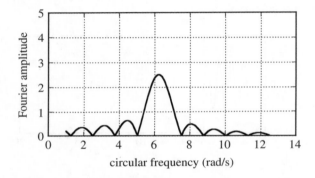

Figure 9.24 Fourier amplitude spectrum of short-duration sinusoidal motion. (© 2008 from *Structural Design Optimization Considering Uncertainties* by Y. Tsompanakis, N.D. Lagaros, and M. Papadrakakis (eds). Reproduced by permission of Taylor and Francis Group, LLC, a division of Informa plc.).

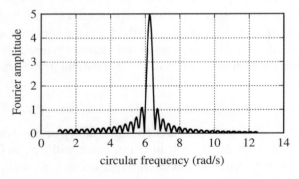

Figure 9.25 Fourier amplitude spectrum of long-duration sinusoidal motion. (© 2008 from *Structural Design Optimization Considering Uncertainties* by Y. Tsompanakis, N.D. Lagaros, and M. Papadrakakis (eds). Reproduced by permission of Taylor and Francis Group, LLC, a division of Informa plc.).

Let the function $f(A, F, \mathbf{k})$ denote the performance requirement including the earthquake input energy by the input with the Fourier amplitude spectrum A based on the energy transfer function F and design $\mathbf{k}$. As in the definition of the robustness in Equations 9.40 and 9.41, the performance requirement may be expressed by

$$f(A, F, \mathbf{k}) \leq f_C \qquad (9.57)$$

The info-gap robustness function (Takewaki and Ben-Haim, 2005) can then be introduced as a measure of robustness for model uncertainty for a given load spectral uncertainty level α_s:

$$\hat{\alpha}_m(\mathbf{k}, f_C, \alpha_s) = \max \left\{ \alpha_m : \left\{ \max_{\substack{F \in \mathcal{F}(\alpha_m, \tilde{F}) \\ A \in \mathcal{A}(\alpha_s, \tilde{A}^2)}} f(A, F, \mathbf{k}) \right\} \leq f_C \right\} \qquad (9.58)$$

Equation 9.58 clearly expresses the robustness (maximum robustness) of the structure under both the load and structural model uncertainties. This enables the simultaneous consideration of both the load and structural model uncertainties. It may be another and future possibility to relate α_s with α_m and define a unified robustness function $\hat{\alpha}_m$ in place of Equation 9.58. This will open a new horizon for truly simultaneous consideration of both uncertainties.

9.9 Numerical Examples II

For a simple and clear presentation of the method explained above, consider the six-story shear building model shown in Figure 9.26(a). Assume that this system has a uniform structural damping of the damping coefficient of 3.76×10^5 N s/m, as shown in Figure 9.27. This structural damping corresponds to a damping ratio of 0.04 in the fundamental vibration mode. Each floor has the same mass $m_i = 32 \times 10^3$ kg and every story stiffness has the same value of 3.76×10^7 N/m. The undamped fundamental natural period of this six-story shear building model is $T_1 = 0.72$ s. A supplemental viscous damper as a passive control system is installed in the first, third, or sixth story, as shown in Figure 9.26(b)–(d). The magnitude of the supplemental viscous damper is shown in Figure 9.28. The nominal damping coefficient $\tilde{c}_d$ of the supplemental viscous damper is determined so as to attain 10 times the damping coefficient of the structural damping in the same story. The uncertain damping coefficient c_d of the supplemental viscous damper is expressed by $c_d = \tilde{c}_d(1 \pm 0.5\alpha_m)$, where α_m is the unknown horizon of uncertainty in the model coefficients. The damping coefficients corresponding to three uncertainty levels $\alpha_m = 0.0, 0.5$, and 1.0 are shown in Figure 9.28.

The degree of uncertainty of the load is assumed to be expressed by the variation $s = \tilde{s}(1 \pm \alpha_s)$ of the squared rectangular Fourier amplitude spectrum of the input ground acceleration where α_s is the unknown horizon of uncertainty in the load. The power

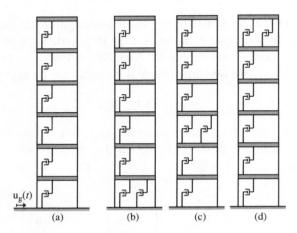

Figure 9.26 Six-story shear building model: (a) bare frame; (b) frame with a supplemental damper in the first story; (c) frame with a supplemental damper in the third story; (d) frame with a supplemental damper in the sixth story. (© 2008 from *Structural Design Optimization Considering Uncertainties* by Y. Tsompanakis, N.D. Lagaros, and M. Papadrakakis (eds). Reproduced by permission of Taylor and Francis Group, LLC, a division of Informa plc.).

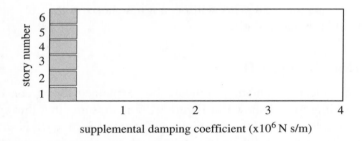

Figure 9.27 Structural damping coefficient. (© 2008 from *Structural Design Optimization Considering Uncertainties* by Y. Tsompanakis, N.D. Lagaros, and M. Papadrakakis (eds). Reproduced by permission of Taylor and Francis Group, LLC, a division of Informa plc.).

of the input defined by Equation 9.54 does not vary and is given by $\overline{C}_A = 11.4\,\mathrm{m}^2/\mathrm{s}^3$. This indicates that the area of squared Fourier amplitude spectrum is constant. The nominal level of the rectangular Fourier amplitude spectrum is $\tilde{A} = 2.91\,\mathrm{m/s}$ and its nominal band-width is $\Delta\tilde{\omega} = 4.21\,\mathrm{rad/s}$.

It is meaningful to note that the worst case (critical case), up to uncertainties α_m and α_s, can be obtained from $c_d = \tilde{c}_d(1 - 0.5\alpha_m)$ and $s = \tilde{s}(1 + \alpha_s)$. Although the problem of finding the worst case is very complicated in general (Kanno and Takewaki, 2007), the present case is almost self-evident. This enables one to discuss the info-gap robustness function directly with respect to the earthquake input energy performance.

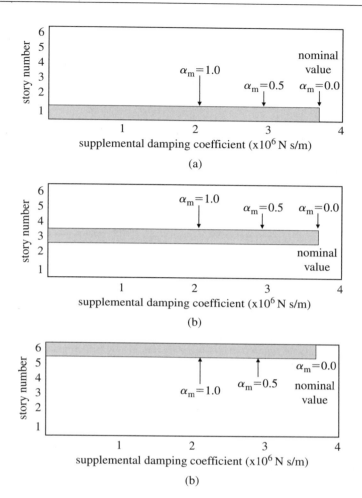

Figure 9.28 Supplemental viscous damping coefficient: (a) first-story allocation; (b) third-story alloca-tion; (c) sixth-story allocation. (© 2008 from *Structural Design Optimization Considering Uncertainties* by Y. Tsompanakis, N.D. Lagaros, and M. Papadrakakis (eds). Reproduced by permission of Taylor and Francis Group, LLC, a division of Informa plc.).

Figure 9.29 shows the energy transfer functions $F_M(\omega)$ defined by Equation 9.52 for three different models: one with a supplemental damper in the first story, one in the third story, and the other in the sixth story. It is observed that the energy transfer functions $F_M(\omega)$ of a passively controlled shear building model with a supplemental damper near the fixed support (base) are smaller than those of the model with a supplemental damper near its tip (top). This means that the allocation of passive dampers into lower stories is effective in reducing the earthquake input energy. It is also observed that the frequencies attaining the peak values are slightly different, but

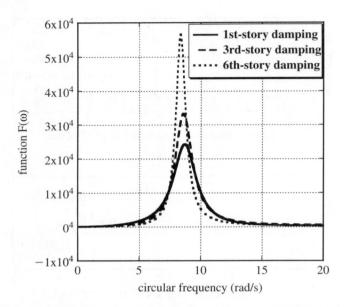

Figure 9.29 Energy transfer functions $F_M(\omega)$ defined by Equation 9.53a for three models: one with an added damper in the first story, one in the third story, and the other in the sixth story. (© 2008 from *Structural Design Optimization Considering Uncertainties* by Y. Tsompanakis, N.D. Lagaros, and M. Papadrakakis (eds). Reproduced by permission of Taylor and Francis Group, LLC, a division of Informa plc.).

the critical rectangular Fourier amplitude of the input acceleration may be almost the same.

Figure 9.30(a) plots the info-gap robustness function $\hat{\alpha}_m$ versus the specified limit value of the earthquake input energy for the null load spectral uncertainty $\alpha_s = 0.0$; that is, no load spectral variation. Figures 9.30(b)–(d) show the plots of $\hat{\alpha}_m$ for smaller load spectral uncertainties $\alpha_s = 0.1$, 0.3, and 0.5; by comparing these plots we can see that robustness to model uncertainty $\hat{\alpha}_m$ decreases as load uncertainty α_s increases. This is because, when the load uncertainty increases, the structural model easily violates the performance requirement for that larger external disturbance. It is also observed that a passively controlled shear building model with a supplemental damper near the fixed support is "more robust" than that with a supplemental damper near its tip (top) in all the cases of the load spectral uncertainties.

Figure 9.31(a) shows the plot of the info-gap robustness function $\hat{\alpha}_m$, of the model with a supplemental damper in the first story, versus the specified limit value of the earthquake input energy for various levels of load uncertainties. It can be understood that, as the level of load uncertainty increases, the info-gap robustness function $\hat{\alpha}_m$ gets smaller; that is, less robust for variation of the structural parameter. Figures 9.31(b) and (c) show the plots of info-gap robustness function $\hat{\alpha}_m$ of the models with a

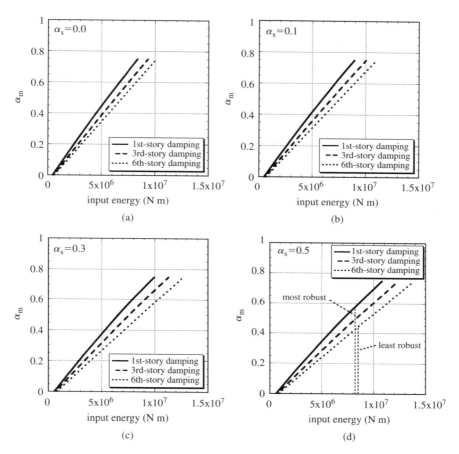

Figure 9.30 Plot of the info-gap robustness function $\hat{\alpha}_m$ versus the specified limit value of the earthquake input energy for various load spectral uncertainties: (a) $\alpha_s = 0.0$; (b) $\alpha_s = 0.1$; (c) $\alpha_s = 0.3$; (d) $\alpha_s = 0.5$. (© 2008 from *Structural Design Optimization Considering Uncertainties* by Y. Tsompanakis, N.D. Lagaros, and M. Papadrakakis (eds). Reproduced by permission of Taylor and Francis Group, LLC, a division of Informa plc.).

supplemental damper in the third and sixth stories respectively, with respect to the specified value of the earthquake input energy for various levels of load uncertainties. A tendency similar to Figure 9.31(a) is observed.

Figure 9.32 is a plot of the info-gap robustness function $\hat{\alpha}_m$ with respect to the level of the load spectral uncertainty α_s for the model with a supplemental damper in the first story. From this figure, the designer can understand the effect of the load spectral uncertainty α_s on the info-gap robustness function. It is also interesting to note that the info-gap robustness function $\hat{\alpha}_m$ and the level of the load spectral uncertainty α_s introduce a new trade-off relationship.

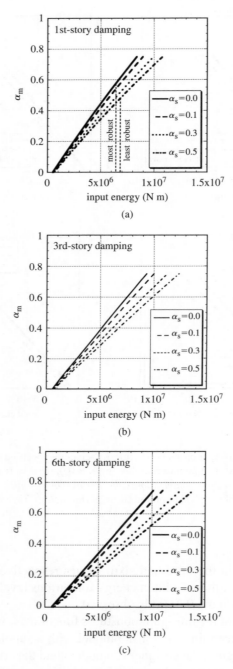

Figure 9.31 Plot of the info-gap robustness function $\hat{\alpha}_m$ versus the specified limit value of the earthquake input energy for various levels of load uncertainties: (a) model with a supplemental damper in the first story; (b) model with a supplemental damper in the third story; (c) model with a supplemental damper in the sixth story. (© 2008 from *Structural Design Optimization Considering Uncertainties* by Y. Tsompanakis, N.D. Lagaros, and M. Papadrakakis (eds). Reproduced by permission of Taylor and Francis Group, LLC, a division of Informa plc.).

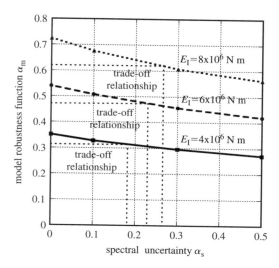

Figure 9.32 Plot of the info-gap robustness function $\hat{\alpha}_m$ with respect to the level of the load spectral uncertainty α_s for various requirements of earthquake input energies $E_I = 4.0 \times 10^6$, 6.0×10^6, 8.0×10^6 N m (first-story damping model). (© 2008 from *Structural Design Optimization Considering Uncertainties* by Y. Tsompanakis, N.D. Lagaros, and M. Papadrakakis (eds). Reproduced by permission of Taylor and Francis Group, LLC, a division of Informa plc.).

9.10 Summary

The results in this chapter may be summarized as follows.

1. A probabilistic critical excitation method is developed by using a stochastic response index as the objective function to be maximized. The energy (area of PSD function) and the intensity (maximum value of PSD function) of the excitations are fixed in the method.
2. Without the restriction on the excitation intensity, the PSD function of the critical excitation is reduced to the Dirac delta function at the frequency with the maximum transfer function. When the restriction on the excitation intensity exists, the critical excitation has a PSD function with the maximum intensity limit in a finite interval (band-limited white noise) for which a function including the squares of the transfer functions is maximized.
3. The explained probabilistic critical excitation method is applicable to MDOF structural systems with nonproportional damping.
4. While the degree of conservativeness of the explained critical excitation is about 2 or 3 for recorded ground motions without a remarkable predominant frequency, it is close to unity for ground motions with a remarkable predominant frequency. The resonant characteristic of such ground motion can be well represented by the critical excitation obtained.

5. Optimal damper placement under uncertain input acceleration is a challenging problem. With the aid of the characteristic that the critical input is insensitive to the damper placement of viscous dampers, the steepest direction search algorithm explained here can be utilized for optimal damper placement in building structures subjected to the critical excitations.

6. Since a transfer function can be obtained in closed form for MDOF systems with nonproportional damping due to the tri-diagonal property of the coefficient matrix, the mean-square responses of the building structure to the random earthquake input and their derivatives with respect to the design variables (damping coefficients of added dampers) can be computed very efficiently.

7. The optimal damper placement obtained from the present formulation can actually reduce the objective function effectively compared with uniform damper placement.

8. The earthquake input energy is an appropriate measure for evaluating the performance level of passively controlled structures. A critical excitation problem can be stated for the earthquake input energy as a criticality measure. The critical excitations depend upon the dynamic properties of the passively controlled mass–spring–damper systems and it is necessary to deal with load and structural model uncertainties simultaneously.

9. Info-gap uncertainty models are very useful in describing both the load and structural model uncertainties. Determination of the critical states in the load and structural parameters is an essential step to the investigation of the robustness of the passively controlled mass–spring–damper systems. In cases where the critical states in the load and structural parameters are not found easily, a sophisticated method for finding those is necessary.

10. A passively controlled mass–spring–damper system with a supplemental damper near the fixed support is more robust than that with an added damper near its tip (top). The increased robustness can be evaluated quantitatively.

11. The simultaneous consideration of the load and structural model uncertainties introduces a new class of trade-off. The robustness to structural model uncertainty increases as the uncertainty level of the load gets smaller.

Appendix 9.A: System Mass, Damping, and Stiffness Matrices for a Two-story Shear Building Model

The system mass, damping, and stiffness matrices for a two-story shear building model may be expressed by

$$\mathbf{M} = \text{diag}(m_1 \quad m_2) \qquad (A9.1)$$

$$\mathbf{C} = \begin{bmatrix} C_1 + C_2 & -C_2 \\ -C_2 & C_2 \end{bmatrix} \qquad (A9.2)$$

$$\mathbf{K} = \begin{bmatrix} k_1 + k_2 & -k_2 \\ -k_2 & k_2 \end{bmatrix} \tag{A9.3}$$

where $C_i = c_i + c_{Vi}$; c_i is an original frame damping coefficient and c_{Vi} is an added damper damping coefficient.

Appendix 9.B: Closed-form Expression of the Inverse of a Tri-diagonal Matrix

Consider the following symmetric tri-diagonal matrix of $M \times M$:

$$\mathbf{A} = \begin{bmatrix} d_M & -e_M & & & & 0 \\ -e_M & \ddots & \ddots & & & \\ & \ddots & \ddots & \ddots & & \\ & & \ddots & \ddots & -e_3 & \\ & & & -e_3 & d_2 & -e_2 \\ 0 & & & & -e_2 & d_1 \end{bmatrix} \tag{B9.1}$$

Let us define the following principal minors:

$$P_0 = 1, \quad P_1 = d_1, \quad P_2 = \begin{vmatrix} d_2 & -e_2 \\ -e_2 & d_1 \end{vmatrix}, \quad \ldots, \quad P_M = \det \mathbf{A} \tag{B9.2}$$

$$P_{R0} = 1, \quad P_{R1} = d_M, \quad P_{R2} = \begin{vmatrix} d_M & -e_M \\ -e_M & d_{M-1} \end{vmatrix}, \quad \ldots, \quad P_{RM} = \det \mathbf{A} \tag{B9.3}$$

The principal minors satisfy the following recurrence formula:

$$P_{j-1} = d_{j-1} P_{j-2} - e_{j-1}^2 P_{j-3} \quad (j = 3, \ldots, M) \tag{B9.4}$$

The jth column of $\mathbf{A}^{-1}$ may be expressed as

$$\frac{1}{\det \mathbf{A}} \left\{ \left(\prod_{i=M-j+2}^{M} e_i \right) P_{M-j} P_{R0} \quad \left(\prod_{i=M-j+2}^{M-1} e_i \right) P_{M-j} P_{R1} \quad \cdots \right.$$

$$\left(\prod_{i=M-j+2}^{M-j+2} e_i \right) P_{M-j} P_{R(j-2)} \quad P_{M-j} P_{R(j-1)} \quad \left(\prod_{i=M-j+1}^{M-j+1} e_i \right) P_{M-j-1} P_{R(j-1)}$$

$$\left. \cdots \left(\prod_{i=3}^{M-j+1} e_i \right) P_1 P_{R(j-1)} \quad \left(\prod_{i=2}^{M-j+1} e_i \right) P_0 P_{R(j-1)} \right\}^{\mathrm{T}} \tag{B9.5}$$

References

Ahmadi, G. (1979) On the application of the critical excitation method to aseismic design. *Journal of Structural Mechanics*, **7**, 55–63.

Akiyama, H. (1985) *Earthquake Resistant Limit-State Design for Buildings*. University of Tokyo Press, Tokyo, Japan.

Arias, A. (1970) A measure of earthquake intensity, in *Seismic Design for Nuclear Power Plants* (ed. R.J. Hansen), MIT Press, Cambridge, MA, 438–469.

Baratta, A., Elishakoff, I., Zuccaro, G., and Shinozuka, M. (1998) A generalization of the Drenick–Shinozuka model for bounds on the seismic response of a single-degree-of-freedom system. *Earthquake Engineering and Structural Dynamics*, **27**, 423–437.

Ben-Haim, Y. (1996) *Robust Reliability in the Mechanical Sciences*, Springer-Verlag, Berlin.

Ben-Haim, Y. (2001). *Info-gap Decision Theory: Decisions under Severe Uncertainty*, Academic Press, London.

Ben-Haim, Y. (2005) Info-gap decision theory for engineering design, in *Engineering Design Reliability Handbook* (eds E. Nikolaide, D. Ghiocel, and S. Singhal), CRC Press.

Ben-Haim, Y. (2006) *Info-gap Decision Theory: Decisions Under Severe Uncertainty*, 2nd edn, Academic Press, London.

Ben-Haim, Y. and Elishakoff, I. (1990) *Convex Models of Uncertainty in Applied Mechanics*, Elsevier, Amsterdam.

Ben-Haim, Y., Chen, G., and Soong, T.T. (1996) Maximum structural response using convex models. *Journal of Engineering Mechanics*, **122**(4), 325–333.

Cherng, R.H. and Wen, Y.K. (1994a) Reliability of uncertain nonlinear trusses under random excitation. I. *Journal of Engineering Mechanics*, **120** (4), 733–747.

Cherng, R.H. and Wen, Y.K. (1994b) Reliability of uncertain nonlinear trusses under random excitation. II. *Journal of Engineering Mechanics*, **120**(4), 748–757.

Drenick, R.F. (1970) Model-free design of aseismic structures. *Journal of the Engineering Mechanics Division*, **96** (EM4), 483–493.

Drenick, R.F. (1973) Aseismic design by way of critical excitation. *Journal of the Engineering Mechanics Division*, **99** (EM4), 649–667.

Drenick, R.F. (1977) The critical excitation of nonlinear systems. *Journal of Applied Mechanics*, **44** (E2), 333–336.

Ghanem, R.G. and Spanos, P.D. (1991) *Stochastic Finite Elements: A Spectral Approach*, Springer-Verlag, Berlin.

Housner, G.W. (1959) Behavior of structures during earthquakes. *Journal of the Engineering Mechanics Division*, **85** (4), 109–129.

Housner, G.W. and Jennings, P.C. (1975) The capacity of extreme earthquake motions to damage structures, in *Structural and Geotechnical Mechanics: A Volume Honoring N.M. Newmark* (ed. W.J. Hall), Prentice-Hall, Englewood Cliff, NJ, pp. 102–116.

Hudson, D.E. (1962) Some problems in the application of spectrum techniques to strong-motion earthquake analysis. *Bulletin of the Seismological Society of America*, **52**, 417–430.

Igusa, T. and Der Kiureghian, A. (1988) Response of uncertain systems to stochastic excitations. *Journal of Engineering Mechanics*, **114**(5), 812–832.

Iyengar, R.N. (1972). Worst inputs and a bound on the highest peak statistics of a class of non-linear systems. *Journal of Sound & Vibration*, **25**, 29–37.

Iyengar, R.N. and Manohar, C.S. (1987) Nonstationary random critical seismic excitations. *Journal of Engineering Mechanics*, **113**(4), 529–541.

Jensen, H. (2000) On the structural synthesis of uncertain systems subjected to environmental loads. *Structural and Multidisciplinary Optimization*, **20**, 37–48.

Jensen, H. and Iwan, W.D. (1992) Response of systems with uncertain parameters to stochastic excitations. *Journal of Engineering Mechanics*, **114**, 1012–1025.

Kanno, Y. and Takewaki, I. (2007) Worst-case plastic limit analysis of trusses under uncertain loads via mixed 0–1 programming. *Journal of Mechanics of Materials and Structures*, **2**(2), 247–273.

Katafygiotis, L.S. and Papadimitriou, C. (1996) Dynamic response variability of structures with uncertain properties. *Earthquake Engineering and Structural Dynamics*, **25**, 775–793.

Koyluoglu, H.U., Cakmak, A.S., and Nielsen, S.R.K. (1995) Interval algebra to deal with pattern loading and structural uncertainties. *Journal of Engineering Mechanics*, **121**(11), 1149–1157.

Lyon, R.H. (1975) *Statistical Energy Analysis of Dynamical Systems*, MIT Press, Cambridge, MA.

Manohar, C.S. and Sarkar, A. (1995) Critical earthquake input power spectral density function models for engineering structures. *Earthquake Engineering and Structural Dynamics*, **24**, 1549–1566.

Ordaz, M., Huerta, B., and Reinoso, E. (2003) Exact computation of input-energy spectra from Fourier amplitude spectra. *Earthquake Engineering and Structural Dynamics*, **32**, 597–605.

Page, C.H. (1952) Instantaneous power spectra. *Journal of Applied Physics*, **23**(1), 103–106.

Pantelides, C.P. and Tzan, S.R. (1996) Convex model for seismic design of structures: I analysis. *Earthquake Engineering and Structural Dynamics*, **25**, 927–944.

Pirasteh, A.A., Cherry, J.L., and Balling, R.J. (1988) The use of optimization to construct critical accelerograms for given structures and sites. *Earthquake Engineering and Structural Dynamics*, **16**, 597–613.

Qiu, Z. and Wang, X. (2003) Comparison of dynamic response of structures with uncertain-but-bounded parameters using non-probabilistic interval analysis method and probabilistic approach. *International Journal of Solids and Structures*, **40**, 5423–5439.

Sarkar, A. and Manohar, C.S. (1998) Critical seismic vector random excitations for multiply supported structures. *Journal of Sound & Vibration*, **212**(3), 525–546.

Schueller, G.I. (ed.) (2008) Computational methods in optimization considering uncertainties (special issue). *Computer Methods in Applied Mechanics and Engineering*, **198**(1).

Shinozuka, M. (1970) Maximum structural response to seismic excitations. *Journal of the Engineering Mechanics Division*, **96**(EM5), 729–738.

Srinivasan, M., Corotis, R., and Ellingwood, B. (1992) Generation of critical stochastic earthquakes. *Earthquake Engineering and Structural Dynamics*, **21**, 275–288.

Srinivasan, M., Ellingwood, B., and Corotis, R. (1991) Critical base excitations of structural systems, *Journal of Engineering Mechanics*, **117**(6), 1403–1422.

Takewaki, I. (1998) Optimal damper positioning in beams for minimum dynamic compliance. *Computer Methods in Applied Mechanics and Engineering*, **156**(1–4), 363–373.

Takewaki, I. (2001a) A new method for nonstationary random critical excitation. *Earthquake Engineering and Structural Dynamics*, **30**(4), 519–535.

Takewaki, I. (2001b) Probabilistic critical excitation for MDOF elastic–plastic structures on compliant ground. *Earthquake Engineering and Structural Dynamics*, **30**(9), 1345–1360.

Takewaki, I. (2002a) Critical excitation method for robust design: a review. *Journal of Structural Engineering*, **128**(5), 665–672.

Takewaki, I. (2002b) Robust building stiffness design for variable critical excitations. *Journal of Structural Engineering*, **128**(12) 1565–1574.

Takewaki, I. (2004) Bound of earthquake input energy. *Journal of Structural Engineering*, **130**(9), 1289–1297.

Takewaki, I. (2006) *Critical Excitation Methods in Earthquake Engineering*, Elsevier, Amsterdam.

Takewaki, I. and Ben-Haim, Y. (2005) Info-gap robust design with load and model uncertainties. *Journal of Sound & Vibration*, **288**(3), 551–570.

Takewaki, I. and Yoshitomi, S. (1998) Effects of support stiffnesses on optimal damper placement for a planar building frame. *The Structural Design of Tall Buildings*, **7**(4), 323–336.

Tsompanakis, Y., Lagaros, N.D., and Papadrakakis, M. (eds) (2008) *Structural Design Optimization Considering Uncertainties*, Taylor & Francis.

Tsuji, M. and Nakamura, T. (1996) Optimum viscous dampers for stiffness design of shear buildings. *The Structural Design of Tall Buildings*, **5**, 217–234.

Uang, C.M. and Bertero, V.V. (1990) Evaluation of seismic energy in structures. *Earthquake Engineering and Structural Dynamics*, **19**, 77–90.

Wang, P.C., Drenick, R.F., and Wang, W.Y.L. (1978) Seismic assessment of high-rise buildings. *Journal of the Engineering Mechanics Division*, **104**(EM2), 441–456.

Wang, P.C. and Yun, C.B. (1979) Site-dependent critical design spectra. *Earthquake Engineering and Structural Dynamics*, **7**, 569–578.

Westermo, B.D. (1985) The critical excitation and response of simple dynamic systems. *Journal of Sound & Vibration*, **100**(2), 233–242.

Yang, J.-N. and Heer, E. (1971) Maximum dynamic response and proof testing. *Journal of the Engineering Mechanics Division*, **97**(EM4), 1307–1313.

Zhang, R.H. and Soong, T.T. (1992) Seismic design of viscoelastic dampers for structural applications. *Journal of Structural Engineering*, **118**(5), 1375–1392.

10

Theoretical Background of Effectiveness of Passive Control System

10.1 Introduction

In the field of passive structural control, it is generally believed that installation of passive dampers upgrades the performance of building structures under earthquake ground motions (or wind disturbances) with much uncertainty (Arias, 1970; Drenick, 1970; Housner 1975; Housner and Jennings, 1975; Abrahamson et al., 1998; Hall et al., 1995; Heaton et al., 1995). However, it seems that there have never been any definite theoretical bases presented for this. The increase of damping with passive control devices may be an intuitive basis, but more theoretical one would be desirable. For this purpose, the earthquake input energy to building structures is treated here as an index to measure the structural performance.

Many fruitful results have been accumulated on the topic of earthquake input energy to structures (e.g., Tanabashi, 1956; Housner, 1959; Zahrah and Hall, 1984; Akiyama, 1985; Uang and Bertero, 1990; Leger and Dussault, 1992; Kuwamura et al., 1994; Riddell and Garcia, 2001; Ordaz et al., 2003; Takewaki, 2004a, 2004b, 2005a, 2005b, 2006, 2007a, 2007b). In some countries the earthquake input energy has been incorporated as an earthquake input demand. The earthquake input energy has usually been computed in the time domain. The time-domain approach has several advantages; for example, the availability in nonlinear structures, the description of time-history response of the input energy, and the possibility of expressing the input energy rate especially for nonlinear structures. On the other hand, the time-domain approach is not necessarily appropriate for probabilistic and bound analysis under uncertainties (Takewaki, 2001a, 2001b, 2002a, 2002b, 2004a, 2005b, 2006; Takewaki and Ben-Haim, 2005, 2008). For that purpose, the frequency-domain approach (Lyon,

Building Control with Passive Dampers   Izuru Takewaki
© 2009 John Wiley & Sons (Asia) Pte Ltd

1975; Ordaz *et al.*, 2003; Takewaki, 2004a, 2004b, 2005a, 2005b, 2006, 2007a, 2007b) is suitable because it uses the Fourier amplitude spectrum of input ground accelerations and the time-invariant energy transfer functions of the structure.

This chapter is aimed at demonstrating that (i) the time-domain and frequency-domain methods have different advantages and can support each other, (ii) the equi-area property of the energy transfer function in general structural models can be derived by the time-domain method for an idealized model of input motions with a constant Fourier amplitude spectrum, and (iii) the dissipated energy by passive control systems can certainly upgrade the performance of building structures under earthquake ground motions.

The remarkable equi-area property of the energy transfer functions supports the property of a nearly constant input energy. This leads to an advantageous feature that, if the energy consumption in the viscous dampers increases, the input energies to buildings can be reduced drastically.

However, it should also be kept in mind that an increase of dampers in the base-isolation story in base-isolated structures does not necessarily lead to favorable situations. Excessive installation of dampers in the base-isolation story may cause a larger earthquake input to the superstructures. This should be discussed in the future.

10.2 Earthquake Input Energy to SDOF model

First of all, the earthquake input energy to a damped linear SDOF system is formulated both in the time and frequency domains.

The SDOF system has mass m, stiffness k, and viscous damping coefficient c. Let $\Omega = \sqrt{k/m}$, $h = c/(2\Omega m)$, and x respectively denote the undamped natural circular frequency, the damping ratio, and the displacement of the mass relative to the ground. The earthquake input energy to this SDOF system by a ground acceleration $\ddot{u}_g(t)$ from $t = 0$ to $t = t_0$ (end of input) can be defined as the work of the ground on the structural system (see Figure 10.1) and is expressed by

$$E_I = \int_0^{t_0} m(\ddot{u}_g + \ddot{x})\dot{u}_g \, dt \tag{10.1}$$

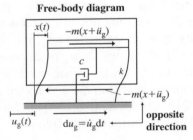

Figure 10.1 Work by ground on structural system. (I. Takewaki, K. Fujita, "Earthquake Input Energy to Tall and Base-isolated Buildings in Time and Frequency Dual Domains," *Journal of the Structural Design of Tall and Special Buildings* © 2009 John Wiley & Sons, Ltd).

Integration by parts of Equation 10.1 provides

$$E_I = \int_0^{t_0} m(\ddot{x} + \ddot{u}_g)\dot{u}_g \, dt = \int_0^{t_0} m\ddot{x}\dot{u}_g \, dt + \left[\frac{1}{2}m\dot{u}_g^2\right]_0^{t_0}$$

$$= \left[m\dot{x}\dot{u}_g\right]_0^{t_0} - \int_0^{t_0} m\dot{x}\ddot{u}_g \, dt + \left[\frac{1}{2}m\dot{u}_g^2\right]_0^{t_0} \tag{10.2}$$

If $\dot{x} = 0$ at $t = 0$ and $\dot{u}_g = 0$ at $t = 0$ and t_0, then the earthquake input energy can be reduced to the following compact expression:

$$E_I = -\int_0^{t_0} m\ddot{u}_g\dot{x} \, dt \tag{10.3}$$

The earthquake input energy per unit mass can also be expressed in the frequency domain (Lyon, 1975; Ordaz et al., 2003; Takewaki, 2004a, 2004b, 2005a, 2005b, 2006, 2007a, 2007b) by applying the Fourier transformation to $\dot{x}$ and $\ddot{u}_g$ in Equation 10.3. The earthquake input energy per unit mass in the frequency domain may be expressed by

$$\frac{E_I}{m} = \int_0^{\infty} |\ddot{U}_g(\omega)|^2 F(\omega) \, d\omega \tag{10.4}$$

where $F(\omega) = -\text{Re}[H_V(\omega; \Omega, h)]/\pi$ is called the energy transfer function and $H_V(\omega; \Omega, h)$ is the transfer function defined by $\dot{X}(\omega) = H_V(\omega; \Omega, h)\ddot{U}_g(\omega)$. $\dot{X}$ and $\ddot{U}_g(\omega)$ are the Fourier transforms of $\dot{x}$ and $\ddot{u}_g(t)$ respectively. The symbol i denotes the imaginary unit. $H_V(\omega; \Omega, h)$ can be expressed explicitly by

$$H_V(\omega; \Omega, h) = -\frac{i\omega}{\Omega^2 - \omega^2 + 2ih\Omega\omega} \tag{10.5}$$

Equation 10.4 indicates that the earthquake input energy to damped linear elastic SDOF systems does not depend on the phase of input motions (Lyon, 1975; Kuwamura et al., 1994; Ordaz et al., 2003; Takewaki, 2004a, 2004b, 2005a, 2005b, 2006, 2007a, 2007b). It can also be understood from Equation 10.4 that the energy transfer function $F(\omega)$ plays an important role in the evaluation of the earthquake input energy and may have some influence on the investigation of constancy of the earthquake input energy in structures with various model parameters.

Assume that the Fourier amplitude spectrum $|\ddot{U}_g(\omega)|$ of the ground acceleration is constant ($= \bar{A}$). Then, Equation 10.4 can be reduced to the following form:

$$\frac{E_I}{m} = \bar{A}^2 \int_0^{\infty} F(\omega) \, d\omega \tag{10.6}$$

With the help of the residue theorem, the relation $\int_0^\infty F(\omega)\,d\omega = 1/2$ holds regardless of Ω and h (Ordaz *et al.*, 2003; Takewaki, 2004a). Figure 10.2 shows the plot of the energy transfer function $F(\omega)$ for various fundamental natural periods and damping ratios of the SDOF system. Equation 10.6 is then reduced to

$$\frac{E_I}{m} = \frac{1}{2}\overline{A}^2 \tag{10.7}$$

Equation 10.7 implies that the constancy criterion of the earthquake input energy to the SDOF system is related directly to the Fourier amplitude spectrum. This is not coincident with the theory due to Housner (1959) and Akiyama (1985) supporting the role of the velocity response spectrum. However, it should be noted that the Fourier amplitude spectrum and the velocity response spectrum have an approximate relation (Hudson, 1962) and further discussion would be necessary.

10.3 Constant Earthquake Input Energy Criterion in Time Domain

In the case where the Fourier amplitude spectrum of a ground acceleration is ideally constant, the discussion on constancy of earthquake input energy can be made in the time domain. The essence of this procedure is shown in this section.

Consider a ground acceleration as the Dirac delta function:

$$\ddot{u}_g(t) = \overline{A}\delta(t) \tag{10.8}$$

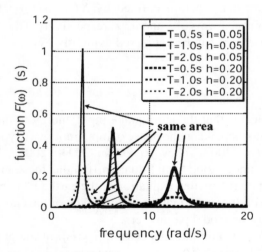

Figure 10.2 Energy transfer functions for various fundamental natural periods and damping ratios. With permission from ASCE. For oBook – With permission from ASCE. This material may be downloaded for personal use only. Any other use requires prior permission of the American Society of Civil Engineers. This material may be found at http://cedb.asce.org/cgi/WWWdisplay.cgi?0411843.

where $\delta(t)$ is the Dirac delta function and the following relation holds:

$$\int_{-\infty}^{\infty} \ddot{u}_g(t)e^{-i\omega t}\,dt = \int_{-\infty}^{\infty} \overline{A}\delta(t)e^{-i\omega t}\,dt = \overline{A} \tag{10.9}$$

Assume that the mass is at rest at first. The velocity change (initial velocity) of the mass may be evaluated by the impulse divided by the mass:

$$\frac{1}{m}\int_{-\infty}^{\infty}\{-m\ddot{u}_g(t)\}\,dt = \int_{-\infty}^{\infty}\{-\overline{A}\delta(t)\}\,dt = -\overline{A} \tag{10.10}$$

Because the constant Fourier amplitude spectrum of a ground acceleration corresponds to the Dirac delta function in the time domain, the initial input energy may be given by

$$E_{\mathrm{I}} = \frac{1}{2}m\overline{A}^2 \tag{10.11}$$

This energy will be dissipated later by the viscous damping system.

10.4 Constant Earthquake Input Energy Criterion to MDOF Model in Frequency Domain

Consider a proportionally damped MDOF structure of the mass matrix $[M]$. Let $\{x\}$ denote the horizontal nodal displacements of masses relative to the ground and let $\{1\}$ denote the influence coefficient vector. The input energy to this MDOF structure may be described as

$$\begin{aligned}
E_{\mathrm{I}} &= -\int_{-\infty}^{\infty}\{\dot{x}\}^{\mathrm{T}}[M]\{1\}\ddot{u}_g\,dt \\
&= -\int_{-\infty}^{\infty}\left[\frac{1}{2\pi}\int_{-\infty}^{\infty}\{\dot{X}\}^{\mathrm{T}}e^{i\omega t}\,d\omega\right][M]\{1\}\ddot{u}_g\,dt \\
&= \int_0^{\infty}|\ddot{U}_g(\omega)|^2 F_{\mathrm{MP}}(\omega)\,d\omega
\end{aligned} \tag{10.12}$$

In Equation 10.12, $F_{\mathrm{MP}}(\omega)$ may be defined by

$$F_{\mathrm{MP}}(\omega) = -\frac{1}{\pi}\{\mathrm{Re}[H_{\mathrm{V}}(\omega;\Omega_i,h_i)]\}^{\mathrm{T}}[\Phi]^{\mathrm{T}}[M]\{1\} \tag{10.13}$$

where $\{H_{\mathrm{V}}(\omega;\Omega_i,h_i)\}$, Ω_i, h_i, and $[\Phi]$ denote the velocity transfer function, the ith undamped natural circular frequency, the ith damping ratio, and the modal matrix. By use of $\{H_{\mathrm{V}}(\omega;\Omega_i,h_i)\}$, the Fourier transform of the nodal velocity vector may be expressed as

$$\{\dot{X}(\omega)\} = [\Phi]\{H_{\mathrm{V}}(\omega;\Omega_i,h_i)\}\ddot{U}_g(\omega) \tag{10.14}$$

If $|\ddot{U}_g(\omega)|$ is constant ($=\bar{A}$) with respect to frequency, then the input energy may be expressed as

$$E_I = \bar{A}^2 \int_0^{\infty} F_{MP}(\omega)\,d\omega \qquad (10.15)$$

Substitution of Equation 10.13 into Equation 10.15 leads to

$$E_I = -\frac{\bar{A}^2}{\pi}\left\{\int_0^{\infty} \mathrm{Re}[H_V(\omega;\Omega_i,h_i)]\,d\omega\right\}^T [\Phi]^T[M]\{1\} \qquad (10.16)$$

With the help of the residue theorem in each mode, the input energy to the proportionally damped MDOF structure may result in

$$E_I = \frac{1}{2}\bar{A}^2\{1\}^T[\Phi]^T[M]\{1\} = \frac{1}{2}\bar{A}^2\sum_{j=1}^{N} m_j \qquad (10.17)$$

N denotes the number of masses and m_j is the mass corresponding to the jth horizontal nodal displacement. Equation 10.17 implies that, if the Fourier amplitude is constant with respect to frequency, the input energy to the proportionally damped MDOF structure depends only on the total mass of the model.

The relation of Equation 10.17 can also be derived by the idea of Equations 10.8–10.11 in the time domain because the initial velocity $-\bar{A}$ is given simultaneously at all the masses by an ideal input with a constant Fourier amplitude spectrum (see Figure 10.3).

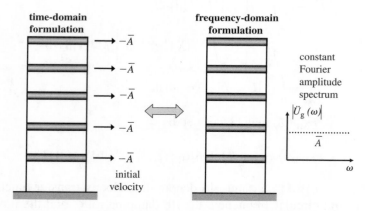

Figure 10.3 Correspondence of time- and frequency-domain dual formulations. (I. Takewaki, K. Fujita, "Earthquake Input Energy to Tall and Base-isolated Buildings in Time and Frequency Dual Domains," *Journal of the Structural Design of Tall and Special Buildings* © 2009 John Wiley & Sons, Ltd).

10.5 Earthquake Input Energy as Sum of Input Energies to Subassemblages

As a representative model of structures including passive dampers, consider the connected N-story MDOF building structure shown in Figure 10.4. Let $\{m_i^{(1)}\}$ and $\{m_i^{(2)}\}$ denote the masses of building FL (flexible) and building ST (stiff) respectively. The story stiffnesses and viscous damping coefficients of building FL are denoted by $\{k_i^{(1)}\}$ and $\{c_i^{(1)}\}$ and those of building ST are denoted by $\{k_i^{(2)}\}$ and $\{c_i^{(2)}\}$. Let $\{c_i^{(3)}\}$ denote the damping coefficients of the connecting viscous dampers. These connected building structures are subjected to a horizontal ground motion $\ddot{u}_g(t)$. The seismic horizontal displacements of the masses in building FL under $\ddot{u}_g(t)$ are denoted by $\{u_i^{(1)}\}$ and those of building ST are denoted by $\{u_i^{(2)}\}$.

The equations of motion of this interconnected building system subjected to the ground acceleration $\ddot{u}_g(t)$ may be expressed by

$$\mathbf{M}\ddot{\mathbf{u}} + (\mathbf{C_F} + \mathbf{C_D})\dot{\mathbf{u}} + \mathbf{K}\mathbf{u} = -\mathbf{M}\mathbf{1}\ddot{u}_g \tag{10.18}$$

where $\mathbf{M}$, $\mathbf{K}$, $\mathbf{C_F}$, and $\mathbf{C_D}$ are the mass, the stiffness, and the damping matrices for the frame and the damping matrix for the connecting viscous dampers. The vector $\mathbf{1}$ indicates $\mathbf{1} = \{1 \cdots 1\}^T$ and $\mathbf{u} = \{u_1^{(1)} \cdots u_N^{(1)} u_1^{(2)} \cdots u_N^{(2)}\}^T$ is the set of displacements including those for both buildings.

Consider the three subassemblages shown in Figure 10.5. The work done by the boundary forces around a subassemblage on the corresponding displacements can be regarded as the input energy to the subassemblage. The earthquake input energy to the overall system in the time domain may be obtained by summing up all the energies; that is, E_I^1 in building FL, E_I^2 in building ST, and E_I^3 in connecting viscous

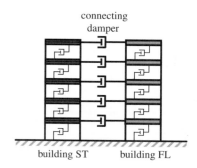

Figure 10.4 Two buildings connected by viscous dampers (ST, stiff; FL, flexible).

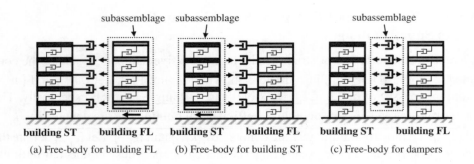

Figure 10.5 Free-body diagram for each structural subsystem.

dampers (see Figure 10.5):

$$E_I^A = E_I^1 + E_I^2 + E_I^3 = \int_0^\infty \left\{ \sum_{i=1}^N m_i^{(1)}(\ddot{u}_g + \ddot{u}_i^{(1)}) + \sum_{i=1}^N m_i^{(2)}(\ddot{u}_g + \ddot{u}_i^{(2)}) \right\} \dot{u}_g \, dt$$

(10.19)

On the other hand, the earthquake input energy to the overall system in the frequency domain may be obtained by summing up all the energies stated above. Actually, the earthquake input energy to the overall system in the frequency domain can be derived as follows by applying the Fourier transformation to $\ddot{u}_g(t)$ and the relative floor velocities:

$$E_I^A = E_I^1 + E_I^2 + E_I^3 = \int_0^\infty \{F_C^{(1)}(\omega) + F_C^{(2)}(\omega) + F_C^{(3)}(\omega)\} |\ddot{U}_g(\omega)|^2 \, d\omega \quad (10.20)$$

In Equation 10.20, $F_C^{(1)}(\omega)$, $F_C^{(2)}(\omega)$, and $F_C^{(3)}(\omega)$ are the energy transfer functions defined for building FL, building ST, and connecting viscous dampers respectively. These functions can be expressed as follows (Takewaki, 2007b):

$$F_C^{(1)}(\omega) = \frac{1}{\pi} \text{Re} \left[\left\{ \sum_{i=1}^N \omega^2 c_i^{(3)} (H_i^{(2)}(\omega) - H_i^{(1)}(\omega)) H_i^{(1)}(\omega)^* \right\} \right.$$
$$\left. + \frac{i}{\omega}(-k_1^{(1)} - i\omega c_1^{(1)}) H_1^{(1)}(\omega) \right]$$

(10.21a)

$$F_C^{(2)}(\omega) = \frac{1}{\pi} \text{Re} \left[-\left\{ \sum_{i=1}^N \omega^2 c_i^{(3)} (H_i^{(2)}(\omega) - H_i^{(1)}(\omega)) H_i^{(2)}(\omega)^* \right\} \right.$$
$$\left. + \frac{i}{\omega}(-k_1^{(2)} - i\omega c_1^{(2)}) H_1^{(2)}(\omega) \right]$$

(10.21b)

$$F_C^{(3)}(\omega) = \frac{1}{\pi} \text{Re} \left[\sum_{i=1}^{N} \omega^2 c_i^{(3)} |H_i^{(2)}(\omega) - H_i^{(1)}(\omega)|^2 \right] \quad (10.21c)$$

$\{H_1^{(1)} \cdots H_N^{(1)} \ H_1^{(2)} \cdots H_N^{(2)}\}$ in Equations 10.21a–10.21c are the displacement transfer functions defined by

$$\mathbf{H} = \{H_1^{(1)} \cdots H_N^{(1)} \ H_1^{(2)} \cdots H_N^{(2)}\}^T = \frac{\mathbf{U}}{\ddot{U}_g} = -[-\omega^2 \mathbf{M} + i\omega(\mathbf{C_F} + \mathbf{C_D}) + \mathbf{K}]^{-1} \mathbf{M1}$$

$$(10.22)$$

It can also be found that, even if the stiffness and damping properties vary in buildings and connecting viscous dampers, $F_C(\omega) = F_C^{(1)}(\omega) + F_C^{(2)}(\omega) + F_C^{(3)}(\omega)$ has an equi-area property under the assumption that the total mass of the buildings does not change. This can be proved by applying the idea in Section 10.3 to the present model.

It should be remarked that the equi-area property exists in general elastic structures when the total mass of the structures does not change. This guarantees the stable characteristic of an approximately constant earthquake input energy in wide-range stiffness and damping parameters.

Numerical examples are presented here. Consider five-story connected building models. As stated before, the right building is called "building FL" and the left building is called "building ST." Note again that building FL is the building with low stiffness and damping and building ST is the building with high stiffness and damping. The given parameters in the two buildings are as follows: $m_i^{(1)} = m_i^{(2)} = 32.0 \times 10^3 \, \text{kg} \ (i = 1, \cdots, 5)$, $k_i^{(1)} = 1.88 \times 10^7 \, \text{N/m}$, $c_i^{(1)} = 1.88 \times 10^5 \, \text{N s/m} \ (i = 1, \cdots, 4)$, $k_5^{(1)} = 3.76 \times 10^7 \, \text{N/m}$, $c_5^{(1)} = 3.76 \times 10^5 \, \text{N s/m}$, $k_i^{(2)} = 3.76 \times 10^7 \, \text{N/m}$, $c_i^{(2)} = 3.76 \times 10^5 \, \text{N s/m} \ (i = 1, \cdots, 5)$. The lowest mode damping ratio of building FL is 0.035 and that of building ST is 0.049. The fundamental natural period of building FL is 0.90 s and that of building ST is 0.64 s.

Two damping design cases are treated first. These two cases have different levels of supplemental viscous dampers (see Figure 10.6); that is, Model DL (distributed/low damping) and Model DH (distributed/high damping). While Model DL indicates the buildings connected by distributed viscous dampers with small damping, Model DH represents the buildings connected by distributed viscous dampers with large damping. The added dampers are equally located at all the floor levels. The damping coefficients of the supplemental viscous dampers in Model DL are $c_i^{(3)} = c_i^{(2)}/100 = 3.76 \times 10^3 \, \text{N s/m} (i = 1, \cdots, 5)$ and those in Model DH are $c_i^{(3)} = c_i^{(2)}/10 = 3.76 \times 10^4 \, \text{N s/m}$. This means that the damping coefficients of the supplemental viscous dampers are 0.01 times the structural damping of building ST in Model DL and 0.1 times the structural damping of building ST in Model DH.

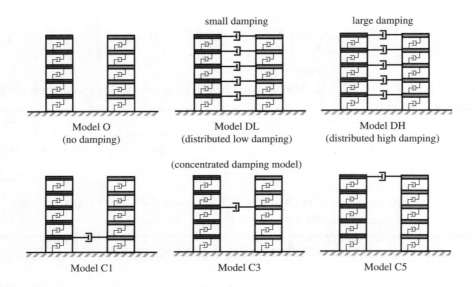

Figure 10.6 Various models (Model O, disconnected; Model DL, distributed/low damping; Model DH, distributed/high damping; Model C1, concentrated at first floor; Model C3, concentrated at third floor; Model C5, concentrated at fifth floor).

   The ground motion of El Centro NS (Imperial Valley 1940) is used here as the input acceleration. Figures 10.7(a)–(c) show the time histories of input energy in Model O (without any connecting viscous damper; see Figure 10.6), Model DL, and Model DH. In Figures 10.7(a)–(c), the time-domain method has been used. It is observed that a lot of energy is absorbed by the connecting viscous dampers in Model DH and the input energies to both buildings are reduced greatly, especially in Model DH. It is interesting to note that the total input energies are almost the same irrespective of difference in the models. In order to use the frequency-domain method in obtaining the time history of input energy, the Fourier transform of a truncated input ground motion is required (see Takewaki, 2005c).

   Figure 10.8 shows the time histories of input energy in Models C1, C3, and C5 (see Figure 10.6 for these models). It is observed that Model C5 is the most effective, as stated just before from the configuration of the energy transfer function. It is also interesting to note that, although the input energies to building FL, building ST, and added connecting viscous dampers are different model by model, the overall input energies in the final state are almost the same. This means that, if it is possible to dissipate a lot of energy in the connecting viscous dampers, the input energies to the buildings can be reduced drastically. The constancy of the area of $F_C^{(1)}(\omega) + F_C^{(2)}(\omega) + F_C^{(3)}(\omega)$ may be proved by introducing the virtual input with a constant Fourier amplitude spectrum and discussing that the energy input in this case is provided only at the initial moment by

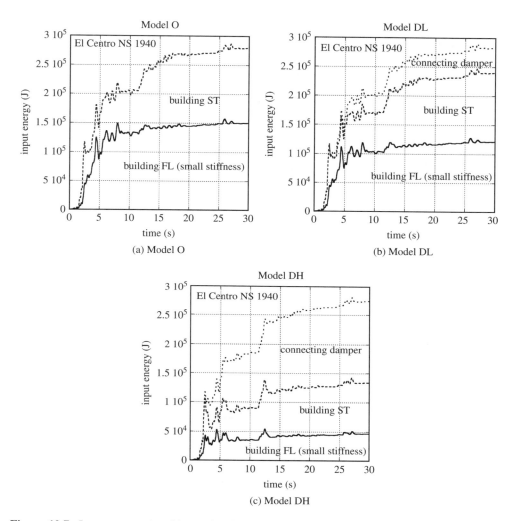

Figure 10.7 Input energy time-history for distributed damping models (El Centro NS 1940). With permission from ASCE. For oBook – With permission from ASCE. This material may be downloaded for personal use only. Any other use requires prior permission of the American Society of Civil Engineers. This material may be found at http://cedb.asce.org/cgi/WWWdisplay.cgi?0702403.

the initial impulse causing the initial velocity and the initial kinetic energy. This initial kinetic energy does not depend on the structural properties except for the total mass.

10.6 Effectiveness of Passive Dampers in Terms of Earthquake Input Energy

In Section 10.3, the constant input energy criterion has been proved in the time domain so long as the input acceleration can be expressed by a constant Fourier amplitude

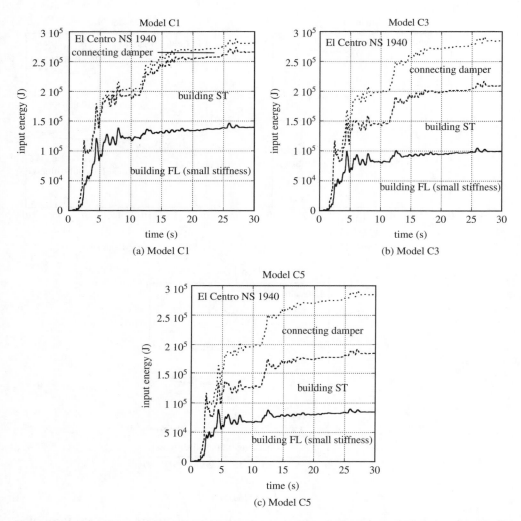

Figure 10.8 Input energy time-history for concentrated damping models (El Centro NS 1940). With permission from ASCE. For oBook – With permission from ASCE. This material may be downloaded for personal use only. Any other use requires prior permission of the American Society of Civil Engineers. This material may be found at http://cedb.asce.org/cgi/WWWdisplay.cgi?0702403.

spectrum. The input acceleration with a constant Fourier amplitude spectrum is equivalent to the Dirac delta function in the time domain. Moreover, in Section 10.4, the constant input energy criterion has been verified for proportionally damped MDOF systems in the frequency domain. With the aid of the proof in the time domain, it can be shown that this proof can be extended to nonproportionally damped MDOF systems (Takewaki and Fujita, 2008). It should be noted that, because actual recorded

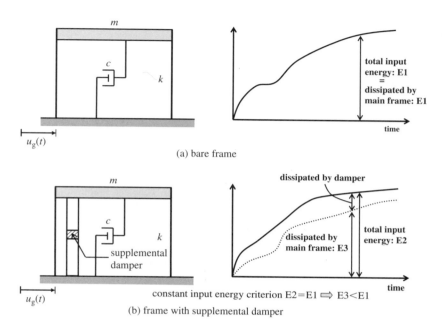

Figure 10.9 Schematic explanation of effectiveness of supplemental dampers.

ground motions do not have the property of a constant Fourier amplitude spectrum, this criterion is approximate. However, the backbone constructed for an idealized ground motion reveals a specific property on the energy transfer function which plays a central role in the formulation of earthquake input energy in the frequency domain.

Figure 10.9 shows a schematic explanation of the effectiveness of supplemental dampers. More specifically, if the supplemental passive dampers can absorb the earthquake input energy as much as possible, then the input energy to the frame can be reduced drastically. Without the constant input energy criterion, the proof of the effectiveness of passive dampers is not straightforward.

10.7 Advantageous Feature of Frequency-domain Method

As shown in Sections 10.2, 10.4, and 10.5, the earthquake input energy is usually formulated in the time domain. The time-domain method has an advantage that the idea explained in Equations 10.8–10.11 can be applied to any elastic structures. It is also interesting to note that this idea can be applied even to inelastic structures. On the other hand, the frequency-domain method has another advantage.

Consider again an SDOF model. Assume that the Fourier amplitude spectrum of the ground acceleration is bounded by $A^{\mathrm{L}} \leq |\ddot{U}_{\mathrm{g}}(\omega)| \leq A^{\mathrm{U}}$ in terms of two constant lines

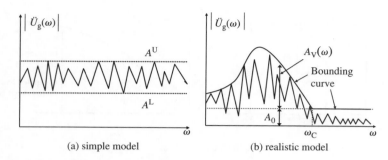

Figure 10.10 Bounding of Fourier amplitude spectrum of ground acceleration. (I. Takewaki, K. Fujita, "Earthquake Input Energy to Tall and Base-isolated Buildings in Time and Frequency Dual Domains," *Journal of the Structural Design of Tall and Special Buildings* © 2009 John Wiley & Sons, Ltd).

(see Figure 10.10(a)). It can then be shown from Equations 10.7 and 10.11 that the earthquake input energy E_I is bounded by the following relation:

$$\frac{1}{2}m(A^L)^2 \leq E_I \leq \frac{1}{2}m(A^U)^2 \tag{10.23}$$

In another case, assume that the upper bound of the Fourier amplitude spectrum $|\ddot{U}_g(\omega)|$ of the ground acceleration is given by the following form (see Figure 10.10(b)):

$$|\ddot{U}_g(\omega)|^U = \begin{cases} A_0 + A_V(\omega) & (0 \leq \omega \leq \omega_C) \\ A_0 & (\omega_C \leq \omega) \end{cases} \tag{10.24}$$

Equation 10.24 implies that the Fourier amplitude spectrum of the ground acceleration in higher frequency ranges is bounded by a constant line, while that in lower frequency ranges is bounded by a nonlinear function. Then the earthquake input energy E_I can be bounded by the following relation:

$$E_I \leq \frac{1}{2}mA_0^2 + \int_0^{\omega_C} F(\omega)\{2A_0A_V(\omega) + A_V(\omega)^2\} \, d\omega \tag{10.25}$$

The second term in the right-hand side of Equation 10.25 can be evaluated numerically. Since the earthquake ground motion has a lot of uncertainties (e.g., the occurrence probability, intensity, frequency content, and duration), it appears difficult to specify its Fourier amplitude uniquely. However, even in such a case, it may be possible to set an upper bound of its Fourier amplitude based on the database or theoretical background. These bounding relations, Equations 10.23 and 10.25, are expected to be useful for estimating the upper bound of the energy demand. It should be noted that this bound estimate is difficult in the time-domain method.

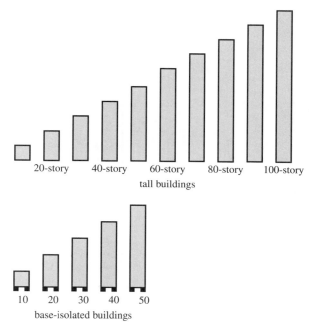

Figure 10.11   Tall buildings of 10–100 stories and base-isolated buildings of 10–50 stories. (I. Takewaki, K. Fujita, "Earthquake Input Energy to Tall and Base-isolated Buildings in Time and Frequency Dual Domains," *Journal of the Structural Design of Tall and Special Buildings* © 2009 John Wiley & Sons, Ltd).

10.8 Numerical Examples for Tall Buildings with Supplemental Viscous Dampers and Base-isolated Tall Buildings

10.8.1 Tall Buildings with Supplemental Viscous Dampers

The 10- to 100-story standard buildings shown in Figure 10.11 are considered. The plan of these buildings is $40\,\text{m} \times 40\,\text{m}$ and the mass per unit floor is 1.28×10^6 kg. The buildings are designed so that the fundamental natural period of the model is given by $T_1 = 0.1N$ (N is the number of stories in the building) and the buildings have a straight-line lowest eigenmode. The structural damping ratio is specified as 0.02. In this section, two kinds of buildings are considered: one without passive viscous dampers and the other with passive viscous dampers. The amount of passive viscous dampers is determined from the condition that the additional damping ratio attains 0.08. It is well known that the passive dampers are not effective in upper stories in tall buildings. This effect is included here by introducing the effective coefficients. These effective coefficients multiplied by the original damping coefficients are set as 0.9 in the first story and 0.5 in the topmost story (linear interpolation). The additional damping ratio 0.08 is evaluated for the case of the effective coefficients equal to 1.0.

The energy transfer functions divided by the total mass for these 20-, 40-, 60-, 80-, and 100-story buildings without and with passive viscous dampers are shown in

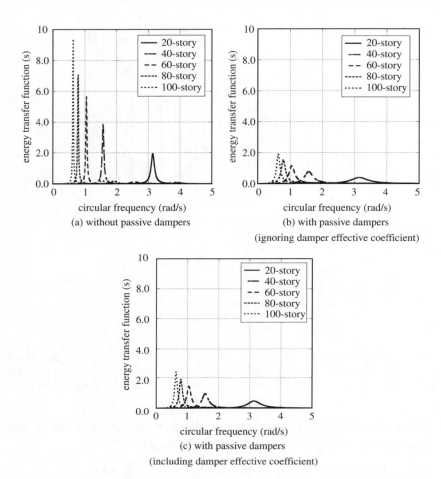

Figure 10.12 Energy transfer functions for tall buildings without and with passive dampers. (I. Takewaki, K. Fujita, "Earthquake Input Energy to Tall and Base-isolated Buildings in Time and Frequency Dual Domains," *Journal of the Structural Design of Tall and Special Buildings* © 2009 John Wiley & Sons, Ltd).

Figures 10.12(a) and (b). Those with passive viscous dampers including consideration of the effective coefficients are shown in Figure 10.12(c). It is observed that, because the passive dampers in lower stories are effective for structural control and the overall bending deformation does not affect those dampers so much, the difference between Figures 10.12(b) and (c) is small. As the amount of damping increases, the amplitude of the energy transfer function at the fundamental natural frequency becomes smaller. However, it should be noted that, for buildings with a specified number of stories, the area of the energy transfer functions is constant and the input energy exhibits a stable characteristic.

Figure 10.13 shows the comparison of energy transfer functions divided by the total mass of the overall models and those due to passive dampers. The energy transfer

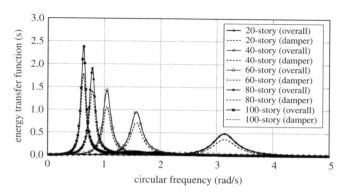

Figure 10.13 Energy transfer functions for overall models and passive dampers in tall buildings. (I. Takewaki, K. Fujita, "Earthquake Input Energy to Tall and Base-isolated Buildings in Time and Frequency Dual Domains," *Journal of the Structural Design of Tall and Special Buildings* © 2009 John Wiley & Sons, Ltd).

functions of subassemblages have been discussed by Takewaki (2007b) and the corresponding expression has been used for computation. It is observed that most parts are governed by the passive dampers in this case. This shows that the expression in terms of the energy transfer functions is appropriate for a clear understanding of the location of the principal energy consumption.

10.8.2 Base-isolated Tall Buildings

Consider next mid-rise and high-rise base-isolated buildings. In Japan, not a few high-rise base-isolated buildings have been constructed to meet the requirements on safety and serviceability. In order to present the input energy characteristics to mid- and high-rise base-isolated buildings, energy transfer functions for these buildings are investigated here. The plan of the buildings is 40 m × 40 m, as in the case of tall buildings stated above, and the mass per unit floor is set to 1.28×10^6 kg. The masses of the super-buildings are shown in Table 10.1 and the mass of the isolation floor is specified as three times the floor mass in the super-building. A natural-rubber isolator of diameter 800 mm is used. The vertical and horizontal stiffnesses of the isolator are assumed to be 4.03×10^9 N/m and 1.42×10^6 N/m respectively.

The super-buildings are designed so that the fundamental natural period of the model with a fixed base is given by $T_1 = 0.1N$ (N is the number of stories in the building) and the lowest eigenmode of the model with fixed base becomes the straight line. On the other hand, the horizontal stiffness of the isolation story is determined from the hybrid inverse formulation (Takewaki, 1998) in which the fundamental natural period of the overall model attains a specified value (see Table 10.2) for a given super-building. The damping coefficient of the viscous-type damper in the isolation story is determined from the requirement that the nominal lowest damping ratio of the overall model attains 0.1. The structural damping ratio of the super-building is given as 0.02 for the fixed-base model (isolation-story is fixed). It should be noted that a

Table 10.1 Parameters determined from the condition under gravity loading. (I. Takewaki, K. Fujita, "Earthquake Input Energy to Tall and Base-isolated Buildings in Time and Frequency Dual Domains," *Journal of the Structural Design of Tall and Special Buildings* © 2009 John Wiley & Sons, Ltd).

No. of stories	Total mass of building (kg)	Stiffness of building (SDOF model) (N/m)	Required minimum number of isolators from compressive stress condition under gravity loading	Horizontal stiffness of isolation story (N/m)
10	1.28×10^7	5.05×10^8	17	2.42×10^7
20	2.56×10^7	2.53×10^8	34	4.83×10^7
30	3.84×10^7	1.68×10^8	51	7.25×10^7
40	5.12×10^7	1.26×10^8	68	9.66×10^7
50	6.42×10^7	1.01×10^8	85	1.21×10^8

Table 10.2 Parameters determined from the natural period condition. (I. Takewaki, K. Fujita, "Earthquake Input Energy to Tall and Base-isolated Buildings in Time and Frequency Dual Domains," *Journal of the Structural Design of Tall and Special Buildings* © 2009 John Wiley & Sons, Ltd).

No. of stories	Fundamental natural period (s) (isolator only)	Allowable maximum number of isolators from natural period condition	Fundamental natural period (s) (isolator + friction damper)
10	5.29	17 (no friction damper)	5.29
20	5.25	25 (26.5 % friction damper)	5.63
30	5.60	32 (37.3 % friction damper)	6.47
40	6.15	40 (41.2 % friction damper)	7.24
50	6.82	47 (44.7 % friction damper)	8.01

two-DOF model is used in this hybrid inverse formulation only. Because the compressive stress of the isolator under gravity loading also has to satisfy a certain constraint (see Table 10.1 for the minimum number of isolators), friction-type bearings are used; that is, a certain part of gravity load is sustained by the friction-type bearings (see Table 10.2 for the necessary number of friction-type bearings). Owing to an extremely small friction coefficient of the friction-type bearings, the dissipation energy in the friction-type bearings has been ignored.

The energy transfer functions divided by the total mass for 10-, 20-, 30-, 40-, and 50-story base-isolated buildings (see Figure 10.11) are shown in Figure 10.14. As stated before, the areas of the energy transfer functions divided by the total mass are constant. In Figure 10.14, both the overall energy transfer function and that corresponding to the damper in the isolation story are plotted. It can be seen that the largest part of the energy transfer function is governed by the quantity due to the damper in the isolation story.

10.8.3 Energy Spectra for Recorded Ground Motions

In order to investigate the property of energy spectra $\sqrt{2E_I/M}$ (M is the total mass of the building) (see Akiyama, 1985; Ordaz *et al.*, 2003) for recorded ground motions,

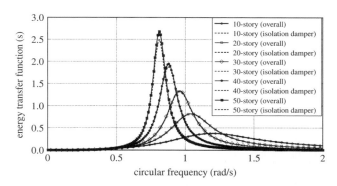

Figure 10.14   Energy transfer functions for overall models and isolation dampers in base-isolated buildings. (I. Takewaki, K. Fujita, "Earthquake Input Energy to Tall and Base-isolated Buildings in Time and Frequency Dual Domains," *Journal of the Structural Design of Tall and Special Buildings* © 2009 John Wiley & Sons, Ltd).

six recorded ground motions are used: El Centro NS (Imperial Valley 1940), Taft EW (Kern County 1952), Hachinohe EW 1968 (Tokachi-oki 1968) (far-field motion), JMA Kobe NS (Hyogoken-Nanbu 1995) (near-field motion), KBU NS (Hyogoken-Nanbu 1995) (near-field motion), Port of Tomokomai EW (Tokachi-oki 2003) (long-period ground motion). The Fourier amplitude spectra of these ground motions are shown in Figure 10.15.

Figure 10.16(a) shows the energy spectra for above-mentioned tall buildings with respect to their fundamental natural periods for El Centro NS 1940. It is observed that the energy spectra exhibit a stable property regardless of the inclusion of passive dampers, except for the models of 10, 20, and 30 stories. After further investigation, it was found that the Fourier amplitude spectra corresponding to the fundamental natural frequencies (natural periods of 1, 2, and 3 seconds) of the models of 10, 20, and 30 stories exhibit larger values than other frequency ranges. On the other hand, Figure 10.16(b) indicates the energy spectra of the above-mentioned base-isolated buildings with respect to their fundamental natural periods for El Centro NS 1940. It is found that the energy spectra for base-isolated buildings exhibit a good correlation with those for tall buildings, as long as the fundamental natural period is the same.

Figures 10.17–10.21 show these energy spectra for Taft EW 1952, Hachinohe EW 1968, JMA Kobe NS 1995, KBU NS 1995, and Port of Tomokomai EW 2003 respectively. It is found that Port of Tomokomai EW 2003 has a strong effect on the response of the models with long fundamental natural periods (Ariga *et al.*, 2006). The irregular phenomena in the models of 10, 20, and 30 stories for Hachinohe EW 1968 and KBU NS 1995 are due to the same reason stated above (small magnitude of Fourier amplitude spectra in the long-period range). It can be concluded that, although the energy spectra depend on the intensity of ground motions, the energy spectra for base-isolated buildings exhibit a good correspondence to some extent with those for tall buildings, as long as the fundamental natural period is the same.

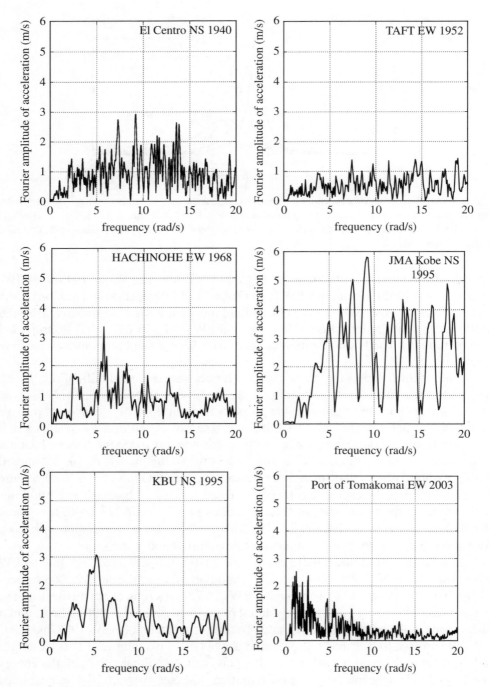

Figure 10.15 Fourier amplitude spectra for El Centro NS 1940, Taft EW 1952, Hachinohe EW 1968, JMA Kobe NS 1995, KBU NS 1995, Port of Tomokomai EW 2003. (I. Takewaki, K. Fujita, "Earthquake Input Energy to Tall and Base-isolated Buildings in Time and Frequency Dual Domains," *Journal of the Structural Design of Tall and Special Buildings* © 2009 John Wiley & Sons, Ltd).

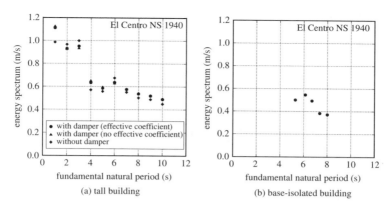

Figure 10.16 Energy spectra for tall buildings and base-isolated buildings (El Centro NS 1940). (I. Takewaki, K. Fujita, "Earthquake Input Energy to Tall and Base-isolated Buildings in Time and Frequency Dual Domains," *Journal of the Structural Design of Tall and Special Buildings* © 2009 John Wiley & Sons, Ltd).

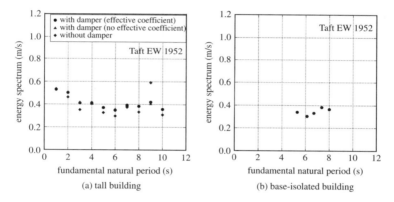

Figure 10.17 Energy spectra for tall buildings and base-isolated buildings (Taft EW 1952). (I. Takewaki, K. Fujita, "Earthquake Input Energy to Tall and Base-isolated Buildings in Time and Frequency Dual Domains," *Journal of the Structural Design of Tall and Special Buildings* © 2009 John Wiley & Sons, Ltd).

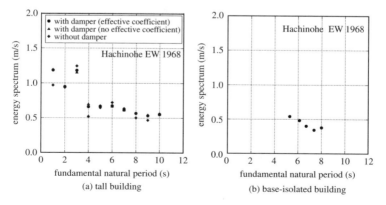

Figure 10.18 Energy spectra for tall buildings and base-isolated buildings (Hachinohe EW 1968). (I. Takewaki, K. Fujita, "Earthquake Input Energy to Tall and Base-isolated Buildings in Time and Frequency Dual Domains," *Journal of the Structural Design of Tall and Special Buildings* © 2009 John Wiley & Sons, Ltd).

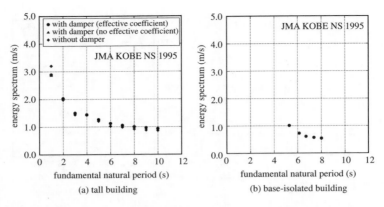

Figure 10.19 Energy spectra for tall buildings and base-isolated buildings (JMA Kobe NS 1995). (I. Takewaki, K. Fujita, "Earthquake Input Energy to Tall and Base-isolated Buildings in Time and Frequency Dual Domains," *Journal of the Structural Design of Tall and Special Buildings* © 2009 John Wiley & Sons, Ltd).

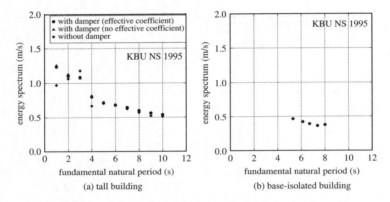

Figure 10.20 Energy spectra for tall buildings and base-isolated buildings (KBU NS 1995). (I. Takewaki, K. Fujita, "Earthquake Input Energy to Tall and Base-isolated Buildings in Time and Frequency Dual Domains," *Journal of the Structural Design of Tall and Special Buildings* © 2009 John Wiley & Sons, Ltd).

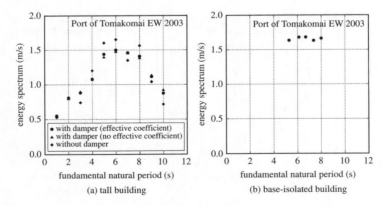

Figure 10.21 Energy spectra for tall buildings and base-isolated buildings (Port of Tomakomai EW 2003). (I. Takewaki, K. Fujita, "Earthquake Input Energy to Tall and Base-isolated Buildings in Time and Frequency Dual Domains," *Journal of the Structural Design of Tall and Special Buildings* © 2009 John Wiley & Sons, Ltd).

10.9 Summary

The results may be summarized as follows.

1. The energy transfer function $F(\omega)$ of an SDOF model can be characterized as the function such that multiplication with the Fourier amplitude of the ground motion squared and the integration of the resulting function in the frequency domain provide the earthquake input energy. The energy transfer function, a function of frequency, has an equi-area property. The residue theorem can be used for evaluating the integration of the energy transfer function in the infinite frequency range. This property guarantees that, if the Fourier amplitude spectrum of a ground acceleration is uniform, the constancy criterion of the earthquake input energy holds strictly. Otherwise, its constancy is not guaranteed. However, this equi-area property of the energy transfer function guarantees the stable characteristic of the earthquake input energy to elastic structures.

2. This equi-area property of the energy transfer function $F(\omega)$ in more general structural models can be proved by using the time-domain method for an idealized model of input ground motions with a constant Fourier amplitude spectrum. This idealized input model represents the Dirac delta function in the time domain and the earthquake input energy can be characterized by the initially given kinetic energy depending only on the total mass.

3. In two buildings connected by viscous dampers, the earthquake input energy to the overall system including these two buildings is nearly constant irrespective of the quantity of connecting viscous dampers. This property is also guaranteed by the equi-area property of the energy transfer function and leads to an advantageous feature that, if the energy consumption in the connecting dampers increases, the earthquake input energies to the buildings can be reduced effectively.

4. The frequency-domain method can provide a theoretical basis for damper effectiveness in view of the equi-area property of the energy transfer function.

5. As for tall buildings including viscous or visco-elastic dampers, the input energy to the buildings is approximately constant regardless of the quantity of those viscous or visco-elastic dampers. This property is also guaranteed by the equi-area property of the energy transfer function and leads to an advantageous feature that, if the energy consumption in the passive dampers increases, the input energies to the buildings can be reduced effectively. The same advantage exists in base-isolated buildings.

6. The frequency-domain method is appropriate for bound analysis of input energy in which the lower and/or upper bounds of the Fourier amplitude spectrum of input motions are specified. Dual use of the frequency-domain and time-domain techniques may be preferable in an advanced seismic analysis for more robust design.

7. Although the energy spectra depend on the intensity of ground motions, the energy spectra for base-isolated buildings exhibit a good correspondence to some extent with those for tall buildings, as long as the fundamental natural period is the same.

References

Abrahamson, N., Ashford, S., Elgamal, A. *et al.* (eds) (1998) Proceedings of 1st PEER Workshop on Characterization of Special Source Effects.

Akiyama, H. (1985) *Earthquake Resistant Limit-State Design for Buildings*, University of Tokyo Press, Tokyo, Japan.

Arias, A. (1970) A measure of earthquake intensity, in *Seismic Design for Nuclear Power Plants* (ed. R.J. Hansen), MIT Press, Cambridge, MA, pp. 438–469.

Ariga, T., Kanno, Y., and Takewaki, I. (2006) Resonant behavior of base-isolated high-rise buildings under long-period ground motions. *The Structural Design of Tall and Special Buildings*, **15** (3), 325–338.

Drenick, R.F. (1970) Model-free design of aseismic structures. *Journal of the Engineering Mechanics Division*, **96** (EM4), 483–493.

Hall, J.H., Heaton, T.H., Halling, M.W., and Wald, D.J. (1995) Near-source ground motion and its effect on flexible buildings. *Earthquake Spectra*, **11**, 569–605.

Heaton, T.H., Hall, J.H., Wald, D.J., and Halling, M.W. (1995) Response of high-rise and base-isolated buildings in a hypothetical MW 7.0 blind thrust earthquake. *Science*, **267**, 206–211.

Housner, G.W. (1959) Behavior of structures during earthquakes. *Journal of the Engineering Mechanics Division*, **85** (4), 109–129.

Housner, G.W. (1975) Measures of severity of earthquake ground shaking. Proceedings of the US National Conference on Earthquake Engineering, Ann Arbor, Michigan, pp. 25–33.

Housner, G.W. and Jennings, P.C. (1975) The capacity of extreme earthquake motions to damage structures, in *Structural and Geotechnical Mechanics: A Volume Honoring N.M. Newmark* (ed. W.J. Hall), Prentice-Hall, Englewood Cliffs, NJ, pp. 102–116.

Hudson, D.E. (1962) Some problems in the application of spectrum techniques to strong-motion earthquake analysis. *Bulletin of the Seismological Society of America*, **52**, 417–430.

Kuwamura, H., Kirino, Y., and Akiyama, H. (1994) Prediction of earthquake energy input from smoothed Fourier amplitude spectrum. *Earthquake Engineering and Structural Dynamics*, **23**, 1125–1137.

Leger, P. and Dussault, S. (1992) Seismic-energy dissipation in MDOF structures. *Journal of Structural Engineering*, **118** (5), 1251–1269.

Lyon, R.H. (1975) *Statistical Energy Analysis of Dynamical Systems*, MIT Press, Cambridge, MA.

Ordaz, M., Huerta, B., and Reinoso, E. (2003) Exact computation of input-energy spectra from Fourier amplitude spectra. *Earthquake Engineering and Structural Dynamics*, **32**, 597–605.

Riddell, R. and Garcia, J.E. (2001) Hysteretic energy spectrum and damage control. *Earthquake Engineering and Structural Dynamics*, **30**, 1791–1816.

Takewaki, I. (1998) Hybrid inverse eigenmode problem for a shear building supporting a finite-element subassemblage. *Journal of Vibration and Control*, **4** (4), 347–360.

Takewaki, I. (2001a) A new method for nonstationary random critical excitation. *Earthquake Engineering and Structural Dynamics*, **30** (4), 519–535.

Takewaki, I. (2001b) Probabilistic critical excitation for MDOF elastic-plastic structures on compliant ground. *Earthquake Engineering and Structural Dynamics*, **30** (9), 1345–1360.

Takewaki, I. (2002a) Critical excitation method for robust design: A review. *Journal of Structural Engineering*, **128** (5), 665–672.

Takewaki, I. (2002b) Robust building stiffness design for variable critical excitations. *Journal of Structural Engineering*, **128** (12), 1565–1574.

Takewaki, I. (2004a) Bound of earthquake input energy. *Journal of Structural Engineering*, **130** (9), 1289–1297.

Takewaki, I. (2004b) Frequency domain modal analysis of earthquake input energy to highly damped passive control structures. *Earthquake Engineering and Structural Dynamics*, **33** (5), 575–590.

Takewaki, I. (2005a) Frequency domain analysis of earthquake input energy to structure–pile systems. *Engineering Structures*, **27** (4), 549–563.

Takewaki, I. (2005b) Bound of earthquake input energy to soil–structure interaction systems. *Soil Dynamics and Earthquake Engineering*, **25** (7–10), 741–752.

Takewaki, I. (2005c) Closure to the discussion by Ali Bakhshi and Hooman Tavallali to "Bound of earthquake input energy" by Izuru Takewaki. *Journal of Structural Engineering*, **131** (10), 1643–1644.

Takewaki, I. (2006) *Critical Excitation Methods in Earthquake Engineering*, Elsevier, Amsterdam.

Takewaki, I. (2007a) Closed-form sensitivity of earthquake input energy to soil-structure interaction system. *Journal of Engineering Mechanics*, **133** (4), 389–399.

Takewaki, I. (2007b) Earthquake input energy to two buildings connected by viscous dampers. *Journal of Structural Engineering*, **133** (5), 620–628.

Takewaki, I. and Ben-Haim, Y. (2005) Info-gap robust design with load and model uncertainties. *Journal of Sound and Vibration*, **288** (3), 551–570.

Takewaki, I. and Ben-Haim, Y. (2008) Info-gap robust design of passively controlled structures with load and model uncertainties, in *Structural Design Optimization Considering Uncertainties* (eds T. Yiannis, L. Nikos, and P. Manolis), Taylor & Francis, Ch 19, pp. 531–548.

Takewaki, I. and Fujita, K. (2008) Earthquake input energy to tall and base-isolated buildings in time and frequency dual domains. *Structural Design of Tall and Special Buildings*, published online.

Tanabashi, R. (1956) Studies on the nonlinear vibrations of structures subjected to destructive earthquakes. Proceedings of the First World Conference on Earthquake Engineering, Berkeley, CA, vol. 6, pp. 1–16.

Uang, C.M. and Bertero, V.V. (1990) Evaluation of seismic energy in structures. *Earthquake Engineering and Structural Dynamics*, **19**, 77–90.

Zahrah, T.F. and Hall, W.J. (1984) Earthquake energy absorption in SDOF structures. *Journal of Structural Engineering*, **110** (8), 1757–1772.

11

Inelastic Dynamic Critical Response of Building Structures with Passive Dampers

11.1 Introduction

There are many investigations on the seismic response of passively controlled structures. Among them, the studies related to the topic explained in this chapter include those by Zhang and Soong (1992), Connor and Klink (1996), Connor et al. (1997), Kasai et al. (1998), Singh and Moreschi (2001), Uetani et al. (2003), Attard (2007), Aydin et al. (2007), and Paola and Navarra (2009). Most of these investigations were carried out for recorded or spectrum-compatible ground motions.

Owing to the inherent irregularities and uncertainties of earthquake occurrence mechanisms and ground properties, it is very difficult to predict the properties of ground motions, the site of occurrence, and the time of occurrence within a reasonable accuracy to be allowed in the seismic-resistant design practice (Drenick, 1970; Takewaki, 2004, 2006). It is desirable, therefore, to develop an approach using the most unfavorable ground motion (critical excitation) among the possible ones for structural design.

The usual intensity normalization method of ground motions using the maximum acceleration and velocity may not necessarily be reasonable from the viewpoint of physical and risk-based evaluation. While some intensity normalization methods of ground motions including structural properties have been proposed (e.g., response spectrum, energy spectrum), there are quite a few treating only the ground motion parameters.

The objective of this chapter is to present the critical excitations for passively controlled building structures with viscous or hysteretic dampers and to disclose the response properties of these structures under critical excitations. A sinusoidal motion,

Building Control with Passive Dampers   Izuru Takewaki
© 2009 John Wiley & Sons (Asia) Pte Ltd

with a variable duration, resonant to the fundamental natural period of these build-ing structures is treated as an approximate critical excitation for such structures. A long-duration far-field ground motion is shown to be a critical excitation for high-rise buildings with longer natural periods, and a near-field ground motion can be a critical excitation for low-rise buildings with shorter natural periods. It is also shown that, as far as the resonant sinusoidal ground motions are concerned and where the usual relation is used between the fundamental natural period of buildings and the number of stories, the maximum elastic interstory drift is directly related to the maximum velocity of the sinusoidal ground motion irrespective of the number of stories and the total input energy to structures in the elastic range is directly related to the velocity power of the input motion irrespective of the number of stories. Because the maximum deformation and the maximum input (or dissipation) energy are two major performance indices in performance-based structural design practice, these properties appear to be very useful in constructing intensity normalization measures of ground motions. It is further shown that the former relationship holds even in the case under spectrum-compatible ground motions.

In the last part, that these relationships may hold approximately even in the inelastic range is discussed. While it is usual in the current structural design practice that passive dampers are installed in order for structures to remain elastic even under severe earthquakes, it may also be true that clarification of the limit state in the inelastic range of such structures is useful and meaningful from the viewpoint of ensuring the true structural safety margin. It may be possible and useful to estimate the upper bound of the total input energy in the inelastic range under a resonant sinusoidal motion in terms of the maximum displacement at the equivalent height of the reduced SDOF model. This possibility will also be discussed.

11.2 Input Ground Motion

11.2.1 Acceleration Power and Velocity Power of Sinusoidal Motion

The acceleration power is defined by $\int_{-\infty}^{\infty} \ddot{u}_g(t)^2 \, dt = \overline{C}_A$ and the velocity power is defined by $\int_{-\infty}^{\infty} \dot{u}_g(t)^2 dt = \overline{C}_V$ (Arias, 1970; Drenick, 1970; Housner and Jennings, 1975; Takewaki, 2004, 2006). It is known that the resonant sinusoidal motion can be an approximate critical excitation to elastic and inelastic structures under the constraint of acceleration power or velocity power (Drenick, 1970; Takewaki, 2004, 2006). Therefore, a resonant sinusoidal motion will be used here as an input motion.

Let $\ddot{u}_g(t) = a_{max} \sin \omega_G t$ denote the acceleration of ground motion, where a_{max} and ω_G are the maximum ground acceleration and the circular frequency of the sinusoidal ground motion. The duration and frequency of the ground motion are denoted by t_0 and $T_G = 2\pi/\omega_G$ respectively. When the duration of the ground motion is given by $t_0 = nT_G/4$ $(n = 1, 2, \cdots)$, the acceleration power and velocity power can be expressed by

$$\overline{C}_A = \int_0^{t_0} \ddot{u}_g(t)^2 dt = \frac{a_{max}^2}{2} t_0 \tag{11.1}$$

$$\overline{C}_V = \int_0^{t_0} \dot{u}_g(t)^2 dt = \frac{v_{max}^2}{2} t_0 \tag{11.2}$$

where $v_{max} = a_{max}/\omega_G$ is the maximum ground velocity.

Let t_A and t_V denote arbitrary times before the ending time t_0 of input motion. The ratios $\overline{a}(t_A)$ and $\overline{v}(t_V)$ are defined by

$$\overline{a}(t_A) = \frac{\int_0^{t_A} \ddot{u}_g(t)^2 dt}{\int_0^{t_0} \ddot{u}_g(t)^2 dt} \tag{11.3}$$

$$\overline{v}(t_V) = \frac{\int_0^{t_V} \dot{u}_g(t)^2 dt}{\int_0^{t_0} \dot{u}_g(t)^2 dt} \tag{11.4}$$

The times t_{A10} and t_{A90} denote the times corresponding to $\overline{a}(t_{A10}) = 0.1$ and $\overline{a}(t_{A90}) = 0.9$ respectively, and the times t_{V10} and t_{V90} denote the times corresponding to $\overline{v}(t_{V10}) = 0.1$ and $\overline{v}(t_{V90}) = 0.9$ respectively. The effective duration of primary (intensive) ground motion is defined by the acceleration point of view ${}_e t_{A0} = t_{A90} - t_{A10}$ or the velocity point of view ${}_e t_{V0} = t_{V90} - t_{V10}$. An example of the effective duration ${}_e t_{A0} = t_{A90} - t_{A10}$ based on the acceleration power is shown in Figure 11.1.

The duration of the sinusoidal ground motion is determined from the natural period of the objective building structure (for treating resonant cases) and the data of the effective durations of actual ground motions. In this chapter, the near-field ground motion is characterized by a period of 0.5 s and a duration of 4 s (this is critical to the five-story building model) and the far-field ground motion is characterized by a period of 2.0 s and a duration of 36 s (this is critical to the 20-story building model). The acceleration power, velocity power, and the effective duration of the representative actual ground motions are shown in Table 11.1 for reference.

11.2.2 Pulse-like Wave and Long-period Ground Motion

The pulse-like velocity wave is simulated (Xu et al., 2007) by a modulated sinusoidal wave which is defined by

$$\dot{u}_p = Ct^n e^{-at} \sin \omega_p t \tag{11.5}$$

$$\ddot{u}_p = Ct^n e^{-at} [(n/t - a) \sin \omega_p t + \omega_p \cos \omega_p t] \tag{11.6}$$

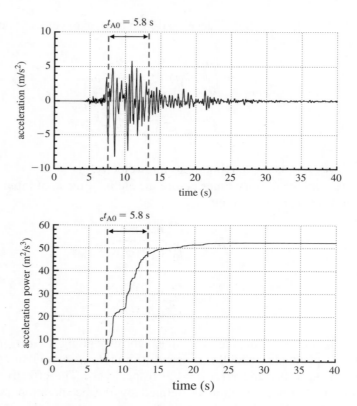

Figure 11.1   An example of effective duration $_et_{A0} = t_{A90} - t_{A10}$ based on acceleration power.

where C, a, n, and ω_p denote the amplitude scaling factor, decay factor, nonnegative integer parameter controlling the skewness of the pulse envelope with respect to time, and the pulse circular frequency. Figure 11.2 indicates the comparison between the pulse-like velocity wave expressed by $T_p = 2\pi/\omega_p = 0.5$ s, $n = 2$, $a = 1.5$, and $C = 2.0$, and the sinusoidal velocity wave expressed by a period of 0.5 s, a duration of 4 s, and a maximum velocity of 0.4 m/s.

It has also been reported that there is a velocity wave similar to a sinusoidal motion in the recorded long-period ground motions and that wave could be resonant to the building structure with a long natural period. Figure 11.3 shows the comparison between the Tomakomai EW and NS velocity waves (Tokachioki Earthquake of 2003) and the sinusoidal motion with a period of 7.0 s.

The representation of near-field ground motions and long-period ground motions by sinusoidal waves enables one to remove uncertainties resulting from ground properties, and so on, and to understand clearly the response characteristics of building structures with passive dampers under critical inputs.

Table 11.1 Acceleration power, velocity power, and effective duration of representative recorded ground motions.

Earthquake	Site and component	$C_A(\mathrm{m^2/s^3})$	$C_V(\mathrm{m^2/s})$	$_et_{A0}$ (s)	$_et_{V0}$ (s)
Near fault motion					
Rock records					
Loma Prieta 1989	Los Gatos NS	49.5	1.49	9.1	5.9
	Los Gatos EW	19.4	0.26	6.1	5.7
Hyogoken–Nanbu	JMA Kobe NS	52.4	0.79	5.8	7.9
1995	JMA Kobe EW	34.0	0.52	7.5	8.5
Soil records					
Cape Mendocino	Petrolia NS	21.5	0.25	16.0	14.8
1992	Petrolia EW	23.9	0.51	13.9	5.6
Northridge 1994	Rinaldi NS	25.0	0.62	5.5	4.2
	Rinaldi EW	46.3	1.13	7.0	6.5
	Sylmar N S	31.3	0.86	4.4	3.9
	Sylmar EW	16.3	0.45	5.2	4.6
Imperial Valley 1979	Meloland NS	5.4	0.36	5.5	16.6
	Meloland EW	6.9	1.06	4.8	23.3
Long-duration motion					
Rock records					
Michoacan 1985	Caleta de Campos NS	4.0	0.08	18.9	14.7
	Caleta de Campos EW	2.9	0.04	23.3	23.5
Miyagiken-oki 1978	Ofunato NS	2.4	0.01	11.8	12.1
	Ofunato EW	4.2	0.03	11.8	25.7
Soil records					
Chile 1985	Vina del Mar NS	34.3	0.46	41.5	43.1
	Vina del Mar EW	18.7	0.20	40.7	43.4
Olympia 1949	Seattle Army Base NS	1.3	2.29	28.0	39.6
	Seattle Army Base EW	0.9	0.02	31.8	40.3

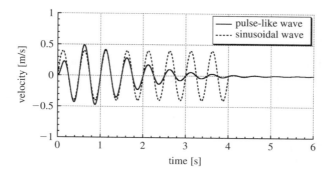

Figure 11.2 Pulse-like velocity wave and sinusoidal velocity wave.

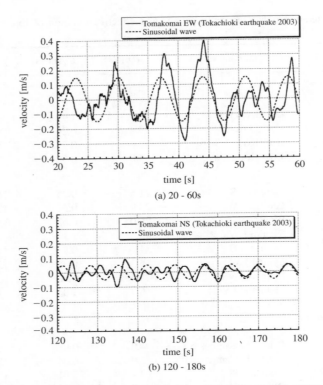

(a) 20 - 60s

(b) 120 - 180s

Figure 11.3 Velocity waves of Tomakomai EW and NS (Tokachioki Earthquake 2003) as a representative long-period ground motion and the corresponding sinusoidal velocity wave.

11.3 Structural Model

11.3.1 Main Frame

Consider four building structures of 5, 10, 20, and 40 stories with the same floor plan, as shown in Figure 11.4(a). These building structures are modeled by shear building models, as shown in Figure 11.4(b). The building parameters are shown in Table 11.2. The yield story displacement of the main frame is given by 3500 mm × 1/150 rad = 23 mm, where 1/150 rad is the yield story deformation angle. The ratio of the post-yield stiffness of the main frame to the initial stiffness is 0.01 and the structural damping ratio is specified as $h_f = 0.02$. The case of a ratio of the post-yield stiffness to the initial stiffness of 0.5 will also be investigated to discover the effect of this ratio. The response analysis is conducted by a reduced SDOF model. The equivalent mass of the reduced SDOF model is calculated by the equivalence of the base shear in the lowest mode vibration component and the equivalent height of the mass of the reduced SDOF model is obtained from the equivalence of the overturning moment at the base in the lowest mode vibration component.

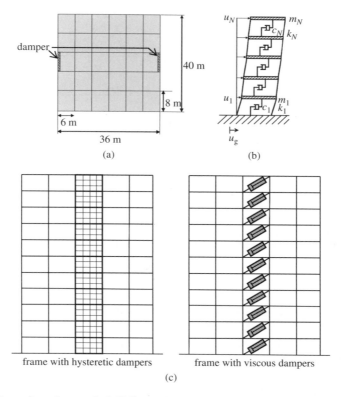

Figure 11.4 Floor plan of example buildings, shear building model, and frames with hysteretic or viscous dampers.

Table 11.2 Parameters of main frame.

Building width	6 m × 6 bays
Building length	8 m × 5 bays
Story height	3.5 m
Mass/unit floor area	800 kg/m^2
Fundamental natural period T	Proportional to number of stories N ($T = 0.1\,N$)
Story mass M	$m_i = 1152 \times 10^3$ kg
Story stiffness k_i	Determined from lowest mode of straight line
Damping coefficient c_i	Stiffness-proportional damping

The fundamental natural period T of the building structure with the number of stories N is assumed to be expressed as $T = 0.1N$.

11.3.2 Building Model with Hysteretic Dampers

The low-yield-point steel LYP100 is used as a hysteretic passive damper and is installed as a wall-type system into the main frame, as shown in Figure 11.4(c). Figure 11.5

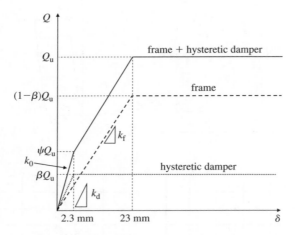

Figure 11.5 Restoring-force characteristic in the representative story of the building including the hysteretic passive damper.

shows the restoring-force characteristic in the representative story of a building including these hysteretic passive dampers. The hysteretic passive dampers are assumed to have an elastic–perfectly plastic restoring-force characteristic. The story shear, the interstory drift, and the ratio of the strength of these hysteretic passive dampers to the total story strength are denoted by Q, δ, and β respectively.

Let ψ denote the trigger-level coefficient, which indicates the ratio of the story shear at the yield displacement of the hysteretic passive dampers to the story shear (total story strength) at the yield displacement of the main frame. The ratio of the elastic damper stiffness k_d to the elastic main frame stiffness k_f is called the stiffness ratio and is expressed by K. If the frame ductility ratio is indicated by γ_F, then the equivalent viscous damping ratio h_e of this hysteretic system (main frame with hysteretic dampers) may be expressed by

$$h_e = \frac{2}{\pi}\left[1 - \frac{\beta^2}{(1-\beta)K\gamma_F} - \frac{1-\beta}{\gamma_F}\right] \tag{11.7}$$

Figure 11.6 shows the $h_e - \beta$ relation for two stiffness ratios K (1 and 3) and two frame ductility ratios γ_F (1 and 2).

In this chapter, the stiffness ratio is given by $K = 1$. In this case, $\beta = 1/11 \doteq 0.091$ and $\psi \doteq 0.182$. The equivalent viscous damping ratio at the frame yield point is given by $h_e \doteq 0.052$ from $\gamma_F = 1$ (see Figure 11.6).

For comparison, the stiffness ratio of $K = 3$ is also investigated additionally. In this case, $\beta = 3/13 \doteq 0.231$ and $\psi \doteq 0.308$. The equivalent damping ratio in this case is turned out to be $h_e \doteq 0.132$ from $\gamma_F = 1$ (see Figure 11.6).

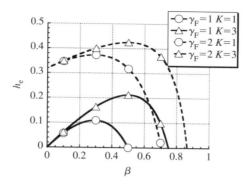

Figure 11.6 The $h_e - \beta$ relation for two stiffness ratios K and two frame ductility ratios γ_F.

11.3.3 Building Model with Viscous Dampers

An oil damper is used as a viscous passive damper (see Figure 11.4(c)). The capacity of the damper is given by the specified damping ratio h_a, which is equal to the equivalent viscous damping ratio at the frame yield point in the frame including the hysteretic passive dampers. In this chapter, the equivalent viscous damping ratio is given by $h_a = 0.052$.

11.3.4 Dynamic Response Evaluation

In this chapter, the dynamic response of building models under a resonant sinusoidal wave and a spectrum-compatible motion is obtained by using a reduced SDOF model. Figure 11.7 shows the schematic diagram for response evaluation by the reduced SDOF model. Once the earthquake response of the reduced SDOF model is computed, the local responses of the MDOF model can be obtained approximately in terms of this response of the reduced SDOF model and the lowest eigenmode (straight line in this chapter). The accuracy of this method has been confirmed by comparison with the results by time-history response analysis for the original MDOF model under a resonant sinusoidal wave and a spectrum-compatible ground motion. An example is shown in Figure 11.7.

11.4 Response Properties of Buildings with Hysteretic or Viscous Dampers

11.4.1 Two-dimensional Sweeping Performance Curves

In order to evaluate the overall performance of building structures with passive dampers, it is effective and useful to sweep out various responses (peak interstory drift, total input energy, etc.) with respect to an appropriate level parameter of

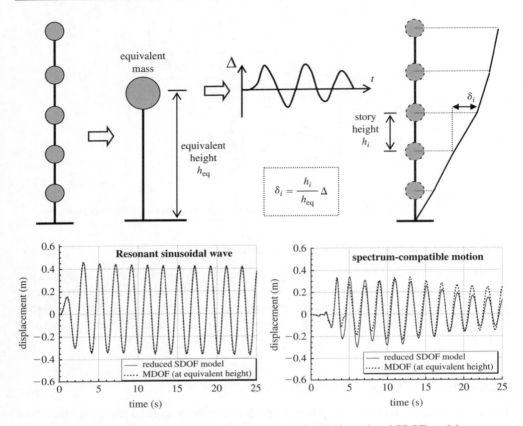

Figure 11.7 Schematic diagram for response evaluation by reduced SDOF model.

ground motion. First, the peak interstory drift and total input energy are swept out with respect to the maximum acceleration of the resonant sinusoidal wave. This figure is called the sweeping performance curve. The sweeping performance curve enables one to express effectively the response performance of structures with passive dampers to the ground motion. In the sweeping performance curve, overall shapes are important.

As stated before, the near-field ground motion is characterized by a period of 0.5 s and a duration of 4 s (this is critical to the five-story building model with viscous dampers) and the far-field ground motion is characterized by a period of 2.0 s and a duration of 36 s (this is critical to the 20-story building model with viscous dampers). For the building models with hysteretic dampers, a slightly shorter equivalent fundamental natural period is introduced by using the frame yielding point as the target point for defining the equivalent stiffness of the building models.

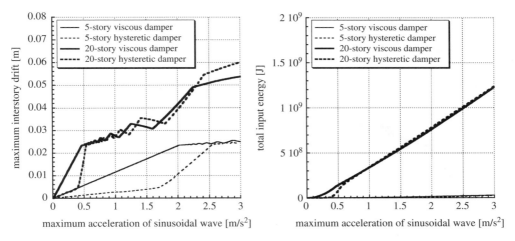

Figure 11.8 Sweeping performance curves of the maximum interstory drift and the total input energy with respect to the maximum acceleration of the sinusoidal motion.

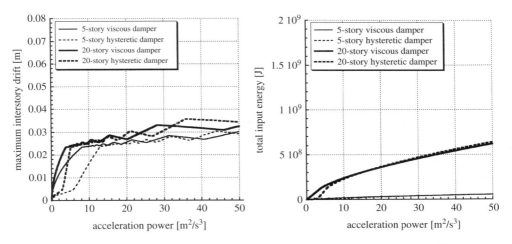

Figure 11.9 Sweeping performance curves of the maximum interstory drift and the total input energy with respect to the acceleration power of the sinusoidal motion.

11.4.2 Two-dimensional Sweeping Performance Curves with Respect to Various Normalization Indices of Ground Motion

The sweeping performance curve is useful in comparing the performances of the structures with viscous dampers and those with hysteretic dampers. Figures 11.8–11.12 plot the sweeping performance curves for structures (5- and 20-story buildings) with hysteretic dampers with a stiffness ratio $K = 1$ and those with viscous dampers with an equivalent viscous damping ratio $h_a \doteq 0.052$.

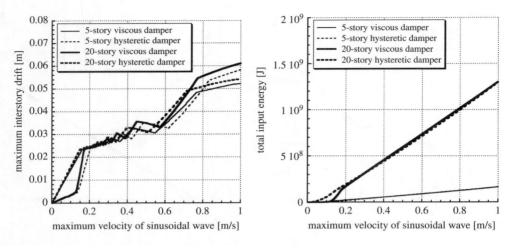

Figure 11.10 Sweeping performance curves of the maximum interstory drift and the total input energy with respect to the maximum velocity of the sinusoidal motion.

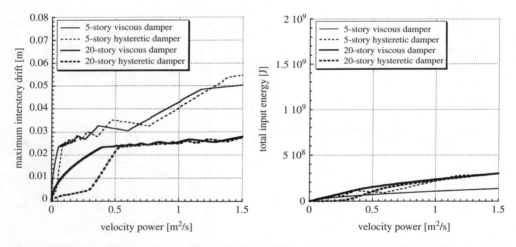

Figure 11.11 Sweeping performance curves of the maximum interstory drift and the total input energy with respect to the velocity power of the sinusoidal motion.

Figures 11.8–11.11 show the sweeping performance curves of the maximum inter-story drift and the total input energy with respect to the maximum acceleration, the acceleration power, the maximum velocity, and the velocity power of the resonant sinusoidal motion. It is seen from Figure 11.8 that hysteretic dampers are effective for reducing the interstory drifts, especially in the elastic range.

It is observed from Figure 11.10 that, when the maximum ground velocity is adopted as the normalization index, the maximum interstory drift exhibits a stable property in

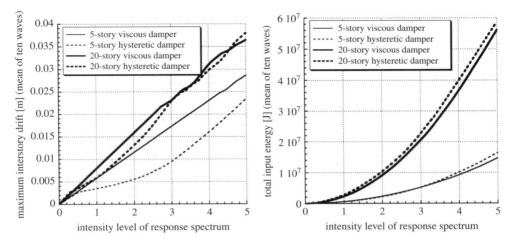

Figure 11.12 Sweeping performance curves of the maximum interstory drift and the total input energy with respect to the intensity level of the response spectrum.

the elastic range irrespective of the number of stories and damper type. This property can be proved theoretically (see Appendix 11.A). Furthermore, this property also holds approximately in the inelastic range.

It is also observed from Figures 11.8–11.11 that, when the velocity power is adopted as the normalization index, the total input energy in the elastic range exhibits a stable property irrespective of the number of stories and damper type (see Appendix 11.B). This property also holds approximately in the inelastic range. Furthermore, a tight upper bound of the total input energy in the inelastic range can be derived from the sweeping performance curve of the maximum displacement of the reduced SDOF model defined before with respect to an intensity normalization index.

On the other hand, Figure 11.12 shows the sweeping performance curves of the maximum interstory drift and the total input energy with respect to the intensity of the response spectrum (Newmark and Hall, 1982; see Appendix 11.C). The value 1 on the horizontal axis corresponds to the response spectrum of level 1 with the maximum ground velocity $v_{max} = 0.25$ m/s. Ten simulated motions compatible with this response spectrum of level 1 have been generated and used in the construction of the sweeping performance curves. Figure 11.12 shows that the hysteretic dampers exhibit a high response reduction property in the frame of five stories.

Figures 11.13–11.15 show the displacement response spectrum due to Newmark and Hall (1982), a sample of acceleration of spectrum-compatible motion, and the velocity response spectra for various damping ratios respectively.

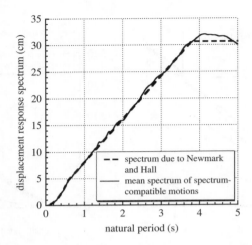

Figure 11.13 Displacement response spectrum due to Newmark and Hall (1982) and mean spectrum of 10 spectrum-compatible ground motions.

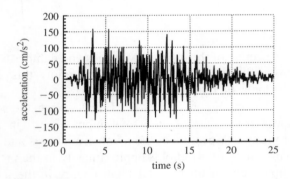

Figure 11.14 A sample of acceleration of spectrum-compatible motion.

11.5 Upper Bound of Total Input Energy to Passively Controlled Inelastic Structures Subjected to Resonant Sinusoidal Motion

In this section, a method is explained for predicting the upper bound of total input energy to passively controlled inelastic structures subjected to a resonant sinusoidal motion. A reduced SDOF model is used here again, as in the previous sections. The original MDOF shear building model has also been used to investigate the accuracy of the present method. Figure 11.16 shows a schematic diagram of the relation between the story shear force and the corresponding displacement at the equivalent height in the reduced SDOF model. The following notation is used:

Δ_{max} :   maximum horizontal displacement of reduced SDOF model

$h_a \doteq 0.052$:   additional equivalent viscous damping ratio of frame as SDOF model with hysteretic damper at frame yield point

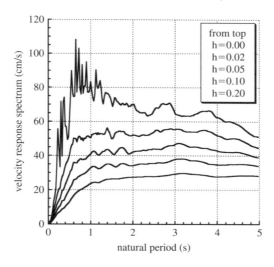

Figure 11.15 Velocity response spectrum for various damping ratios.

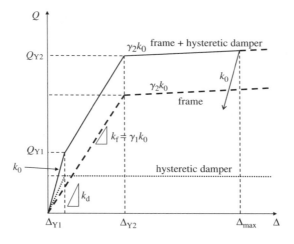

Figure 11.16 Schematic diagram of relation between story shear force and drift in reduced SDOF model.

n :   number of cycles of resonant sinusoidal motion
M :   equivalent mass of reduced SDOF model
k_f: elastic stiffness of frame as SDOF model
c_f: structural damping coefficient of frame as SDOF model
h_f: structural damping ratio of frame as SDOF model
k_0:   sum of elastic stiffness of frame as SDOF model and elastic stiffness of
         hysteretic damper in SDOF model
γ_1:   ratio of elastic stiffness of frame as SDOF model to k_0
γ_2:   ratio of post-yield stiffness of frame as SDOF model to k_0

Δ_{Y1} :   yield interstory drift of hysteretic damper in SDOF model
Δ_{Y2} :   yield interstory drift of frame as SDOF model
Q_{Y1} :   story shear force at yield point of hysteretic damper in SDOF model
Q_{Y2} :   story shear force at yield point of frame as SDOF model
ω_G: circular frequency of sinusoidal input motion
k_e: equivalent stiffness of frame as SDOF model at maximum deformation
$\quad\Delta_{max}$.

11.5.1 Structure with Supplemental Viscous Dampers

The dissipation energy due to plastic deformation of the main frame may be expressed by

$$W_p = 4(1 - \gamma_2)(\Delta_{max} - \Delta_{Y2})k_f \Delta_{Y2} n \tag{11.8}$$

The dissipation energy due to structural damping of the main fame may be described by

$$W_{hf} = \pi c_f \omega_G \Delta_{max}^2 n = 2\pi h_f M \omega_G^2 \Delta_{max}^2 n \tag{11.9}$$

The dissipation energy by supplemental viscous dampers may be obtained using

$$W_{hd} = \pi c_d \omega_G \Delta_{max}^2 n = 2\pi h_d M \omega_G^2 \Delta_{max}^2 n \tag{11.10}$$

The strain energy just after the input of sinusoidal motion can be evaluated approximately by

$$\frac{1}{2}k_e \Delta_{max}^2 \tag{11.11}$$

The total input energy E to the frame with supplemental viscous dampers can then be expressed as follows, depending on the magnitude of Δ_{max}.

1. When $\Delta_{max} \leq \Delta_{Y2}$:

$$\begin{aligned} E &= W_{hf} + W_{hd} + \frac{1}{2}k_e \Delta_{max}^2 \\ &= [2\pi(h_f + h_d)M \omega_G^2 \Delta_{max}^2]n + \frac{1}{2}k_e \Delta_{max}^2 \end{aligned} \tag{11.12}$$

2. When $\Delta_{Y2} < \Delta_{max}$:

$$\begin{aligned} E &= W_p + W_{hf} + W_{hd} + \frac{1}{2}k_e \Delta_{max}^2 \\ &= [4(1 - \gamma_2)(\Delta_{max} - \Delta_{Y2})k_f \Delta_{Y2} + 2\pi(h_f + h_d)M \omega_G^2 \Delta_{max}^2]n + \frac{1}{2}k_e \Delta_{max}^2 \end{aligned} \tag{11.13}$$

11.5.2 Structure with Supplemental Hysteretic Dampers

The total input energy E to the frame with supplemental hysteretic dampers can be expressed as follows, depending on the magnitude of Δ_{max}.

1. When $\Delta_{max} \leq \Delta_{Y1}$:

$$E = W_{hf} + \frac{1}{2}k_e\Delta^2_{max}$$

$$= (2\pi h_f M \omega^2_G \Delta^2_{max})n + \frac{1}{2}k_e\Delta^2_{max} \tag{11.14}$$

2. When $\Delta_{Y1} < \Delta_{max} \leq \Delta_{Y2}$:

$$E = W_{hf} + W_{hd} + \frac{1}{2}k_e\Delta^2_{max}$$

$$= [4(1-\gamma_1)(\Delta_{max} - \Delta_{Y1})k_0\Delta_{Y1} + 2\pi h_f M \omega^2_G \Delta^2_{max}]n + \frac{1}{2}k_e\Delta^2_{max} \tag{11.15}$$

3. When $\Delta_{Y2} < \Delta_{max}$:

$$E = W_p + W_{hf} + W_{hd} + \frac{1}{2}k_e\Delta^2_{max}$$

$$= \{4(1-\gamma_2)\{\Delta_{max} - (\Delta_{Y1} + l_0)\}k_0(\Delta_{Y1} + l_0) - 4[(\Delta_{Y2} - \Delta_{Y1})$$

$$\times k_0 - Q_{Y2} - Q_{Y1}]l_0 + 2\pi h_f M \omega^2_G \Delta^2_{max}\}n + \frac{1}{2}k_e\Delta^2_{max} \tag{11.16}$$

In Equation 11.16, l_0 denotes the following quantity:

$$l_0 = \frac{Q_{Y2} - Q_{Y1} - \gamma_2 k_0(\Delta_{Y2} - \Delta_{Y1})}{k_0(1-\gamma_2)} \tag{11.17}$$

Figure 11.17(a) shows the two-dimensional sweeping performance curve (maximum drift) for the corresponding inelastic SDOF model (post-yield stiffness ratio $\gamma_2 = 0.01$) including supplemental viscous dampers with respect to the maximum acceleration of the resonant sinusoidal ground motion. Figure 11.17(b) illustrates the corresponding diagram for post-yield stiffness ratio $\gamma_2 = 0.5$. It is observed that the model with a ratio of 0.5 exhibits a stable property even in the inelastic range. On the other hand, Figure 11.18(a) presents the corresponding diagram for the inelastic SDOF model including supplemental hysteretic dampers. These figures were obtained by time-history response analysis for the SDOF model. Figure 11.18(b) illustrates the corresponding diagram for post-yield stiffness ratio $\gamma_2 = 0.5$.

The maximum drifts Δ_{max} obtained in Figures 11.17 and 11.18 are used in Equations 11.12–11.16. Figure 11.19(a) and Figure 11.20(a) show the comparison between the predicted upper bound of total input energy and the result by the time-history response analysis (post-yield stiffness ratio of 0.01). The elements of dissipation energy consisting of the upper bound are also illustrated in Figures 11.19(a) and

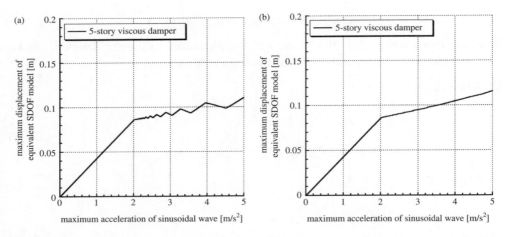

Figure 11.17 Two-dimensional sweeping performance curve (maximum drift) for the corresponding inelastic SDOF system including supplemental viscous dampers with respect to the maximum acceleration of the resonant sinusoidal ground motion: (a) post-yield stiffness ratio of 0.01; (b) post-yield stiffness ratio of 0.5.

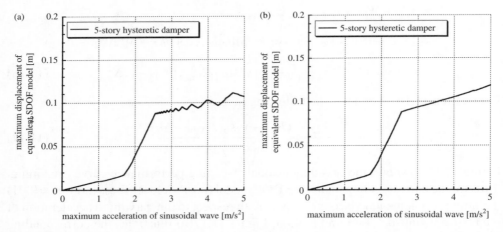

Figure 11.18 Two-dimensional sweeping performance curve (maximum drift) for the corresponding inelastic SDOF system including supplemental hysteretic dampers with respect to the maximum acceleration of the resonant sinusoidal ground motion: (a) post-yield stiffness ratio of 0.01; (b) post-yield stiffness ratio of 0.5.

11.20(a). Figures 11.19(b) and 11.20(b) illustrate the corresponding diagram for post-yield stiffness ratio $\gamma_2 = 0.5$. This shows that the method of predicting the upper bound in terms of the two-dimensional sweeping performance curve for the corresponding inelastic SDOF model including supplemental dampers can provide a reasonably accurate upper bound and can be used in the preliminary design stage of supplemental dampers.

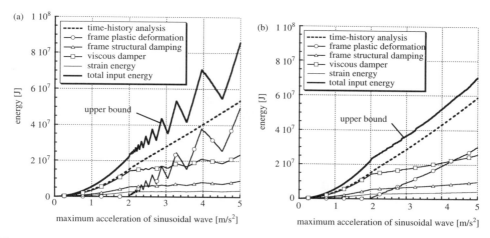

Figure 11.19 Comparison between the predicted upper bound of total input energy and the result by the time-history response analysis and the elements of dissipation energy consisting of the upper bound: (a) supplemental viscous dampers, post-yield stiffness ratio of 0.01; (b) supplemental viscous dampers, post-yield stiffness ratio of 0.5.

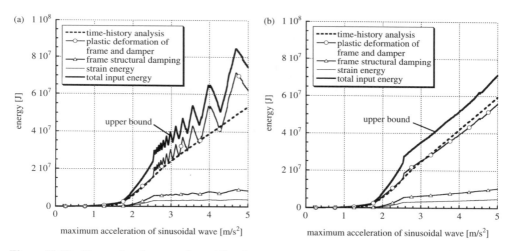

Figure 11.20 Comparison between the predicted upper bound of total input energy and the result by the time-history response analysis and the elements of dissipation energy consisting of the upper bound: (a) supplemental hysteretic dampers, post-yield stiffness ratio of 0.01; (b) supplemental hysteretic dampers, post-yield stiffness ratio of 0.5.

11.6 Relationship of Maximum Interstory Drift of Uncontrolled Structures with Maximum Velocity of Ground Motion

Figure 11.21 shows the two-dimensional sweeping performance curve (maximum interstory drift) for the original inelastic MDOF model without supplemental dampers

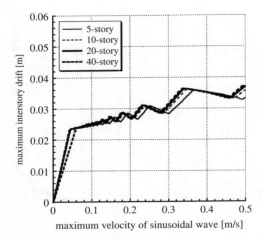

Figure 11.21 Two-dimensional sweeping performance curve (maximum interstory drift) for the corresponding inelastic SDOF system without supplemental dampers with respect to the maximum velocity of the resonant sinusoidal ground motion (post-yield stiffness ratio of 0.01).

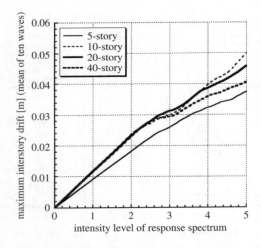

Figure 11.22 Two-dimensional sweeping performance curve (mean value of the maximum interstory drifts to 10 spectrum-compatible motions) for the corresponding inelastic SDOF system without supplemental dampers with respect to the intensity level of the spectrum-compatible ground motions (post-yield stiffness ratio of 0.01).

with respect to the maximum velocity of the resonant sinusoidal ground motion. The interstory drift has been transformed from the response of the corresponding reduced SDOF model, as in the previous sections. Furthermore, Figure 11.22 illustrates the two-dimensional sweeping performance curve (mean value of the maximum interstory drifts to 10 spectrum-compatible motions) for the original inelastic MDOF model

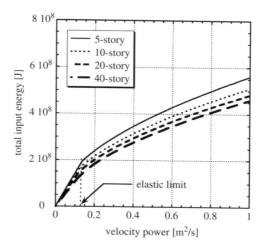

Figure 11.23 Relation of total input energy with velocity power of resonant sinusoidal ground motion (post-yield stiffness ratio of 0.01).

without supplemental dampers with respect to the maximum velocity of the spectrum-compatible ground motions. These figures were obtained by time-history response analysis. Figures 11.21 and 11.22 support the validity of the fact that, when the maximum ground velocity is adopted as the normalization index, the maximum interstory drift exhibits a stable property irrespective of the number of stories.

These properties also hold for the spectrum-compatible ground motions. When the response amplification factor in the velocity-sensitive region is denoted by $A_V(h)$ in terms of the damping ratio h and the lowest mode vibration component only is employed, then the maximum interstory drift is obtained by

$$\delta_{i\,max} \cong \frac{3N}{20\pi(2N+1)} v_{max} A_V(h_1) \tag{11.18}$$

where h_1 is the lowest-mode damping ratio. The expression $3N/[20\pi(2N+1)]$ in Equation 11.18 exhibits an almost constant value irrespective of the number of stories N (see Appendix 11.A), and this indicates clearly the one-to-one correspondence of the maximum interstory drift with the maximum ground velocity irrespective of the number of stories.

11.7 Relationship of Total Input Energy to Uncontrolled Structures with Velocity Power of Ground Motion

Figure 11.23 shows the relation of the total input energy with the velocity power. This figure has been obtained by time-history response analysis. It is observed that the total input energy to elastic buildings is strongly related to the velocity power regardless of the number of stories so far as the resonant sinusoidal ground motions are concerned. This can be proved by noting that the coefficient $30\pi(N+1)/[2(2N+1)]$ becomes almost constant for $N \geq 5$(see Appendix 11.B). Figure 11.23 shows that this property can also hold approximately in the elastic–plastic buildings.

11.8 Summary

The results are summarized as follows.

1. Since a resonant sinusoidal motion can be an approximate critical excitation to elastic and inelastic structures under the constraint of acceleration or velocity power, a resonant sinusoidal motion with variable period and duration has been used as an input wave of the near-field and far-field ground motions. This enables one to understand clearly the relation of the intensity normalization index of ground motion (maximum acceleration, maximum velocity, acceleration power, velocity power) with the response performance (peak interstory drift, total input energy).

2. In order to evaluate the overall performance of building structures with passive dampers, it is effective to sweep out various responses (peak interstory drift, total input energy, etc.) with respect to the maximum acceleration of the sinusoidal wave. This figure is called the sweeping performance curve. The sweeping performance curve enables one to express effectively the response performance of structures with passive dampers to the ground motion represented by a resonant sinusoidal wave or a suite of response spectrum-compatible motions. The sweeping performance curve is useful in comparing the performances of the structures with viscous dampers and those with hysteretic dampers.

3. When the fundamental natural period T of the building structure with the number of stories N is assumed to be expressed as $T = 0.1N$, the maximum ground velocity plays an important role in the sweeping performance curve of the maximum interstory drift. More specifically, when the maximum ground velocity is adopted as the normalization index, the maximum interstory drift exhibits a stable property irrespective of the number of stories and damper type. This fact can be proved by introducing the relationship of the fundamental natural period T of the building structure with the number of stories N in the response evaluation. This property also holds approximately in the inelastic range.

4. When the fundamental natural period T of the building structure with the number of stories N is assumed to be expressed as $T = 0.1N$, the velocity power plays an important role in the sweeping performance curve of the total input energy. More specifically, when the velocity power is adopted as the normalization index, the total input energy exhibits a stable property irrespective of the number of stories and damper type. This fact can be proved in the time and frequency dual domains. This property also holds approximately in the inelastic range.

5. The upper bound of total input energy to passively controlled inelastic structures subjected to resonant sinusoidal ground motions can be derived in terms of the two-dimensional sweeping performance curve (maximum interstory drift) for the corresponding inelastic SDOF model with respect to the maximum velocity of the resonant sinusoidal ground motion.

Appendix 11.A: Relationship of the Maximum Interstory Drift with the Maximum Velocity of the Resonant Sinusoidal Ground Motion

Consider the following resonant sinusoidal ground motion:

$$\ddot{u}_g(t) = a_{\max} \sin \omega_G t \tag{A11.1a}$$

$$\dot{u}_g(t) = -v_{\max} \cos \omega_G t \tag{A11.1b}$$

In Equations A11.1a and A11.1b, $a_{\max}$ and $v_{\max}$ denote the acceleration amplitude and velocity amplitude of the sinusoidal ground motion respectively, and ω_G indicates the circular frequency of that ground motion. In this case, the equations of motion for an MDOF shear building model may be expressed by

$$\mathbf{M}\ddot{\mathbf{u}} + \mathbf{C}\dot{\mathbf{u}} + \mathbf{K}\mathbf{u} = -\mathbf{M1}a_{\max} \sin \omega_G t \tag{A11.2}$$

Assume that the present MDOF model of N stories has an equal mass m in all stories and a straight-line lowest eigenmode. Assume also proportional damping for structural damping. Then, the response of the sth normal coordinate system (ω_s: sth natural circular frequency, h_s: sth damping ratio) can be described as

$$\ddot{q}_{0s}(t) + 2h_s\omega_s\dot{q}_{0s}(t) + \omega_s^2 q_{0s}(t) = -a_{\max} \sin \omega_G t \tag{A11.3}$$

Consider the resonant case, which is expressed by

$$\omega_1 = \omega_G \tag{A11.4}$$

If the damping ratio is small, then the first normal coordinate can be expressed as

$$q_{01} \cong (q_{01st})_0 \frac{1}{2h_1}(e^{-h_1\omega_G t} - 1)\cos \omega_G t \tag{A11.5}$$

$$(q_{01st})_0 = \frac{-a_{\max}}{\omega_G^2} \tag{A11.6}$$

Assume that the higher order components can be neglected in the resonant situation with the first natural vibration. Then, the response is described by

$$\mathbf{u} \cong \mathbf{u}_1\gamma_1 q_{01} \cong \mathbf{u}_1\gamma_1(q_{01st})_0 \frac{1}{2h_1}(e^{-h_1\omega_G t} - 1)\cos \omega_G t \tag{A11.7}$$

where γ_1 is the lowest mode participation factor. As a normalization condition of the lowest eigenmode, put the first component of $\mathbf{u}_1$ as unity. If we take the limit $t \to \infty$, the maximum interstory drift can be reduced to

$$\delta_{j\,\mathrm{max}} \cong -\frac{1}{2h_1}\gamma_1(q_{01st})_0 \tag{A11.8}$$

The straight-line lowest eigenmode with the above-mentioned normalization condition leads to the following lowest mode participation factor γ_1:

$$\gamma_1 = \frac{3}{2N+1} \tag{A11.9}$$

Substitution of Equations A11.6 and A11.9 into Equation A11.8 provides

$$\delta_{j\,\mathrm{max}} \cong \frac{3}{2N+1}\frac{a_{\mathrm{max}}}{\omega_{\mathrm{G}}^2}\frac{1}{2h_1} = \frac{3}{2N+1}\frac{v_{\mathrm{max}}}{\omega_{\mathrm{G}}}\frac{1}{2h_1} \tag{A11.10}$$

Assume that the fundamental natural period of the shear building model is given by the relation $T_1 = 0.1N$ (N: number of stories). Then, Equation A11.10 can be reduced to

$$\delta_{j\,\mathrm{max}} \cong \frac{3N}{20\pi(2N+1)}\frac{1}{2h_1}v_{\mathrm{max}} \tag{A11.11}$$

The number $3N/[20\pi(2N+1)]$ in Equation A11.11 can be shown to be almost constant (0.0217 for $N=5$, 0.0227 for $N=10$, 0.0233 for $N=20$, 0.0236 for $N=40$), irrespective of the number of stories. Therefore, Equation A11.11 implies that, if the lowest mode damping ratio h_1 is specified, the maximum interstory drift $\delta_{j\,\mathrm{max}}$ is proportional to the maximum ground velocity v_{max}.

Appendix 11.B: Relationship of the Total Input Energy with the Velocity Power of the Resonant Sinusoidal Ground Motion

Consider the same model as in Appendix 11.A. The equivalent mass M as a reduced SDOF model can be expressed by $M = [3N(N+1)m]/[2(2N+1)]$ in terms of the floor mass m and the number of stories N. In addition, the displacement Δ of this reduced SDOF model can be described in terms of the corresponding interstory drift δ (equal in every story) as

$$\Delta = \frac{2N+1}{3}\delta \tag{B11.1}$$

For the amplitude, $\Delta_{\mathrm{max}} = (2N+1)\delta_{\mathrm{max}}/3$ holds.

When the number of cycles n of the sinusoidal ground motion is large, the elastic strain energy just after the end of the sinusoidal input is negligible. In this case, the total input energy during n cycles ($=t_0/0.1N$) is equal to the energy dissipated by the viscous damping. By using Equations 11.2 and A11.11, the total input energy can be expressed by

$$
\begin{aligned}
E \cong W_{h} \\
&= (2\pi h_1 M \omega_G^2 \Delta_{\max}^2)n \\
&= (2\pi h_1 \Delta_{\max}^2)M \left(\frac{20\pi}{N}\right)^2 \left(\frac{t_0}{0.1N}\right) \\
&= \frac{\pi}{2h_1} \left[2h_1 \delta_{\max} \frac{20\pi(2N+1)}{3N}\right]^2 \frac{3N(N+1)m}{2(2N+1)} \left(\frac{t_0}{0.1N}\right). \qquad \text{(B11.2)} \\
&= \frac{\pi}{2h_1} v_{\max}^2 \frac{3N(N+1)m}{2(2N+1)} \left(\frac{t_0}{0.1N}\right) \\
&= \frac{30\pi(N+1)}{2h_1(2N+1)} mC_V
\end{aligned}
$$

The number $2\pi h_1 M \omega_G^2 \Delta_{\max}^2$ in Equation B11.2 indicates the energy dissipated in the reduced SDOF model by the viscous damping in one cycle. Furthermore, the number $[30\pi(N+1)]/[2(2N+1)]$ in Equation B11.2 can be shown to be almost constant (25.69 for $N = 5$, 24.67 for $N = 10$, 24.12 for $N = 20$, and 23.84 for $N = 40$) irrespective of the number of stories. Therefore, Equation B11.2 implies that, if the lowest mode damping ratio h_1 is specified, the total input energy E is proportional to the velocity power C_V of ground motion.

This fact can also be proved in the frequency domain. The displacement Δ of the reduced SDOF model is denoted by u hereafter. The total input energy per unit mass to the reduced SDOF model can be expressed in the frequency domain by

$$
\begin{aligned}
\frac{E}{M} &= -\int_{-\infty}^{\infty} \dot{u}\ddot{u}_g dt \\
&= -\int_{-\infty}^{\infty} \left(\frac{1}{2\pi}\int_{-\infty}^{\infty} \dot{U}e^{i\omega t} d\omega\right) \ddot{u}_g dt \\
&= -\frac{1}{2\pi}\int_{-\infty}^{\infty} \dot{U} \left(\int_{-\infty}^{\infty} \ddot{u}_g e^{i\omega t} dt\right) d\omega \qquad \text{(B11.3)} \\
&= -\frac{1}{2\pi}\int_{-\infty}^{\infty} \ddot{U}_g(-\omega)\dot{U} d\omega \\
&= -\frac{1}{2\pi}\int_{-\infty}^{\infty} \ddot{U}_g(-\omega)[H_V(\omega; \Omega, h)\ddot{U}_g(\omega)] d\omega
\end{aligned}
$$

where $H_V(\omega; \Omega, h) = -i\omega/(\Omega^2 - \omega^2 + 2ih\Omega\omega)$ is the velocity transfer function defined by $\dot{U}(\omega) = H_V(\omega; \Omega, h)\ddot{U}_g(\omega)$. $\ddot{U}_g(\omega)$ is the Fourier transform of $\ddot{u}_g(t)$.

By using the relations $\ddot{U}_g(-\omega) = \ddot{U}_g^*(\omega)$ and $\ddot{U}_g(\omega)\ddot{U}_g^*(\omega) = |\ddot{U}_g(\omega)|^2$, Equation B11.3 can be reduced to

$$\begin{aligned}\frac{E}{M} &= \frac{1}{2\pi}\int_{-\infty}^{\infty}|\ddot{U}_g(\omega)|^2\{-\text{Re}[H_V(\omega;\Omega,h)]\}\mathrm{d}\omega \\ &= \int_0^{\infty}|\ddot{U}_g(\omega)|^2\left\{-\frac{1}{\pi}\text{Re}[H_V(\omega;\Omega,h)]\right\}\mathrm{d}\omega\end{aligned}$$

(B11.4)

In Equation B11.4, $\text{Re}[H_V(\omega;\Omega,h)] = -2h\Omega\omega^2/[(\Omega^2-\omega^2)^2 + 4h^2\Omega^2\omega^2]$.

It is interesting to note that the long-duration sinusoidal motion corresponds to the Dirac delta function in the frequency domain with the infinite peak at the frequency ω_G of the sinusoidal ground motion. Thus, substitute $\omega = \Omega = \omega_G$ in Equation B11.4. By using the relation

$$C_V = \int_{-\infty}^{\infty}v(t)^2\,\mathrm{d}t = \frac{1}{\pi}\int_0^{\infty}|\dot{U}_g(\omega)|^2\,\mathrm{d}\omega = \frac{1}{\pi}\int_0^{\infty}\frac{1}{\omega^2}|\ddot{U}_g(\omega)|^2\mathrm{d}\omega$$

(B11.5)

Equation B11.4 is arranged to

$$\frac{E}{M} = \int_0^{\infty}|\ddot{U}_g(\omega)|^2\frac{1}{2h\omega_G\pi}\,\mathrm{d}\omega = \frac{\omega_G C_V}{2h}$$

(B11.6)

Let us rewrite the lowest mode damping ratio by h_1 in place of h. Since the mass M of the reduced SDOF model can be expressed by $M = 3N(N+1)m/[2(2N+1)]$ in terms of floor mass m as stated before, the total input energy to the equivalent SDOF model can be reduced to

$$E = \frac{\omega_G C_V}{2h_1}M = \frac{30\pi(N+1)}{2h_1(2N+1)}mC_V$$

(B11.7)

Equation B11.7 is equivalent to Equation B11.2.

Appendix 11.C: Design Response Spectrum by Newmark and Hall (1982)

A simplified version of the design displacement response spectrum by Newmark and Hall (1982) can be expressed by

$$S_D(\omega;h) = S_D^A(\omega;h) = \frac{\ddot{u}_{g\,\text{max}}[3.21 - 0.68\ln(100h)]}{\omega^2} \qquad (\omega_U \le \omega) \qquad \text{(C11.1a)}$$

$$S_D(\omega;h) = S_D^V(\omega;h) = \frac{\dot{u}_{g\,\text{max}}[2.31 - 0.41\ln(100h)]}{\omega} \qquad (\omega_L \le \omega \le \omega_U)$$

(C11.1b)

$$S_D(\omega;h) = S_D^D(\omega;h) = u_{g\,\text{max}}[1.82 - 0.27\ln(100h)] \qquad (\omega \le \omega_L) \qquad \text{(C11.1c)}$$

In Equations C11.1a–C11.1c, $u_{g\,max}$, $\dot{u}_{g\,max}$, and $\ddot{u}_{g\,max}$ are the maximum ground displacement, maximum ground velocity, and maximum ground acceleration respectively, and the circular frequencies ω_U and ω_L are derived from the relations $S_D^A(\omega_U; h) = S_D^V(\omega_U; h)$ and $S_D^V(\omega_L; h) = S_D^D(\omega_L; h)$.

References

Arias, A. (1970) A measure of earthquake intensity, in *Seismic Design for Nuclear Power Plants* (ed. R.J. Hansen), MIT Press, Cambridge, MA, pp. 438–469.

Attard, T.L. (2007) Controlling all interstory displacements in highly nonlinear steel buildings using optimal viscous damping. *Journal of Structural Engineering*, **133** (9), 1331–1340.

Aydin, E., Boduroglub, M.H., and Guney, D. (2007) Optimal damper distribution for seismic rehabilitation of planar building structures. *Engineering Structures*, **29**, 176–185.

Connor, J.J. and Klink, B.S.A. (1996) *Introduction to Motion-Based Design*, WIT Press.

Connor, J.J., Wada, A., Iwata, M., and Huang, Y.H. (1997) Damage-controlled structures, I: Preliminary design methodology for seismically active regions. *Journal of Structural Engineering*, **123** (4), 423–431.

Drenick, R.F. (1970) Model-free design of aseismic structures. *Journal of the Engineering Mechanics Division*, **96** (EM4), 483–493.

Housner, G.W. and Jennings, P. (1975) The capacity of extreme earthquake motions to damage structures, in *Structural and Geotechnical Mechanics: A Volume Honoring N.M. Newmark* (ed. W.J. Hall), Prentice-Hall, Englewood Cliffs, NJ, pp. 102–116.

Kasai, K., Fu, Y., and Watanebe, A. (1998) Passive control systems for seismic damage mitigation. *Journal of Structural Engineering*, **124** (5), 501–512.

Newmark, N.M. and Hall, W.J. (1982) *Earthquake Spectrum and Design*, EERI, Oakland, CA.

Paola, M.D. and Navarra, G. (2009) Stochastic seismic analysis of MDOF structures with nonlinear viscous dampers. *Structural Control and Health Monitoring*, **16** (3), 303–318.

Singh, M.P. and Moreschi, L.M. (2001) Optimal seismic response control with dampers. *Earthquake Engineering & Structural Dynamics*, **30** (4), 553–572.

Takewaki, I. (2004) Bound of earthquake input energy. *Journal of Structural Engineering*, **130** (9), 1289–1297.

Takewaki, I. (2006) *Critical Excitation Methods in Earthquake Engineering*, Elsevier, Amsterdam.

Uetani, K., Tsuji, M., and Takewaki, I. (2003) Application of optimum design method to practical building frames with viscous dampers and hysteretic dampers. *Engineering Structures*, **25** (5), 579–592.

Xu, Z., Agrawal, A.K., He, W.-L., and Tan, P. (2007) Performance of passive energy dissipation systems during near-field ground motion type pulses. *Engineering Structures*, **29**, 224–236.

Zhang, R.H. and Soong, T.T. (1992) Seismic design of viscoelastic dampers for structural applications. *Journal of Structural Engineering*, **118** (5), 1375–1392.

Index

Date Due
